America's Food

America's Food
What You Don't Know About What You Eat

Harvey Blatt

The MIT Press
Cambridge, Massachusetts
London, England

For information about special quantity discounts, please e-mail special_sales@mitpress.mit .edu

This book was set in Sabon on 3B2 by Asco Typesetters, Hong Kong.
Printed on recycled paper and bound in the United States of America.

Library of Congress Cataloging-in-Publication Data

Blatt, Harvey.
America's food : what you don't know about what you eat / Harvey Blatt.
    p.   cm.
Includes bibliographical references and index.
ISBN: 978-0-262-02652-9 (hardcover : alk. paper)
1. Food industry and trade—Health aspects—United States. 2. Agriculture—Environmental aspects—United States. I. Title. II. Title: What you don't know about what you eat.

HD9005.B628   2008
363.19′20973—dc22   2008013826

10 9 8 7 6 5 4 3 2

Contents

Introduction

There are no substitutes.
Sometimes I say there are so I can live.
But I know better. Only food can feed;
Not air, not dust, not water through a sieve.
—Abbie Huston Evans, *The Poet*

Child: Mommy, why do they grow green peas at the North Pole?
Mother: Why do you think green peas grow there?
Child: Because they're still frozen when we get them.

The conversation between mother and child makes us smile. Ah, the innocence of youth. But vegetables coated with ice are only one of the many unnatural conditions of the food we see in the local supermarket. Down various aisles we find strangely uniform and incredibly shiny red tomatoes and picture-perfect peaches. The apparent perfection reflects the fact that perhaps one-third of the farm's fruits and vegetables have been discarded by the farmer or the supermarket for aesthetic reasons.[1] How many Americans have ever seen the varied external appearances of fruits and vegetables on a farm? One in five Britons do not know that sausages come from farms.[2] For most Americans food is pretty much an abstract idea, something they do not know or imagine until it appears on the grocery shelf or the table.

Most urban shoppers know that food is produced on farms. But most of them do not know what farms, or what kind of farms, or where the farms are, what knowledge or skills are needed in farming, or how farming today bears little resemblance to farming as practiced a hundred years ago. Farming in 1908 was what we today call organic farming, and most farms raised both crops and animals. We produce much more food today than we did a century ago, and humankind is better off because of it. The proportion of the earth's people who live with hunger has been consistently reduced, in large part because of American food production. We can be proud of what our farmers have accomplished.

But there have been many unsustainable negative side effects of the large-scale industrial agriculture that has developed since our grandfather's day. Soil erosion has carried away a significant portion of our topsoil, and the remaining soil has been depleted of its nutrients. The fertilizers used to replace the nutrients are made from ever more expensive petroleum and replace only some of the lost nutrients. In addition, these artificial fertilizers commonly contain poisonous heavy metals. Pesticides are sprayed on crops on nearly all farms today, and residues are often still present in small amounts on the produce when they reach the supermarket, though they are invisible to us as we pick up and inspect them. Most farmers consider pesticides essential for crop productivity, but there is a growing suspicion among food scientists that the small amounts of pesticides that we swallow with our grains, fruits, and vegetables every day may have a harmful cumulative effect on health.

The gigantic tomatoes and large, tasty ears of corn we appreciate are the result of cross-breeding varieties of tomatoes and corn with desirable characteristics. But recently this hybridization has been replaced by genetic manipulation. Today genes from one type of plant can be replaced with genes from another type of plant. Corn no longer has to mate only with corn. It can receive genes from a tomato, a sunflower, or any other type of plant or bacterium. This type of manipulation is perhaps the most powerful agricultural tool ever developed. One hundred years ago, Darwin's concept of natural selection had already been around for fifty years, but no one would have wanted to interfere with nature's, or God's, plan for living things by altering the arrangement of their atoms. And, in any event, it was not possible to do this in 1859. Twentieth-century science and technology have changed that. We all now eat food whose atoms have been partially rearranged, that is, genetically modified. Many people are concerned about this, and more than half of Americans say they do not want to eat genetically modified food. But genetic modification is not required to be labeled on supermarket boxes, and we are all eating food every day that has been so modified. Is it wise to drift away from "what nature intended"? Is it necessary to improve on the arrangement of atoms (DNA) provided by Mother Nature? How do we balance the benefits and risks?

The production of meat, like the production of grains, has changed drastically since the Pilgrims landed almost four hundred years ago. The number of cattle in the United States is now 105 million,[3] and only a tiny percentage of them spend their brief lives in open pastures grazing on grass. Instead they are herded into gigantic pens, fattened on the grain the crop farmers produce, and routinely fed antibiotics to prevent disease and stimulate faster growth. More than one-third of our grain and nearly three-quarters of our corn is used to fatten livestock.[4] Some scientists concerned with food production have warned that our taste for the flesh of cattle, swine, and chickens may soon cause a shortage of grain for us humans.

Modern commercial cattle pens contain thousands—and sometimes even hundreds of thousands—of confined animals that produce enormous quantities of manure. A dairy cow produces twenty-three times the waste of a human and a hog four times as much.[5] In total, cattle and hogs generate 220 billion gallons of waste each year, much of it stored in gigantic and localized pits. In the United States, 130 times more animal waste is produced than human waste.[6] What can we do with these ever-increasing mountains of waste?

Fish and other types of seafood are an important part of our diet, but modern technology and factory-sized fishing vessels have made it possible to deplete the oceans of their fish. Only 10 percent of all large fish, including tuna, swordfish, marlin, cod, halibut, and flounder are now left in the sea.[7] We catch fish faster than the rate at which they can reproduce in the wild. Fish farming is now a large and growing source of seafood, but carnivorous fish eat smaller fish, so producing the large carnivores Americans enjoy is not sustainable. Commercial food processors are perhaps the closest thing to miracle workers in the food industry. They have turned soybeans into nuts, milk, yogurt, and tofu, a cheese-like substance. Pigs have been turned into pork chops and Spam. Cattle have morphed into hamburger and hot dogs. The types of plants and animals from which our food was derived is not apparent as we push our shopping cart past various food products. Someone has to tell us whether the packaged meat we see originated as an anemic calf (veal), an obese steer (hamburger), an obese hog (pork sausage), or a grotesquely large-breasted turkey (bologna).

The food processing industry is a goliath we have erected as a necessary intermediary between us and the farm. And these manufacturers know that our taste buds are drawn to products that taste sweet. Hence, boxed breakfast cereals and a multitude of other commercial products contain lots of sugar or the subsidized, and thus less expensive, high-fructose corn syrup. Unfortunately, all this sweetness has contributed to overeating and has made the average American fat. We are ultimately responsible for what we eat, but it is hard to resist the taste-tempting products turned out by the food processing companies. Do these companies have a responsibility to protect our health?

It is clear that the world of food production has changed greatly since America was founded. Some of these changes have been beneficial, but many others have not and have generated serious social and environmental problems. This book describes how food is produced in the United States and how the process, driven by the dual motivation for corporate profits and efficiency in food production, has seriously harmed human health and the environment. I hope this story contributes to a better understanding and appreciation for the food we eat and how its production affects our lives and health. I hope it will enable you to make better choices in the marketplace.

1

Old MacDonald Has No Farm: He Dies, She Dies, Sold

There are two spiritual dangers in not owning a farm. One is the danger of supposing that breakfast comes from the grocery, and the other that heat comes from the furnace.
—Aldo Leopold, ecologist

Farmers are the only indispensable people on the face of the earth.
—Li Zhaoxing, Chinese Ambassador to the United States, 1998

What do Americans think of when the words *farm family* are heard? For most of us, raised in cities and on films starring John Wayne, a vision of a rutted dirt road, a farmhouse in need of paint and with a sloping front porch, hard-working pa, pious ma, the young 'uns, a tired old plow horse, and the faithful family dog come to mind. Depending on the film, cattle barons, rustlers, foreclosures, and sympathetic but unyielding bankers may be in the picture. Does this farm family still exist? Did it ever? Well, maybe it did in 1807 or even 1907 but no longer.

Agrarianism and Nostalgia: Those Were the Good Old Days

To own a bit of ground, to scratch it with a hoe, to plant seeds, and watch the renewal of life—this is the commonest delight of the race, the most satisfactory thing a man can do.
—Charles Dudley Warner, *My Summer in a Garden*, 1870

A prevailing philosophy in colonial America was agrarianism, which sprang from the writings of Thomas Jefferson. He believed that country people were hard-working, self-sufficient, morally virtuous, and superior to city dwellers, a point of view no longer dominant in American culture but not extinct either. It survives in the near-worshipful attitude that many city dwellers and suburbanites still have toward the family farm as a romantic enterprise. Hollywood has had a large part in encouraging this notion in many cowboy movies over many decades. The image of the farmer as the salt of the earth, independent son of the soil, and child of nature is a sort of caricature covering over the image of the farmer as a rustic simpleton, uneducated hick, or uncouth redneck. Both images serve to obliterate any concept of farming as an ancient, useful, honorable vocation, requiring intelligence, skill, great patience, and endurance, to say nothing of money. The colonial farmer needed little money to begin farming, only enough to buy a crude plow, a few hand tools, and some seed. A farmer wealthy enough to buy a draft animal might be the envy of the other settlers in the area. But times have changed. Expensive implements are now required for successful farming; modern tractors to cultivate the soil can cost $200,000, and a planter $70,000. Combines, machines that cut, thresh (separate grain from husk), and clean grain, cost $250,000, as much as or more than a new home.

Successful crop farmers must be expert at selecting the varieties of plants that are adapted to their soils and climate. They must be skilled in preparing soil and in planting, growing, protecting, harvesting, and storing crops. They must be able to control weeds, insects, and diseases. In addition to these farming skills, today's farmer must know governmental farm policies, marketing strategies, and environmental laws and be a mechanic, electrician, and accountant, among many other roles. Few other occupations require such a diverse assortment of abilities. This is reflected in educational attainment: Since the early 1980s, farmers and farm managers have more formal education than the average American.[1]

An additional variable, and one that is uncontrollable and unpredictable by the American farmer, is the federal government's foreign policy. If the government wants to punish a foreign nation, a convenient export item to restrict is food, a basic necessity that a hungry country cannot do without. Farmers rely on exports to help maintain food prices and therefore bear the brunt of this restriction. Agricultural exports were relatively unimportant to American farmers until 1955, when they began to increase rapidly, and today about 22 percent of farm production is exported.[2] Indexed for inflation, the value of farm exports increased by a factor of 8 between 1955 and 2001 (figure 1.1).

In 1776, 90 percent of the new Americans were farmers. Farms, like everything else, were located along the eastern seaboard. Of course, there were fewer in frigid Maine and New Hampshire than in the warmer southeastern states, but small farms were everywhere. The food everyone ate was grown and consumed locally. Today's domestically grown food travels thousands of miles and changes hands up to six times before reaching the table. In Iowa, the typical carrot has traveled 1,600 miles from California, a potato 1,200 miles from Idaho, and a chuck roast 600 miles from Colorado. Three-quarters of the apples sold in New York City come from the West Coast or overseas, even though the state produces far more apples than city residents consume.[3] In 1776 there were no railroads or eighteen-wheelers, and long-distance transport by horse or mule was slow, difficult, and hazardous on rutted

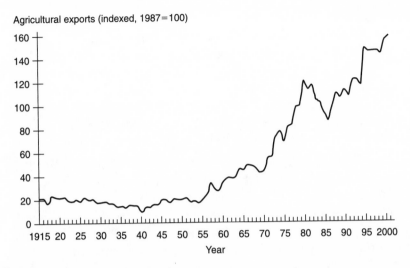

Agricultural exports (indexed, 1987=100)

Figure 1.1
Growth of U.S. agricultural exports since 1915, indexed for inflation. *Source*: U.S. Department of Agriculture.

roads, ancient Indian paths, or animal trails. And the only food preservative available was salt. Calcium propionate, disodium EDTA, and BHA had yet to be invented. Flavor enhancers such as monosodium glutamate were unnecessary. Except in the northeastern United States during the winter months, food was fresh, and chances are that it was grown and harvested by a nearby family, probably the morning or the day before it was consumed.

As America expanded westward beyond the Appalachians during the 1800s, the view was unobstructed as far as the eye could see (no pollutant haze or smog). The ground was flat, the climate was mild for most of the year, and the abundant flowing Ohio, Mississippi, and Missouri rivers had sizable tributaries extending for a thousand miles west to the Rocky Mountains. The flatland covered almost half of the future forty-eight states, described by novelist Willa Cather as "nothing but land, not a country at all, but the material out of which countries are made." There was more high-quality land available for agriculture in the United States than in any other country—more than 800 million acres.[4] When this land was combined with the independent spirit, background knowledge, and the work ethic of the pioneers, agricultural abundance was ensured. Many of the potential farmers in this new nation had either been farmers in Europe or were the children of these immigrants and had learned crop farming working with their fathers. They had a fund of background knowledge and knew what they were doing. They were not totally divorced from agriculture like most of today's Americans, who live in cities or nearby suburbs. Large factories were unknown, and most people were farmers. Operating a farm was a productive and profitable enterprise.

Machines Invade the Farm

The technology of mass production is inherently violent, ecologically damaging, self-defeating in terms of non-renewable resources, and stultifying for the human person.
—E. F. Schumacher, *Small Is Beautiful*, 1973

Most crop farmers in colonial times did their work with hand tools. Only a few had a draft animal, normally an ox or a horse, to pull a crude plow. In 1830 it took 250 to 300 hours of back-breaking labor to produce 100 bushels (the production from 5 acres) of wheat.[5] During the early 1800s, hundreds of inventors, many of them farmers weary of the grinding toil of subsistence farming, built, experimented, and tinkered with machines they hoped would make their work easier. The wildly successful ones like Cyrus McCormick and John Deere founded manufacturing concerns that are still in business (McCormick's firm is now International Harvester). The reliability of their products is forcefully expressed by a sign posted in a John Deere sales office: "The only machine we don't stand behind is our manure spreader." A modern harvester, which costs about $125,000, cuts a swath 16 feet wide with each pass, measuring the amount of grain and its moisture content as it moves, and it can harvest 900 bushels of corn per hour. In 1998 Japanese researchers developed a tractor that can till, seed, and fertilize fields by itself. It uses global positioning satellites to make its way around the field and can find its way with a margin of error of 2 inches.[6]

By the late 1800s, sophisticated farm machines became available that embodied the principles of many of today's modern implements, and fewer farmers were needed: only about half of the population were farmers. The amount of labor needed to produce 100 bushels of wheat in 1890 was 40 to 50 hours, less than one-fifth of the time needed sixty years earlier.[7] Nearly all of the new machines were animal powered, as internal combustion engines were rare and the large and heavy steam engines were too expensive for most farmers. Even after the introduction of smaller gasoline-powered tractors during the first decade of the 1900s, the horse, donkey, and mule population in the United States continued to grow, reaching an all-time high of 26.4 million animals in 1918.[8] These beasts of burden were gradually retired, and by the mid-1900s had all but disappeared as essentials in crop farming. Farmers now use 5 million tractors in place of the horses and mules of yesteryear.[9] Current agricultural technology enables one person to be fed from the food grown on about 21,000 square feet, a plot 145 feet square.[10] Two hundred years ago the area needed was ten times greater.

With farm mechanization, the time required to produce 100 bushels of wheat decreased from about a week in 1890 to 5 hours in 1965 to about 2 hours today; the time required to cultivate an acre of corn decreased from 35 to 40 hours in 1890 to 12 hours in 1945 to less than 3 hours today.[11] Farms were able to increase in size, and the more successful farmers gobbled their neighbors' land. The percentage of Americans who farmed dropped precipitously from about 40 percent in 1900 to 12 percent in 1950. Only 1.9 percent of America's employed labor force works in agriculture today. Like the white rhino and the mountain gorilla, farmers have become an endangered species. Contributing to the endangerment is increased mechanization. According to the U.S. Bureau of Labor Statistics, farmers are more than twice as likely to die on the job as police officers and nearly four times as likely to be killed at work than firefighters.

The typical American farm is sold, probably to become part of a larger farm, at least every generation. Today 2 percent of American farmers own 36 percent of the land, 10 percent own 62 percent of the land, and the bottom 70 percent of farmers own just 16 percent.[12] It is clear that agricultural land is being concentrated in fewer and fewer hands. The number of farms has decreased sharply from a high of 6.8 million in 1935 to only 2.1 million in 2005,[13] although the total number of acres farmed (at a maximum in 1950) is about the same today as it was in 1925 and has been decreasing for decades.[14] Average farm size in America has skyrocketed from about 146 acres in 1900 to 449 acres in 2007 (figure 1.2). (As a point of reference, 640 acres equals 1 square mile.)

Because of the large size and favorable topography of the United States, our capitalist economic system, favorable climate, mechanization, intensive fertilization, and genetic manipulations of crops, American farmers are the backbone of the world's agricultural productivity. Total U.S. agricultural output more than doubled between 1948 and 2004, increasing at an average annual rate of 1.74 percent. Gains in productivity account for all of the growth in output.[15] The United States is the world's major food exporter, as dominant in the world's agriculture as the Organization of Petroleum Exporting Countries is in oil production. Two hundred years ago, typical American farmers would feed their family and perhaps a neighboring family when times were hard. In 1960, the farmer's largesse would feed 26 people; today one farmer feeds about 212 people.[16] Although 98 percent of all farms are still family farms, they are not the mom-and-pop operations of yesteryear, and they may be organized as proprietorships, partnerships, or family corporations.[17] In terms of productivity or profitability, very large operations with their many economies of scale have replaced the 40–acres-and-a-mule concept of cowboy movies. One percent of American farmers account for over 50 percent of farm income; 9 percent account for 73 percent.[18] We are in the age of industrial agriculture with its destructive and dangerous reliance on a few high-yield crops with limited species variation,

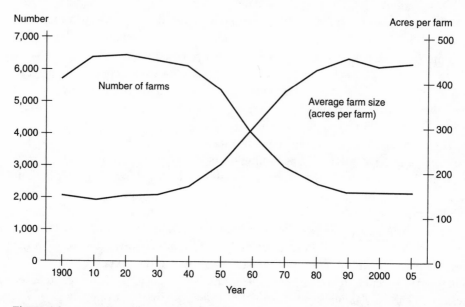

Figure 1.2
Change in number and average size of farms in the United States, 1900–2005. *Source*: U.S. Department of Agriculture.

enormous amounts of artificial fertilizers, and intensive use of ever more lethal pesticides. Traditionally the farmer relied on natural predators to control pests, animal manures for fertilizer, crop rotation (planting different crops in alternate years) to discourage pests and restore soil nutrients, and fallowing (allowing land to remain unplanted every so often to rejuvenate the soil). The degree of specialization in farming is revealed by the fact that in 1900, the average farm produced five commodities;[19] today the average is only one, the mathematical lower limit.

A small number of farm operations produce the majority of agricultural products consumed in America today. This has occurred because since the 1950s, American agricultural policies have been grounded in the belief that farms should produce as much food as possible for the least cost. These policies have led to a landscape of fewer but bigger farms that specialize in a decreasing number of commodities destined for fewer processors and packers. Eighteen percent of all farms, those larger than 500 acres, produce 75 percent of America's harvest.[20] There are relatively few farmers and few farms, and the most successful ones are highly mechanized and enormous. The possible future of farming was all too vividly described by a farmer in Minnesota: "The way things are going now, I foresee the day when there's one farmer on the east side of the Mississippi and one on the west side. They'll be plowing and they'll meet at the river. There'll be a discussion, and shortly thereafter,

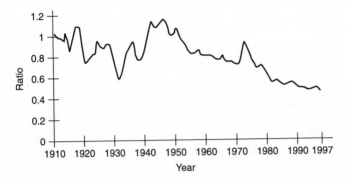

Figure 1.3
Index of average farm income: Ratio of prices received by farmers to prices paid, 1910–1997.
Source: U.S. Department of Agriculture, National Agricultural Statistical Service.

there'll be one hell of a tiling project [to drain the river] and then there'll be only one farmer."[21]

The same observation was made about 3,000 years ago by the Israelite prophet Isaiah (5:8).

Woe to those who add house to house
and join field to field
until everywhere belongs to them
and they are the sole inhabitants of the land.

Where Are the Profits?

How pleasant it is to have money, heigh ho!
How pleasant it is to have money.
—Arthur Hugh Clough, *Dipsychus*, 1850

Despite mechanization and enormous farm sizes, farm incomes have plummeted more or less continuously since 1950 (figure 1.3). Real commodity prices have fallen by about two-thirds over the past fifty years. The share of the food dollar received by American farmers is only 25 cents and has been decreasing for the past thirty years.[22] Adjusted for inflation, consumer food prices increased 3 percent from 1984 to 1998 while prices paid to farmers dropped 36 percent.[23] The vast majority of the money now goes to food processors, food marketers, and agricultural input (i.e., chemicals, seed, and fuel for agricultural machinery) suppliers.

As the prices paid to farmers decreased, the cost of running a farm increased. Between 2001 and 2005 the increase was 5.7 percent. In many areas of the United States, farm families depend on food donations from social service agencies, church pantries, and soup kitchens.[24] The loss of income has been particularly disastrous for small farmers. Eighty percent of farmers on small acreages have farm incomes below the poverty line,[25] and 59 percent of farms have less than $10,000 in sales annually. Most of the families on small farms survive only by supplementing their farm income with off-farm work.[26] In 2001 three out of every four farm households earned the majority of their income from off-farm sources. In 2004 nonfarm jobs accounted for 91 percent of the income of farm households.[27] Of the 956,000 farm operators who indicated that their primary job was off-farm work, 725,000 (76 percent) said that off-farm work was now their career choice.[28] They no longer find farming a rewarding occupation.

Where Are the Children?

The larger our great cities grow, the more irresistible becomes the attraction which they exert on the children of the country, who are fascinated by them, as the birds are fascinated by the lighthouse or the moths by the candle.
—Havelock Ellis, *The Task of Social Hygiene*, 1913

Young people today rarely consider farming as an occupation. The result is that farms are becoming homes for the chronologically challenged. The average age of full-time farmers in 2002 was 55.3; the average age of the nonfarm labor force was only 40.[29] In the ninety-nine U.S. counties with the highest percentage of residents older than 85, all but two are in the Great Plains agricultural belt.[30] Cemeteries in rural farming communities have so many fresh mounds that it looks as if badgers have dug there all winter. One North Dakota farmer in a town that recorded more deaths than births in 2000 joked that the few remaining residents may have to start importing pallbearers.

The children of farmers are leaving the family business and moving to the cities and suburbs. Only 17 percent of the U.S. population lived in rural areas in 2000.[31] In nearly 70 percent of the counties on the agricultural Great Plains, there are fewer people now than there were in 1950, and population decrease has accelerated since 2000.[32] Upon retirement or death, many veteran farmers pass the farm on to children who live elsewhere and have no interest in farming. Although membership in the National FFA Organization, known until 1988 as the Future Farmers of America, has swelled by 100,000 since 1990, very few of these new members plan to be farmers. Their career plans are to be food industry scientists, seed bioengineers, turf grass managers, food economists, nutritionists, florists, landscapers, and renewable fuels engineers.[33] Many of their farmer parents will sell the farmland to commercial developers, who are likely to use the extensive acres of flat land for more suburban housing.

In an era of unparalleled affluence and leisure, the American farmer on a small to moderately sized farm is harder pressed and harder worked than ever before. The farm's margin of profit is small to nonexistent, working hours are long, expenses for equipment and maintenance are increasing rapidly, the farm's labor force is being lost to higher-paying industrial jobs, and the farmer is being forced to spray noxious chemicals on the crops to compete with the megasized industrial farms. The average farmer is now nearing retirement, and the family's children have moved away. A farmer's work has low status in the societal pecking order and is considered marginal to the nation's economy, although farming accounts for about one-tenth of America's gross domestic product. The owner's place is being taken by absentee owners, large corporations, and machines.

For a long time, the news from everywhere in rural America has been almost unrelievedly bad: bankruptcy, foreclosure, depression, suicide, the departure of the young, the loneliness of the old. Between 1980 and 1997, the difference in suicide rates between men in the most rural and most urban counties grew from 21 percent to 54 percent. An astonishing 330 farm operators leave their land every week.[34] With the loss of hereditary farmers who felt an integral part of the land they served has come industrial farming and accelerated soil loss, soil degradation, chemical pollution, loss of genetic diversity, depletion of aquifers, and stream degradation.

Farm Subsidies: The Rich Get Richer

To those who have, more will be given;
from those who have not, what little they have will be taken away.
—Mark 4:25

We have all heard of federal government subsidies, cash given to certain groups to help them survive bad economic times. And there is no doubt that most American farmers are in bad economic times and have been for decades. In some ways, subsidies are analogous to the minimum wage guaranteed to industrial and service workers. In 1940 direct payments to farmers were $3 billion; they have risen cyclically since then, reaching an all-time high of $24 billion in 2005 before declining to $18 billion in 2006 (figure 1.4). When other federal supports such as subsidized water for irrigation and subsidized crop insurance are included, total support for farmers is roughly three times as much as for direct cash subsidies alone.[35] As a percentage of farm profits, government subsidies have ranged from 2 percent in 1974 to 47 percent in 2000. In 2005 they accounted for about 30 percent.[36] The government plans to significantly reduce subsidy payments in the coming years.

Lest we feel farmers are particularly privileged in benefiting from government assistance, we should note that all Americans get federal largesse, although they do not call it a subsidy. Businesses get rapid depreciation allowances on new

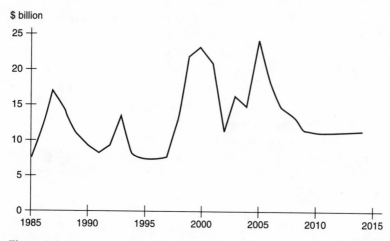

Figure 1.4
Size of direct payments to farmers by the federal government. *Source*: USDA Agricultural Baseline Projections to 2014, February 2005. Economic Research Service, USDA.

equipment; oil companies get depletion allowances as reserves decline; home owners get mortgage interest deductions from taxable income; parents get tax deductions for children; and poor families get federal and state assistance, commonly called welfare. But a subsidy by any other name is still a subsidy.

Agricultural subsidies were created during the Great Depression in the 1930s to promote a rural middle class when much of the population still worked in the farm sector. These subsidies are anachronistic now that agribusiness in developed countries employs only a tiny percentage of the population. But for the past seventy years, the government has paid farmers to grow food. The original system guaranteed price supports: grain would be sold for a minimum price no matter who grew it. But in the 1960s, Congress slowly switched to supporting farmers' incomes, not crop prices. For the purpose of determining who will receive subsidies, a farmer was defined as a person, partnership, or corporation that owns farmland, not as a person who actually farms.[37] Under this definition, farmers who have been receiving large amounts of federal money (which comes from taxes) include former professional basketball star Scottie Pippen, pornographer Larry Flynt, stock brokerage mogul Charles Schwab, more than a dozen senators and congressional representatives (some on agricultural committees), and billionaires David Rockefeller and Ted Turner. Other beneficiaries are well-known Fortune 500 farmers such as International Paper, Chevron, DuPont, and John Hancock Mutual Life Insurance.[38] Other beneficiaries include more than 1,200 universities, government farms (including state prisons), and real estate developers.[39] Between 1999 and 2005, the U.S. De-

Table 1.1
The 16 most costly direct agricultural crop subsidies in 2004 (* = program crops). There also are subsidies for the conservation reserve program ($1.8 billion), disaster payments ($548 million), environmental quality incentive program ($224 million), and the wetlands reserve program ($14 million). (Environmental Working Group, 2005)

Subsidy program	Subsidy	Percent
Corn*	$4,501,951,045	45.40
Cotton*	$1,649,366,720	16.63
Wheat*	$1,215,411,553	12.26
Soybean*	$913,345,172	9.21
Rice*	$636,205,504	6.42
Sorghum*	$313,220,331	3.16
Peanut	$213,046,953	2.15
Dairy Program	$206,530,250	2.08
Barley*	$166,949,308	1.68
Dry Pea	$30,461,699	0.31
Livestock	$27,041,523	0.27
Sunflower	$13,324,195	0.13
Fish	$11,248,791	0.11
Wool	$6,716,264	0.07
Oat*	$5,838,919	0.06
Canola	$5,238,135	0.05
TOTAL	$9,915,896,362	99.99

partment of Agriculture (USDA) paid $1.1 billion to the estates or companies of deceased farmers.[40] These farmers, alive or dead, have received payments based on the amount of land they owned and the acres of crops they plant that are covered by subsidies, called "program crops."

There are eight program crops. In 2002, corn, wheat, soybeans, rice, and cotton growers received 90 percent of the subsidies. Growers of the other favored crops, barley, sorghum, and oats, received 5 percent.[41] Other subsidized crops include apples, peanuts, peas, sunflower, and canola (table 1.1). Growers of most of the 400 other domestic food crops, the 66 percent of farmers who produce products such as such as eggs, poultry, cattle, nuts, tomatoes, strawberries, cantaloupes, and most vegetables, do not qualify for farm subsidies. Recently, however, competition from nations with low labor costs has caused American vegetable growers to call for government assistance.[42] The effect of designating certain crops as program crops is shown by the effect in Iowa. In 1945, Iowa's farmers grew seventeen commercial crops, including potatoes, cherries, peaches, plums, pears, strawberries, raspberries, and wheat. Now the commercial crops are down to four: corn, soybeans, hay, and wheat. Three of these four are subsidized program crops, and the fourth is hay, which farmers need to feed livestock. In 2005, 82 percent of harvested acreage in the United States was in these four crops.[43]

In an attempt to reduce farm welfare payments to the very rich, the farm bill passed in 2002 disqualified farm owners whose annual income exceeds $2.5 million or who do not actually make a living from farming from receiving federal subsidies. The change was largely a public relations ploy, as very few farmers were affected. In the 2005 federal budget, the annual ceiling on subsidy payments was lowered by 30 percent, from $360,000 to $250,000.

Although agricultural subsidies are typically touted as an attempt to provide a safety net for small farmers, a politically useful claim ("we need to save family farms" has a who-can-argue-with-that ring to it), the fact is that the rich benefit most from the subsidy program. According to the Environmental Working Group, between 2003 and 2005, the top 10 percent of American "farmers" received 66 percent of federal subsidy money, with an average payment of $148,077 over the three years.[44] The bottom 80 percent of recipients received 16 percent, with an average payment of $4,508 over the period. Two-thirds of America's farmers do not qualify for any assistance at all. The program is of little help to small farmers and functions mainly as a corporate welfare program. Small farmers say that the annual agricultural subsidies help big agribusinesses by lowering the prices of corn, soybeans, and other grains. These firms then make their profits by selling goods overseas in a system geared toward exports. According to the USDA, subsidy payments induce farmers to grow 25 million acres more corn and soybeans than the country needs. At the low subsidized prices, farmers have to grow larger quantities to make any profit, a cycle that eventually undermines small family farmers. In addition, large farms use their massive government subsidies to buy out smaller farms and increase consolidation in the agriculture industry. This process feeds on itself to eliminate small farmers. Fewer and fewer monster corporations control agricultural production in the United States. John Ikerd of the University of Missouri observed, "Every farm bill since the 1930s has had as its stated objective the preservation of family farms. But the reality has been greater support of specialized agriculture as a means of increasing efficiency and reducing the cost of food to the consumer."[45]

The problem with the American subsidy system is that it emphasizes output. The subsidy system encourages high-output industrial farming that will yield lots of product in the short term instead of sustainable agriculture that has high long-term productivity with very low input. Because subsidies are distributed based around particular crops, they also promote monoculture (growing of a single crop rather than several complementary crops). The externalized costs of environmental degradation also add to the hidden profit of industrial agribusiness. When corporations move in with conventional agriculture, they deplete the nutrients from the soil, trailing runoff pollution from fertilizers (see chapter 2). They do not have to pay for any of the damage they have done or the resources they have stolen; all of it converts into profit. Studies in Great Britain have revealed that a "conservative estimate" of the costs of cleaning up pollution, repairing habitats, and coping with sickness caused by conventional farming almost equals the industry's income.[46]

A large number of economists and groups interested in the welfare of the American farmer continually clamor for major revisions in federal farm policy or for the total elimination of farm subsidies. So far their voices have fallen on deaf congressional ears. There are at least two politically important reasons for congressional inaction. First, the subsidies (paid from tax dollars) allow America's major exported farm products to be sold overseas at low prices, enabling them to flood the world market with inexpensive program food crops. In effect, American tax dollars are paying part of the cost of farm products bought by overseas customers. Robert Zoellick, U.S. trade representative, boasted in 2002 that 1 out of every 3 acres in the United States is planted for export.[47] The United States has a 55 percent share of world corn exports, sold at prices 20 to 30 percent below the cost of production. It exports 60 percent of the wheat crop, selling it at 40 to 46 percent below cost, and exports 30 percent of soybean production, selling it at 30 percent below cost; rice goes at 20 percent below cost.[48] As a result, through 2004, agriculture was one of the few sectors of the American economy where the United States consistently had a trade surplus.[49] But in 2005, reports the USDA, the nation had an agricultural trade deficit for the first time since 1959. After averaging over 40 percent in the 1990s, the surplus dropped to 25 percent in 2000 and 0 percent in 2005. Despite subsidies, U.S. agricultural trade surplus has fallen victim to dramatically increased agricultural output elsewhere, particularly in Brazil.[50]

American businesses that thrive on subsidized global trade want to keep the subsidy system in place. But the subsidy system practiced by wealthy nations is a disaster for poor countries, whose economies depend heavily on agriculture. They are unable to sell their agricultural products. In many poor countries, it is cheaper to buy imported European or American farm products than to grow their own.

Second, the United States is not the major offender in the subsidy scandal. Its major overseas trading partners have even greater farm subsidies, and they refuse to anger their farmers and agribusiness constituents by dropping them. The European Union's support for its producers in 2004 was 33 percent of the value of production; Japan's totaled 56 percent. America's was only 18 percent.[51] Governments in the developed world hand out more than $300 billion in agricultural subsidies each year to their "farmers."[52] Given the enormous political clout of farm and agribusiness lobbies, change will be slow in coming.

Cotton is one of the world's most heavily subsidized crops, and in January 2004, Brazil formally challenged the legality of this most egregious American subsidy in a case submitted to the World Trade Organization (WTO). American subsidies permit exporters to sell cotton at an astonishing 57 percent below the cost of production.[53] The cotton was exported at 37 cents a pound in 2002 but cost agricultural companies 86 cents to produce. In April 2004, the WTO ruled that the U.S. cotton subsidy was indeed violating global trade rules. The U.S. scrapped the cotton subsidy early in 2006.

North American Free Trade Agreement

Governments never learn. Only people learn.
—Milton Friedman, economist, 1980

The North American Free Trade Agreement (NAFTA), which took effect on January 1, 1994, calls for the gradual removal of tariffs and other trade barriers on most goods produced and sold in North America: the United States, Canada, and Mexico. It was designed to increase trade among the three nations involved, benefiting all.[54] Agriculture is one of the areas covered by the pact, and Canada and Mexico are the second and third largest export markets for U.S. agricultural products. In 2000, slightly more than one out of every four dollars earned through U.S. agricultural exports was earned in North America.[55] How has the American farmer fared under NAFTA?

When NAFTA was enacted, it was predicted that exports of agricultural products, such as grain, oilseeds, corn, and livestock, from the United States would increase. Decreases in exports were predicted for products such as melons, cucumbers, tomatoes, orange juice, and green peppers because of cheaper labor in Mexico and ideal growing conditions for these products south of the border. The United States would do better with grains and cattle, Mexico with produce. Some of these expectations were correct. U.S. soybean exports to Mexico have doubled since NAFTA was enacted, but many of the predicted benefits of NAFTA for the American farmer have not materialized. Farm incomes for small- and middle-income farmers have continued to decline and consumer prices have risen, while agribusinesses on both sides of the Rio Grande have prospered. The U.S. trade surplus in agricultural products with partners Canada and Mexico has declined significantly since NAFTA.[56] Imports of agricultural products from these neighbors have increased much faster than exports to them. Tomato farmers in Florida have been particularly hard hit, as imports from Mexico rose by 67 percent between 1994 and 2001 and drove two-thirds of the state's tomato growers out of business. Under NAFTA, the U.S. balance of agricultural trade with Canada went from a $300 million surplus in

1994 to a $1.7 billion deficit in 2002. The trade surplus with Mexico contracted by over $1 billion under NAFTA, to $1.7 billion in 2002. Canada and Mexico are America's largest trading partners, accounting for one-third of the total of $71 billion worth of exports.[57]

In addition to NAFTA's effects on the U.S. trade balance with Mexico, U.S. consumers have been placed at greater risk from contaminated produce. In 2004, the Food and Drug Administration inspected about 100,000 of the nearly 5 million shipments of food crossing our borders, 2 percent of the imports.[58] Imported Mexican strawberries caused a massive hepatitis outbreak among Michigan schoolchildren in 1998, and in 2001 two people died from salmonella poisoning from cantaloupes imported from Mexico. Most recently imports of green onions from Mexico have been suspected in hepatitis A outbreaks that have killed three and sickened more than nine hundred in four states.[59]

In May 2004, Congress passed an extension of NAFTA-CAFTA, the Central American Free Trade Agreement—and President Bush wants to extend the agreement to South America as well to create a Free Trade Area of the Americas (FTAA). Based on their experience with NAFTA, American farmers with small holdings have reason to be concerned.

In free market nonsubsidized competition with less developed countries, the United States is often at a disadvantage. One prominent example is Brazil, the world's biggest exporter of chickens, orange juice, sugar, coffee, and tobacco.[60] It hopes to add soybeans to this list soon. With low labor costs and a climate that varies little the year round, it is not unusual to have two or even three harvests a year and to see combines clearing fields as planters sow another crop in their wake. Brazil in 2003 passed the United States as the world's largest exporter of beef and has 175 million cattle (as compared to 105 million in the United States).[61] In 2005 the United States had an agricultural trade deficit for the first time in nearly fifty years, demonstrating rising dependence on foreign agricultural production and distribution systems whose safety is questionable.

Shrinking Farmland

If people destroy something replaceable made by mankind, they are called vandals; if they destroy something irreplaceable made by God, they are called developers.
—Joseph Wood Krutch, quoted in *Mother Earth News*, 1990

The U.S. population, in 2007 just over 300 million, continues to increase at a rate of about 3 million each year, with most of the growth occurring in and around urban centers. One result of this growth is loss of arable land to development. A study in 2002 found that the United States is losing 2 acres of mostly prime farmland every minute to development—more than 1 million acres per year or 1 percent of our cropland every four years, the fastest such decline in the country's history.[62] The loss of farmland caused by urban sprawl reflects our growing affluence at least as much as the need for new housing. Over the past two decades, the U.S. population increased by 17 percent, while the amount of farmland and green space wrapped into urban areas increased by 50 percent.[63]

Urban sprawl has been occurring for centuries.[64] It is a sign of economic health and a democratizing process that gives people more choice over where they live. Sprawl is now the preferred settlement pattern anywhere there is any measure of affluence and where citizens can choose how they live. The difficulty of stopping urban sprawl was clearly illustrated by Measure 37, passed in 2004 by Oregon voters. Since the early 1970s Oregon has had "smart growth laws" that define living patterns, set land prices, and protect open space. These laws attempt to direct development to areas served by existing roads and utilities and curtail new housing and business construction that will sprawl out to rural areas that lack infrastructure development. Oregon has had the best record in the nation of reining in sprawl, according to state officials and national planning experts,[65] but its record is now crumbling.

Measure 37 compels the government to pay cash to long-time property owners when land use restrictions reduce the value of their property. If the government cannot pay, owners must be allowed to develop their land as they see fit. Because Oregon's local and state governments have almost no money to pay landowners, Measure 37 has unraveled smart-growth laws. Although voters tend to favor protection of farmland and open space, they vote down these protections if they perceive them as restrictions on their own property rights. Preserving farmland often draws a fine line between private property rights and the obligation of a community to protect and preserve land resources for future generations. Who has the right to decide what land will be developed, preserved, or utilized?[66] Should irreplaceable farmland be taxed differently and treated differently from other property? How should the environmental benefits of farmland, such as floodplain protection, groundwater recharge, and wildlife habitat, be factored into evaluating the determination of "value"? These are contentious and highly charged political issues. In November 2006, voters in at least twelve states considered ballot measures to extend protections on property rights.

Suburban housing developments are typically termed "estates" by land developers, and not only for advertising reasons. Residential lots and house sizes are increasing in size, in the extreme tending toward the vastness of estates owned by British nobility. In 1950 the average single-family home was 983 square feet. By 1970 it was 1,500. Today it is more than 2,300. This has occurred even as the average family size has decreased by 20 percent. Houses on 10-acre or larger lots are responsible for 55 percent of the sprawl onto farmlands since 1994.[67] The amount of impervious land surface (an area where water cannot penetrate the soil) owing to human construction is now 43,479 square miles, an area slightly smaller than Ohio (44,994 square miles).[68] Clearly, rates of farmland loss are high. There is 20 percent less farmland in the United States today than in the 1950 because of commercial development.[69]

It would make more sense for the government to use federal money to protect valuable and irreplaceable farmland than to subsidize preferred crops and large landholders. A federal program does exist for this purpose, the Farmland and Ranch Lands Protection Program, which helps purchase development rights to halt sprawl. The program provides matching funds to state, tribal, or local governments and nongovernmental organizations to purchase conservation easements, in which landowners agree not to convert their land to nonagricultural uses and to develop and implement a conservation program for any highly erodable land. In 2006 $73.5 million was distributed to applicants under this program.[70] Requests far exceed available funds.

Most of the population growth is still in cities and surrounding suburbs, and there has been substantial commercial development in suburban areas since World War II. Cities and suburbs are growing at the expense of rural areas. According to the USDA, 3 million acres of American croplands, wetlands, and forests were gobbled up by suburban development in 1997, two-thirds of which were cropland.[71] And the rate is increasing. Conversion of agricultural land to other purposes such as subdivisions and industrial areas is traditionally thought of as happening only around major metropolitan areas, but growing numbers of small and midsized cities are also contributing to farmland loss.[72]

Cities in the United States are compact, dense environments that maximize the use of land per capita and hence minimize the threat to agricultural areas. Suburban development is more damaging to farmland because it tends to be low density, using more land to serve fewer people, and leapfrogs over patches of agricultural areas, making it harder to have a critical mass of farms. Managing a farm surrounded by residential development is fraught with day-to-day operational perils, including nuisance lawsuits by neighbors who build homes in bucolic surroundings but eventually resent the farm sounds and smells.

Land is classified by the USDA in categories of excellence according to its suitability for agriculture. In the best category is land that is nearly level, has ground that is easily worked and favors root penetration, is well drained, is rich in nutrients, retains moisture, and is not easily eroded. Less than one-fifth of existing agricultural land in the United States is rated in this category. Such land, often located in low-lying, fertile valleys, is a farmer's dream. Unfortunately it is also a land developer's dream, so that as cities and suburbs expand, they swallow up the most productive farmland first. Satellite surveys indicate that land with the most productive soils is being paved over 30 percent faster than less productive land.[73] In California, where half the nation's fruits and vegetables are grown, 16 percent of the best soils now underlie urban areas, as do 9 percent of the next best soils.[74] This trend is national. The area in the United States devoted to roads alone is 8.2 million lane miles, and 10,000 miles of new roads are added each year.[75] The area being blacktopped each year is 1.3 million acres, an area equal to the size of Delaware. As environmentalist Rupert Cutler once noted, "Asphalt is the land's last crop."

To provide governmental policymakers with information useful for projecting future changes in land use, the Economic Research Service of the USDA created a system to classify remaining farmland into population interaction zones for agriculture (PIZA).[76] These zones represent areas of agricultural land use in which urban-related activities affect the economic and social environment of agriculture. In these zones, population interaction with farm production activities increases farmland value, changes farm enterprises, and elevates the probability of conversion to urban-related uses.

The growth of cities at the expense of productive farmland not only reduces crop-growing area but also increases air pollution, which decreases crop yields for many miles around.[77] Prior to the advent of the automobile about a hundred years ago, smog was unheard of. In addition, the smoggy haze that now blankets cities decreases the amount of sunlight available for photosynthesis. And the noxious chemicals in the haze reduce plant growth and crop yield and increase plants' susceptibility to disease and insect attack. For many reasons, urbanization can be regarded as a growing and metastasizing cancer in the agricultural community.

The Price of Land: Location, Location, Location

Buy land. They ain't makin' any more of the stuff.
—Humorist Will Rogers, 1930s

Despite the repeated losses from farming operations, the net worth of small to moderately sized family farm operations has risen more or less continuously since the end of World War II, probably mostly because of land speculation but also because of federal crop subsidies. Subsidies are estimated to inflate the price of land by 25 percent.[78] Landowners calculate their land's value based on projected income, so because subsidies make the land more profitable, the owner can charge more money, making the land more expensive for future farmers who want to get started. From this perspective, subsidies are essentially a redistribution of wealth, with the money going mainly to already wealthy landowners.

Farm real estate values increased from about $50 an acre in 1945 to $1,900 in 2006, more than triple the increase because of inflation alone. Land constituted 79 percent of farm business assets in 2000[79] and is likely even greater today because of the increasingly rapid increase in land values, which leaped 21 percent from 2004 to 2005 and another 15 percent from 2005 to 2006 (figure 1.5A). Cropland is even more valuable than farmland in general (figure 1.5B).

What Is the Real Value of Land?

I think it inappropriate to call land a "resource" because that term is tied so closely to economics. We can call gold or chrome or coal a resource, but land and people transcend a one-dimensional economic consideration.

—Wes Jackson, *American Land Forum*, 1986

In twenty-first-century capitalism, the worth of things is valued by its cost. Is a new house worth $290,000 or only $260,000? Is a used car worth $8,000 or $6,000? How about a new suit: $250 or $200? A sirloin steak in a restaurant: $17 or only $8.95? We all tend to evaluate worth in terms of price. But is there another valid way of evaluating things? And if there is, how and when should it be used?

Land suitable for raising food has a value beyond calculation. We may treasure mountains, deserts, glaciers, and wetlands, but their value can hardly be compared with that of cropland. Yet we preserve wetlands, are saddened by shrinking glaciers, worry about desert ecology, and wax ecstatic about mountain beauty while we ignore the welfare of that part of the earth's surface on which our food is grown. The explanation for this lack of public concern results in part from the surpluses of food Americans have come to accept as normal. But as we will see later in this book, our apparently ever-increasing bounty may be reaching its limit. We may need to develop a new land ethic. As author E. F. Schumacher noted thirty-five years ago, "Economics, as currently constituted and practiced, acts as a most effective barrier against the understanding of these natural resource problems, owing to its addiction to purely quantitative analysis and its timorous refusal to look into the real nature of things."[80]

The economic dimensions of farmland protection are important, but farmland protection is not an economic issue. It is a conservation issue with an economic dimension. This difference is real and important. Natural resources are not man-made resources that have value only insofar as they can be converted into dollars. In practical terms, such thinking has too often led economists to conclude that land is best used by being destroyed, that foreclosing forever the possibility for people in the future to harvest food from the land is of no consequence so long as the current owner nets the maximum profit today. Is $10,000 worth of prime farmland, which humans

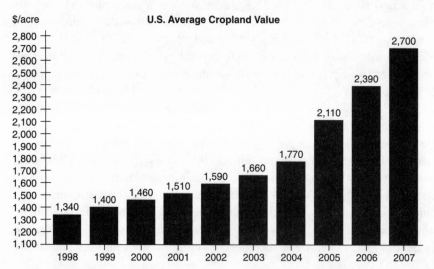

Figure 1.5
Average land values, 1998–2007 (dollars per acre). *Source*: U.S. Department of Agriculture, National Agricultural Statistical Service.

did not make and cannot replace once it is lost, to be given the same value as a $10,000 car? Surely there is something fundamentally wrong with treating the earth as if it were a business in liquidation. Land is the living, dynamic bridge where crops convert solar energy and atmospheric gases to human food. What is more important than that?

America needs a new land ethic, one that treasures the prime farmlands that have made the United States the breadbasket of the world. We must keep this land for agricultural use only, help farmers survive economically and environmentally so that they can profitably produce from them, and insist that farmlands be used in a way that maximizes their long-term health and preservation. As a recent environmental television ad says, "We only have one planet. We only get one chance."

The Reliability of Forecasting

Government-to-government assistance is only as good as the recipient government.... Hunger is not caused by scarcity of land, nor scarcity of food; it's caused by a scarcity of democracy.
—Frances Moore Lappé, *Rain*, 1985

In the terrible history of famines in the world, there is hardly any case in which a famine has occurred in a country that is independent and democratic with an uncensored press.
—Amartya Sen, Nobel laureate in economics, 1998

Will the United States always be the world's major breadbasket? Will federal crop subsidies ever end? Will the fortunes of the small farmer continue to deteriorate? Forecasting the future is almost impossible because of the unpredictability of new discoveries and inventions. In the agricultural realm, I need only cite the famous forecast of Thomas Malthus in his 1798 publication, *An Essay on the Principle of Population*, in which he "proved" mathematically the inevitability of mass starvation because population would inevitably outrun the food supply. Malthus noted that population increases geometrically (2, 4, 8, 16, 32, ...) but the food supply can increase only arithmetically (5, 10, 15, 20, 25, ...) so that starvation is inevitable. His logic seemed irrefutable. Who can argue with mathematics? As Sally Brown expressed in a *Peanuts* comic strip, "People are everywhere. Some people say there are too many of us, but no one wants to leave."

In 1900, Sir William Crookes of Great Britain predicted, "It is almost certain that within a generation the ever-increasing population of the United States will consume all the wheat grown within its borders, and will be driven to import, and like ourselves, will scramble for the lion's share of the wheat crop of the world."[79]

More recently, in 1968, latter-day Malthusian Paul Ehrlich of Stanford University said in his book *The Population Bomb* that "the battle to feed humanity is over. In the 1970s the world will undergo famines ... hundreds of millions of people (including Americans) are going to starve to death." In 1969 Ehrlich was more specific: "By 1985 enough millions will have died to reduce the earth's population to some acceptable level, like 1.5 billion people" Since 1969 the earth's population has continued to rise, from 3.6 billion to 6.7 billion today. Ehrlich also predicted that "by 1980 the United States would see its life expectancy drop to 42 because of pesticides, and by 1999 its population would drop to 22.6 million." Life expectancy in America has continued to rise since Ehrlich's 1969 prediction, from 70.5 years to 77.6 years in 2003, with the population increasing 50 percent, from 200 million to 300 million. In his 1975 book, *The End of Affluence*, he envisioned the president dissolving Congress "during the food riots of the 1980s," followed by the United States suffering a nuclear attack for its mass use of insecticides. Insecticides are now widely used throughout the world.[81]

Lester Brown of the Worldwatch Institute is another modern Malthusian who has echoed the concerns of Thomas Malthus and Sir William Crookes in two books, *Who Will Feed China?* and *Tough Choices: Facing the Challenge of Food Scarcity*. Brown envisioned China's population growth rising to 1.7 billion and commandeering all of the world's grain. This population estimate has turned out to be much too high. The latest forecasts indicate that because of its one-child policy, China's population in 2050 will plateau at 1.5 billion, only 200 million more than its population of 1.3 billion today.[82]

The prognostications of Malthus, Crookes, and Ehrlich were well meaning and based on reasonable assumptions, but they were wrong. Forecasts of world population in the 1990s envisioned stabilization at 12 billion, but the latest predictions of world population growth foresee stabilization at a number only about one-third larger than today: 8.9 billion.[83] Experts agree that food production now is more than adequate to feed today's 6.7 billion people and believe that there will not be a food shortage with 9 billion either.[84] Food shortages in Third World nations now are not due to a worldwide inability to produce food but rather to political instabilities such as wars, inadequacies of transportation, dictatorships, government subsidies to food producers in wealthy countries, poverty in developing countries, and other factors unrelated to photosynthesis and soil fertility. As the quotations at the start of this section say, the problem is a shortage of democratic, representative governments, not a shortage of food.

Who could have predicted the explosive development of pesticides in the post–World War II era, or the development of genetically modified crops in the 1990s? One does not have to be a wild-eyed optimist to feel comfortable with mankind's ability to feed itself if political factors do not intervene. Like current oil shortages, food shortages are political problems and do not originate in chemistry, biology, or a lack of solar radiation for photosynthesis. As a governmental group of agricultural experts noted in 2002.

> Over the next few decades, there are no obvious biological limits on yields that would prevent continued increase. In the longer term, far greater changes are possible. Industrialization of agriculture could mean that raw biomass [crops] is processed into livestock feed and processed food products, using biotechnology-generated microbial organisms—greatly reducing the need for conventional crop production as we now recognize it. As we try to look forward 50 and 100 years, it is not clear whether the crops that will be grown then will resemble the crops grown today....
>
> Biotechnology and precision agriculture are likely to revolutionize agriculture over the next few decades—much as mechanization, chemicals, and plant breeding revolutionized agriculture over the past century.... Biotechnology has the potential to improve adaptability, increase resistance to heat and drought, and change crop maturation schedules."[85]

Today's apparent barriers are tomorrow's accomplishments.

2

Soil Character and Plant Growth: Nature's Magic

The history of every nation is eventually written in the way in which it cares for its soil.
—President Franklin Delano Roosevelt, 1936

Three things are essential to sustain human life: air to breathe, water to drink, and soil to grow crops. All of these were in better condition a few hundred years ago than they are today. Fortunately Americans have awakened to the fact that the air we breathe is dirty and harmful to human health, and there have been many successes with cleanup efforts in recent decades. Similarly, we have realized that the water we drink contains many other substances besides H_2O, and there have been some successes in purification efforts.

But the deterioration in quality and the erosion of the soil in which food grows is a different story. Few Americans realize that our agricultural soil is in as bad condition as our air and water.[1] How often do you encounter in the media a presentation and discussion of soil? It certainly is not a soul-stirring issue to mobilize the troops in national political campaigns. It has an image problem. The word *soil* is synonymous in peoples' minds with "dirt," and bumper stickers asking us to save the soil, while more appealing than calls to save the cockroaches, mosquitoes, or head lice, do not have the same effect on the human psyche as do calls to save whales or bald eagles. But is soil any less essential to our existence than air or water? Clearly the only difference is in the time it takes for its loss to do us in. The loss of air will suffocate anyone within a minute. The absence of water will release you from this world in a few days. But the effects of soil destruction are gradual, reminiscent of the story about the frog in a pot of warming water: As the frog's environment slowly changes, the frog cooks, but without realizing what is happening. Like the frog in the heated water, we can also be unaware of our impending demise until it is too late. The National Research Council wrote in 1993, "Protecting soil quality, like protecting air and water quality, should be a fundamental goal of national environmental policy."[2]

Early in the nation's history, farms were small, they grew grains and raised animals, and alternating the crop grown from year to year was common practice. And every few years the ground would be left fallow so it could regenerate its nutritional capabilities. These had been the successful characteristics of farms for thousands of years. But as all sections of the economy were industrialized throughout the twentieth century, highly mechanized and corporate-run models of crop production were considered more efficient than the traditional small farm models. This was seen as a natural and necessary progression toward greater efficiency and productivity. Particularly after World War II, farms increased greatly in size, adopted monoculture (planting the same crop in the same field year after year), forgot about fallowing, developed a dependence on massive chemical inputs of fertilizers and pesticides, and specialized in either grains or animals. Farms are now factories that generate huge profits for the corporations that run them. Industrial agriculture produces most crops today.

The change from small and diverse family farms that used to produce our grains to monoculture farms covering hundreds of acres has resulted in a large number of undesirable side effects. We now live with poisoned surface water and groundwater, massive fish kills, incredible cruelty to the animals we have chosen to eat, toxic chemicals in food, loss of topsoil and increased soil erosion, increases in food-borne illnesses, loss of biodiversity and wildlife habitat, savaging of ocean life, and the destruction of once-thriving small communities. Today's agribusiness has the worldview that artificial is better than natural, and that because we humans have large brains, we can do better than evolution has done over hundreds of millions of years. As the Bible says (Proverbs 16:18), "Pride goeth before destruction and an haughty spirit before a fall."

The object of industrial agriculture is to produce the most food for the lowest cost. Environmental damage, or damage that develops gradually, is not factored into the definition of "lowest cost." Items such as the depletion of soil nutrients, the cost of purifying water polluted by farm runoff of fertilizers and pesticides, pollution of groundwater by downward percolation of animal urine and manure extracts from gigantic feedlots, and the loss of soil into adjacent streams because of tilling practices are "external costs." They are not a financial burden for the industrial farmer or corporation despite the fact that industrial agriculture is now the biggest polluter in the United States. The public pays for these damages through their tax dollars, a large additional subsidy to industrial farmers. In fact, industrial agriculture is unnecessary, as organic farmers have shown (chapter 4). It persists because of the political clout of the agribusiness conglomerates that now rule food production.

Knowledge of the harmful effects of industrial agriculture first entered American consciousness with the publication of Rachel Carson's classic book, *Silent Spring,* in

1962. In it she documented how the death of the nation's birds was being caused by the use of DDT and other artificial pesticides that had replaced time-honored natural methods of controlling harmful insects. It soon became evident that many of our growing environmental problems, such as water and air pollution, loss of topsoil, and habitat destruction, were direct results of our system of food production.

Many health experts believe that the rising incidence of some cancers can be traced to the massive amounts of pesticides used in the industrial method of producing food. They believe we have built fatal illnesses into our food supply.

Agricultural Soil

Soil is the mother of all things.

—Chinese proverb

Soil forms very slowly from the decay of rocks over hundreds or thousands of years. In the humid, temperate climate and rocks of the American Midwest, the rate is no faster than 1 inch per thirty years. Some authors estimate the rate to be only 1 inch per five hundred years.[3] Because at least 6 inches of topsoil are needed for successful farming, it takes at least two hundred, and possibly a few thousand, years for a usable agricultural soil to form. From the point of view of human survival, people who deliberately damage agricultural soil should be prosecuted. However, using this criterion for responsible stewardship, most of America's farmers would be indicted, as would most of the nation's makers of agricultural policy in Washington.

Fertilizer

O Lord, grant ... that once a week thin liquid manure and guano [bat excretions] may fall from heaven.

—Karel Capek, *The Gardener's Prayer*, 1931

Historically the addition of nutrients to a plant's diet was done the natural way, with animal manure, which provides not only the major and minor nutrients that crops need but also adds organic matter. More than 3,000 years ago the Greek writer Homer made reference to the use of animal manure as fertilizer, as does the Bible. Human feces, known euphemistically as night soil, has also been used since antiquity as fertilizer. In 1649 Tokyo, toilets that emptied into streams or canals were banned so as to maximize the collection of human excrement.[4] However, unlike animal manure, human feces can contain harmful microorganisms, so its use on crops is not recommended.

Less than a century ago, farms contained grass, amber waves of grain, and farm animals, and the relationship between them was symbiotic. The naturally occurring grass fed the animals, and the animals produced manure to fertilize the grain and grass. In nature there is no waste. This symbiotic relationship had another benefit: insurance against the risk of disease. Farm diseases are usually quite specific and attack one type of livestock or crop. The best way to prevent them is to avoid keeping too many of the same animals together in one place and to rotate them so that the cycle of diseases and parasites is broken. The advent of industrial monoculture and animal confinement cancelled this insurance policy.

The farmer maintained a balance of the number of animals on the farm, the amount of manure produced, and the amount needed for fertilizer. But following the devastation of World War II, there was a need for massive increases in food production to feed a hungry world, which resulted in farm specialization and the separation of grain growing from animal farming. Farm animals were confined in pens remote from their source of natural food and regarded as machines, with their excrement regarded not as fertilizer but first a waste and then as a pollutant.

The places where farm animals are kept today are more like factories than farms (in fact, they are called factory farms), with single farms holding thousands to hundreds of thousands of animals in confined spaces. Their manure and urine are no longer dispersed widely over the farmer's fields but are deposited in the confined areas where the animals are penned.

The amount of manure produced annually by the millions of farm animals in the United States is staggering: 1.5 billion tons, 5 tons for each of America's 300 million people, 130 times more manure than Americans produce themselves.[5] A single 1,000-pound cow produces 15 tons of manure per year. One hog yields 1.7 tons of feces and urine in a year. This staggering amount of urine and manure is not easily disposed of and is typically stored in shallow pits called lagoons. The waste solids sink to the bottom of the pit and are broken down by bacteria, and the liquids at the top of the lagoon are sprayed onto fields. However, it is not uncommon for major spillage of animal waste to occur from the lagoons; more than 1,000 such spills were documented between 1995 and 1998, with disastrous effects in nearby streams and on the olfactory senses of local residents. In a well-publicized occurrence in 1995, an 8-acre lagoon holding 25 million gallons of putrefying hog urine and feces burst and spilled into the ironically named New River in North Carolina, immediately killing over 10 million fish. As noted in a 1995 study in *Brain Research Bulletin*, "The amount of anger people [in North Carolina] have over odor pollution is tremendous. The smell gets into bedding, carpets, and drapes. People can't sell their houses because no one wants to live near a hog farm."[6] There are no federal or state standards for odor pollution.

Manure is nutritious for plants. Excluding water and carbon dioxide, the nutrients that crops need in the greatest amounts are nitrogen (N), phosphorous (P), and potassium (K). A ton of average animal manure contains about as much

Table 2.1
Loss of nutrient content of some types of meat and milk products, 1940–2002

	Calcium	Magnesium	Iron
Beef rump steak	−4%	−7%	−55%
Corned beef	−45	−48	−76
Streaky bacon	−87	−16	−78
Chicken roast	−31	No change	−69
Turkey	−71	−4	−79
Milk	−2	−21	ND
Cheddar cheese	−9	−38	ND
Vegetables	−46	−24	−27

Note: ND = not determined.
Source: *Food Magazine*, January–March 2006, p. 10.

NPK as a 100-pound bag of 10-5-10 fertilizer.[7] The manure also contributes other nutrients, such as calcium, magnesium, and sulfur, needed in small amounts (micronutrients) and provides a bonus in the form of organic matter added to the soil. Healthy agricultural soil contains 3 to 6 percent organic matter.

The change from using manure to using fertilizers containing only nitrogen, phosphorous, and potassium has depleted crops of micronutrients that used to be present and that humans need for health.[8] Studies of 64 foods made between 1940 and 1991 reveal that since 1940, vegetables have lost 76 percent of their copper, 46 percent of the calcium, 27 percent of the iron, 24 percent of the magnesium, and 16 percent of their potassium. From 1978 to 1991 they lost 57 percent of their zinc. Other nutrients have suffered similar declines. The loss of mineral nourishment from grains is also reflected in the nutrient content of the flesh of the animals that eat the grain (table 2.1). It should not be surprising that when the soil is deficient, so also will be the grasses that grow in that soil, the cattle that eat the grasses, and the people who eat the cattle and drink the milk. These data say that you must eat more food today to get the same amount of nourishment people got from their food a few decades ago.

Artificial Fertilizer

The loss of animal manure as a natural fertilizer is harmful to crops in many ways. Because artificial fertilizers are designed to release the nutrient immediately, removal of the nutrient by rainwater can remove much of it before the plant roots can grab it. This is most serious for nitrogen. Almost half of the artificial nitrogen fertilizer spread onto fields is not taken up by crops but instead washes away.[9] In contrast, an organic fertilizer such as manure is more beneficial because it releases its nutrient chemicals slowly; the microorganisms in the soil must first break down the organic material and convert it into inorganic chemicals that the plant can use.

The United States is considered a mature market for NPK fertilizer, with usage varying only 2 to 3 percent from year to year. A major reason for the stability is that the saturation level has been reached. Additional fertilizer has little effect on production because most crops are physiologically incapable of absorbing more N, P, or K.[10] Unlike people, a plant cannot overeat. Its food intake is biologically controlled. American farmers are now getting the maximum possible crop production per acre, barring new developments in genetic engineering. With fertilizer sales stagnant, agribusiness companies are pioneering genetic modification of crops to increase their profits (chapter 5).

Of the major nutrients needed for crops, nitrogen is most commonly the limiting one. Nationwide, nitrogen averages about 57 percent of the three elements in NPK applications; phosphorous, 21 percent, and potassium, 22 percent.[11] The major change in NPK percentages over the past few decades has been in the application of nitrogen, which has increased from 37 percent in 1960 to 57 percent today, a change that has had terrible environmental effects.

Air consists of 78 percent nitrogen, but plants cannot use nitrogen gas; it must be converted into a form soluble in water to be accessible to the plant. This is accomplished by several types of bacteria that live in agricultural soil. Plants such as alfalfa, clover, soybeans, peas, and beans send out a chemical signal that binds to a receptor inside the bacterium, drawing the microbe to the plant. The bacteria set up house in nodules along the roots, where they convert the nitrogen gas in the air into ammonia for the alfalfa in exchange for energy. Legumes have the ability to attract the nitrogen-fixing microbes. In the past, when the crop in a field was changed yearly, alfalfa and clover, the most effective legumes, were commonly alternated with the corn or wheat to reinvigorate the nitrogen content of the soil.

Humans do not eat alfalfa and clover, so they are not the monocrop planted yearly in agricultural fields. Monocropping with corn, wheat, or other popular grains requires adding massive amounts of inorganic nitrogen fertilizer, which can be deadly to soil microbiota. High-nitrogen chemical fertilizers can be so deadly to earthworms that they have been recommended as a way to rid golf courses of night crawlers while greening up the lawn at the same time. The monocropping practiced by industrial agriculture decreases the number of living organisms in the soil and also decreases the diversity of soil animals (species) compared to native vegetation.[12] The huge amounts of nitrogen fertilizer, coupled with repeated dowsing with poisonous artificial pesticides, deplete agricultural soil of its essential animal life.

Killing earthworms by excessive nitrogen is particularly harmful to agricultural soil. The microscopic animals in the soil need oxygen to survive, which they obtain from the air located a few inches above their heads. The soil must be churned continually to maintain this aeration, a service performed in temperate regions largely by the 3,000 species of earthworms, one of which grows up to 13 feet long. It is humbling to think that this service, essential for human existence, is performed by a minibrained, blind, deaf, speechless, spineless, toothless, slimy vegan creature 2 or 3 inches long and a few tenths of an inch wide that weighs only a small fraction of an ounce. An acre of soil may contain 1 million earthworms busily preparing the soil for farming by burrowing through the soil, creating pore networks, and depositing tons of plant nutrients in their castings (feces). Researchers have found that the infiltration rate of water into an agricultural soil depends almost entirely on the number of earthworms.[13]

In addition to their burrowing activities, earthworms deposit their excreta as solid casts, a balanced selection of minerals and plant nutrients in a form accessible for root uptake. Earthworm casts are five times richer in available nitrogen, seven times richer in available phosphates, and eleven times richer in available potassium than the surrounding soil. A single worm can produce 10 pounds of casts per year. As bruising as it may be to our egos, in the grand scheme of things an earthworm is more important to the maintenance of civilization than you or I.

Nitrate Pollution

Humans have approximately doubled the amount of fixed nitrogen in circulation in the quest for more and better crops. The amount of nitrogen fertilizer applied to crops in the 1980s alone was greater than the amount applied in all previous human history.[14] Because nitrogen applications have increased by 50 percent over the past fifty years and because almost half of the nitrogen spread on fields is not taken up by crops but washes away, severe nitrate pollution has developed in streams and in the near-shore Gulf of Mexico.

The excess nitrate fuels the rampant growth of algae, and when they die, their decomposition by bacteria uses most of the oxygen in the water, reducing the dissolved oxygen concentration by two-thirds and suffocating fish and other marine life. The nitrogen and other nutrients pouring out of the mouth of the Mississippi River have caused a dead zone in the near-shore Gulf of Mexico that lasts from February to October and varies in size from 3,000 to 8,000 square miles, the latter area about the size of New Jersey or Israel. The size of the dead zone in the summer of 2007 was one of the largest ever recorded. There are twenty-two such zones in U.S. coastal waters.[15] There are no mandatory controls on applications of nitrogen by farmers, and nitrogen pollution in coastal waters is increasing about 1 percent per year.[16] One-third of America's coastal rivers and bays show similar effects on a smaller scale.[17] Internationally there are more than 200 near-shore dead zones, and their number and size have risen every decade since 1970 and seem to be accelerating.

The number jumped by more than a third between 2004 and 2006.[18] The largest one covers 38,000 square miles, an area just slightly smaller than Ohio. According to the National Oceanographic and Atmospheric Administration, nutrient pollution (mostly nitrogen and phosphorous) has degraded more than half of U.S. estuaries.[19] In 2001 the National Research Council named nutrient pollution and the sustainability of fisheries as the most important problems facing the nation's coastal waters in the next decade.

Another unwanted by-product of artificial fertilizers has been heavy metal pollution. An analysis by the U.S. Public Interest Research Group in 2001 revealed that each of twenty-nine commercial fertilizers examined contained twenty-two toxic heavy metals. In twenty of the products, the levels exceeded the limits set for waste sent to public landfills, with especially high amounts of arsenic, lead, mercury, cadmium, and chromium. Dioxin was also exceptionally high.[20]

Sewage Sludge

'Tis a sordid profit that's accompanied by the destruction of health.
—Bernardino Ramazzini, *Treatise on the Diseases of Tradesmen*, 1705

Sludge, also known as biosolids, accumulates as waste products in sewage treatment plants. There is a lot of it because each American generates almost 200 gallons of sewage per day. We use a lot of water to dispose of a small volume of waste. From this input, treatment plants create 8 million tons of sludge each year, two-thirds of it used by crop farmers as fertilizer in crop farming.[21] Compared to animal manure, human sludge contains only 3 percent as much nitrogen, 11 percent as much phosphorous, and 0.2 percent as much potassium,[22] but farmers can get sludge free, and in some cases they may even get paid to take it. However, the idea of using material containing human toilet waste, processed or not, to grow the food we eat does not sit well with most Americans.

Sewage sludge contains pathogens and toxic heavy metals, but despite eighty years of its use as fertilizer in the United States, there have been no documented cases of illness attributed to sewage sludge.[23] This negative result may be due to inadequate investigation by the Environmental Protection Agency (EPA) and the difficulty of isolating the effects of using sewage sludge from other aspects of environmental pollution.[24]

Pesticides

Under the philosophy that now seems to guide our destinies, nothing must get in the way of the man with the spray gun.
—Rachel Carson, *Silent Spring*, 1962

The crops we grow are enjoyed not only by us humans but also by insects and other small creatures that live near the ground surface. We call them pests, but they are part of the natural ecology and no doubt appreciate our efforts to increase and improve their food supply. Our competitors at the dinner table go by such names as corn earworm (the most widespread pest in North America, which also eats tomatoes and green vegetables), European corn borer, corn rootworm, rusty grain beetle, Colorado potato beetle, the familiar grasshopper, and a host of other insects, bacteria, fungi, and viruses.[25] And, of course, there are weeds, defined as plants you do not want. In traditional farming, before the introduction of pesticides, crops were rotated every couple of years and then a field was left fallow after seven years (a biblical command; Leviticus 25:4), effectively killing off pests, bacteria, and viruses by removing their food supply. Most pathogens are specialized to infect a single type of crop plant.[26] Therefore, their populations and the risk of disease can be decreased by crop rotation. However, a few diseases are able to infect more than one plant type, so crop rotation does not affect them. With today's intensive agriculture, continuous monocultures of a single crop are the order of the day, and pesticides are now widely considered essential for pest control.

American farmers spent more than $9 billion in 2005 for more than 900 million pounds of pesticides applied to nearly 100 crops,[27] and the amount of pesticides used increased thirty-three-fold between 1980 and 2000 as farmers waged war against more than 10,000 different kinds of insects, 80,000 plant diseases, and 30,000 species of weeds. Two-thirds of the expenditure was for herbicides, 20 percent for insecticides, 8 percent for fungicides, and 6 percent for other formulations. With increased pesticide use comes heightened resistance, in accord with the familiar concept of survival of the fittest. The enemies with some slight variations in their genetic makeup survive and reproduce, and their descendants require either ever stronger old poisons or new poisons.[28]

Pesticides are classified according to their intended target organism: insecticides (for bugs), herbicides (for weeds), fungicides (for fungi), nematicides (for worms), rodenticides, and miticides (for mites). Birds are serious consumers of crops too, but they are protected by federal laws, unlike rats or grasshoppers.[29] Insecticides and herbicides dominate pesticide applications. Tens of thousands of different pesticides are used in American agriculture. They are used on nearly all of America's 2.2 million farms, and most farmers consider them essential to maintain productivity.

The U.S. Food and Drug Administration has determined that at least fifty-three carcinogenic pesticides are applied in large amounts to major food crops and that in 2005, 73 percent of fresh fruit and vegetables contained pesticide residues, as did 61 percent of processed fruits and vegetables, 22 percent of soybean samples, 75 percent of the wheat samples, 99 percent of milk and cream samples, 8 percent of pork samples, and 16 percent of the bottled water samples.[30] In its study, multiple

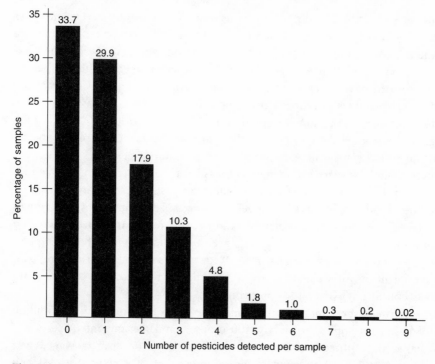

Figure 2.1
Percentage of pesticides detected in food samples. *Source*: U.S. Department of Agriculture, 2006. *Note*: The study examined 13,208 samples of fresh and processed fruit and vegetables, soybeans, wheat flour, milk, and drinking water for a variety of pesticides and growth regulators. Seventy-eight percent of the samples were fruit and vegetables, 10 percent were grains, 6 percent were milk, and 6 percent were drinking water. Eighty-four percent of the samples were of domestic origin.

pesticides were found in more than one-third of the samples (figure 2.1). Eighty-four percent of the samples were produced in the United States, and the remainder imported. Few samples contained pesticide levels higher than permissible federal limits, but many environmentalists believe the tolerances set by the government are too high. If a highway speed limit is set at 100 miles per hour, for example, few people will be classed as speeding. Some scientists have warned of "a pandemic of subclinical neurotoxicity" resulting from the low doses of the multitude of artificial chemicals now present in our bodies. The Centers for Disease Control looked for and found 116 pesticides and other artificial chemicals in human blood and urine, an industrial "body burden" that is passed on to children through prenatal exposure and breast milk. The effects of these chemicals may be difficult to separate from normal mildly debilitating aspects of the aging process.

A study in 2006 by the Environmental Working Group found the "dirty dozen"—the most contaminated fruits and vegetables—to be peaches, apples, bell peppers, celery, nectarines, strawberries, cherries, pears, imported grapes, spinach, lettuce, and potatoes (the most pesticide-intensive U.S. crop).[31] The data were accumulated after the produce was washed or peeled, as typical consumers would do. Safest were onions, avocado, frozen sweet corn, pineapple, mango, asparagus, frozen sweet peas, kiwi, bananas, cabbage, broccoli, and papaya. We have come a long way in our concerns since 1774, when farmers resisted using the cast-iron plow because they feared the metal would contaminate the soil.

Washing lettuce and other produce in water alone will remove only a small amount of the harmful substances; washing them in a dilute solution of vinegar or dishwashing detergent will remove most of these substances without harming the taste of the food. But some pesticides are formulated to bind to the surface of the crop and do not easily wash off. Others are taken internally into the plant (strawberries, for example), and therefore are in the fruit and cannot be washed off. As agricultural philosopher Wendell Berry noted, the germs that used to be in our food have been replaced by poisons. Actually, as we have become aware in recent years with salmonella and hepatitis outbreaks, plus the human form of mad cow disease, harmful bacteria and viruses remain in the food supply.

The EPA estimates that there are 10,000 to 20,000 physician-diagnosed cases of pesticide poisoning of agricultural workers each year in the United States. Because many cases are undiagnosed, the true number of cases may well be higher. Clearly, working with and around these chemicals is hazardous to human health.[32]

About twenty ingredients in pesticides have been found to cause cancer in animals. People with high exposure to pesticides, such as farmers, crop duster pilots, and pesticide manufacturers, have higher rates of blood and lymphatic system cancers, melanoma, and cancers of the lip, stomach, brain, lung, and prostate.[33]

Of the five herbicides used most commonly on corn, atrazine has been identified as the one of most concern to humans. It has been banned by the European Union but is one of the most commonly used pesticides in the United States. Amounts of the chemical that exceed federal drinking water standards have been found in 27 percent of samples from smaller tributaries of the lower Mississippi and Missouri rivers. It was detected in all 147 streams sampled.[34] Atrazine is an endocrine disrupter, an artificial chemical that affects glands and hormones that regulate many bodily functions. It is produced by Syngenta and has been shown to feminize fish, amphibians, reptiles, birds, and mammals and is associated with an increased risk of breast cancer and prostate cancer.[35] The lack of regulatory action by the EPA may be related to the fact that atrazine has a world market worth of over $400 million.

Recent comparison of the mental abilities of children of farmers in Mexico gives cause for concern.[36] The children of farmers who have adopted modern farming techniques, such as the use of fertilizers and pesticides, have noticeably decreased conceptual and physical abilities. These include decreased memory and problem-solving skills, lack of physical stamina in exercise and play, and a threefold increase in illnesses and allergies. Other studies have shown that pesticide poisoning reduces IQ levels.

Why Are Pesticides Needed?

In the United States, about 62 percent of planted acreage is treated at least annually with some kind of pesticide (93 percent of row crops are). Corn, soybeans, and sorghum are the major recipients. The untreated areas are largely forage crops (food for domestic animals such as cattle) that are not planted in rows but cover the ground completely, such as grass, clover, and alfalfa (sod crops). Herbicides are used on nearly four times more acreage than insecticides. But even with the current intensive use of pesticides, 13 percent of attainable production in the United States is currently lost to pests compared to 7 percent in the early 1940s despite the fact that we spend more than $4 billion per year on pest control and pesticides are now much stronger than in the 1940s.[37] The losses are subequally attributed to insects, weeds, and pathogens (microorganisms and viruses). The worldwide average loss is 42 percent because of total loss of crops in Africa and elsewhere.[38] Rice is the crop most severely affected, with 51 percent of the worldwide crop lost, mostly to insects. How much worse might it be without pesticides?

Over the past sixty years, the pesticide industry has tried, with some success, to develop pesticides that are less toxic and more selective in their targets, require lower dosage per acre, and have less persistence, all of which would reduce the many cases of pesticide poisoning of farmers and contamination of the soil. Nevertheless, an estimated 300,000 farm workers suffer from pesticide-related illnesses yearly. Farmers who work with pesticides get Parkinson's disease, have damaged immune systems, have lower quantity and quality of sperm in men, and develop several kinds of cancer more often than the general public does.[39] A British study concluded that more than 10 percent of those who are regularly exposed to organophosphate pesticides (chemical compounds composed of phosphorous and an organic, or carbon-containing, molecule; commercial examples are malathion and parathion) will suffer irreversible physical and mental damage.[40] These pesticides account for about half of all insecticides used in the United States, on crops such as wheat, corn, and many important minor crops.[41] There may be other health risks associated with pesticides, but the lengthy interval between exposure and the observation of chronic effects makes risk assessment for these outcomes more difficult to evaluate than acute effects.

In 2005 scientists discovered that exposure to toxic chemicals while a woman is pregnant affects not only her children but her grandchildren and great-grandchildren as well.[42] This means that if your great-grandmother was exposed to an environmental toxin at a critical point in her pregnancy, you may have inherited the effect of the toxic chemical.

Soluble pesticides drain from farms and are found in aquifers and nearby waterways, where they kill aquatic insects, small invertebrates that are food for fish, and the fish themselves. Agriculture is responsible for 70 percent of the stream pollution in the United States, and many major sources of drinking water are polluted with pesticides.

A 1995 study in the midwestern United States found herbicides in the tap water in twenty-eight of twenty-nine cities tested, and in more than half of them, herbicide levels exceeded governmental safety standards.[43] The source of the herbicides is not clear. A report by the U.S. Geological Survey in 2005 found that pesticides are typically present throughout the year in most streams and urban and agricultural areas of the United States but are less common in groundwater.[44] But the amounts of individual pesticides were seldom at concentrations likely to affect humans. Whether the mixtures of the pesticides found are harmful is not known.

Pesticide use in the United States is increasing at about 5 percent per year.[45] However, the increase in consumption does not accurately reflect the amount actually applied to pests. Pesticide delivery systems are varied, but all are very inefficient. For example, two-thirds of all pesticides applied in the United States are sprayed by aircraft, and even under ideal conditions, much of the pesticide released by aerial spraying does not reach its target because it drifts out of the target area. According to some ecologists, less than 1 percent of applied pesticides actually reaches the target pests.[46]

There are no health data for many pesticides for toxicity, cancer-causing potential, or endocrine disruption. More than 2,000 chemicals enter the market every year, and most of them do not go through even the simplest tests to determine toxicity.[47] And when testing is done, the chemicals are tested individually, despite the fact that chemicals ingested in combination can be much more harmful than the chemicals are individually.[48] The large number of chemicals precludes testing combinations of pesticides. To test just the commonest 1,000 toxic chemicals in combinations of three at a standardized dosage would require at least 166 million different experiments. Using different dosages would greatly increase the number of needed experiments. There is no solution to the testing problem. Even the ingredients listed on pesticide labels as inert may not be safe for humans. According to a survey by the Northwest Coalition for Alternatives to Pesticides, about a quarter of substances labeled as inert are not, because they are classified as hazardous under federal statutes.[49]

Endocrine Disrupters

High on the list of dangerous pesticides are the endocrine disrupters.[50] The human endocrine system is a complex network of glands and hormones that regulates many bodily functions, including growth, development and maturation, as well as the way various organs operate. The endocrine glands, which include the pituitary, thyroid, adrenal, thymus, pancreas, ovaries, and testes, release carefully measured amounts of hormones into the bloodstream that act as chemical messengers, traveling to different parts of the body in order to control and adjust many life functions. At least forty chemical compounds used in pesticides are known endocrine disrupters, based on animal studies with rats, mice, fish, dogs, reptiles, and birds. Disruption can occur by halting or stimulating excess production of hormones or changing the way the hormones travel through the body, thus affecting their function.

Sexual and reproductive abnormalities are among the harmful effects endocrine disrupters are known to cause. Although there is no proof that the harmful effects the chemicals are known to cause in other vertebrates cause them in humans as well, many scientists believe it is a strong possibility.

Unfortunately, even some pesticides that have been tested for toxicity and found to be harmful to humans and other organisms can be legally used. One well-known example is Paraquat, which appeared in 1959 and is widely used to control weeds in fruit orchards, and fields of alfalfa, onion, leeks, sugar beets, and asparagus. The EPA classifies Paraquat as a possible human carcinogen but has concluded that the risks posed to individual applicators are minimal and of no concern. It is classed as "moderately hazardous" by the World Health Organization. Poisoning is possible by skin exposure or inhalation, and the chemical attacks the lungs, liver, and kidneys. Experiments have shown that Paraquat is fatal to frogs and tadpoles at the lowest dose tested and kills honeybees at doses lower than those used for weed control. Nevertheless, the producer of this chemical in the UK has published brochures claiming that the herbicide works "in perfect harmony" with nature and is "environmentally friendly."[51]

In 2005, Japanese researchers developed a class of pesticides that immobilize and eventually kill more than half a dozen species of destructive insects, including some that are resistant to other insecticides.[52] Even at concentrations more than a hundred times greater than needed to poison these pests, the pesticide does not appear to harm rats, honeybees, spiders, or any of several beneficial predatory insect species.

The Effectiveness of Agricultural Pesticides

The yields of most crops have increased dramatically in recent decades (figure 2.2), which most farmers attribute in large part to the use of pesticides. There are hundreds of documented cases where chemical pesticides have reduced losses of crops. For example, without chemical control of weeds in monocultured wheat production, yields in the United States would fall by 30 percent, and without herbicides and fungicides, wheat yields would fall by 5 percent. Global losses in monocultured fields

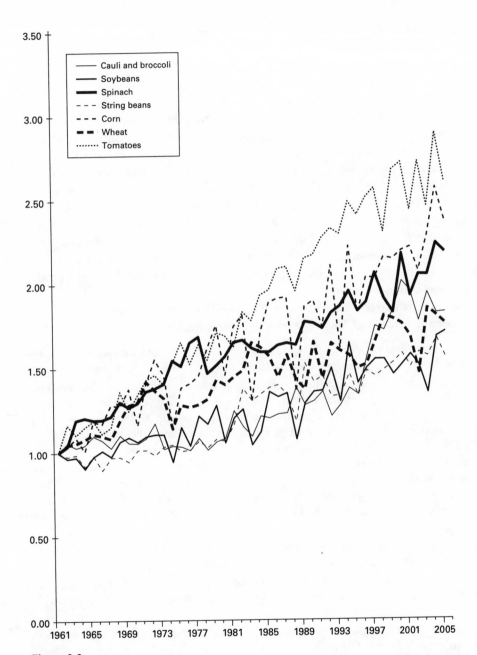

Figure 2.2
Increases in the yields of assorted crops in the United States since 1961. *Source*: Organic Center, 2007. *Note*: The base year is 1961.

would rise from today's levels of around 42 percent to nearly 70 percent in the absence of chemical pesticides.[53]

Data from the U.S. Department of Agriculture reveal a tenfold increase in both the amount and toxicity of insecticide use in the United States from the early 1940s to the 1990s. Paradoxically, during the same period, crop losses from insects rose from 7 percent to 13 percent[54]; losses to plant pathogens grew from 10 percent to 12 percent, and losses from weeds decreased from about 14 percent to 12 percent. Increases in losses from pests in the corn crop confirm this perverse correlation between increased use of ever more potent pesticides and an increased proportion of pest-induced losses. In 1945, when very little insecticide was used, losses were estimated to be around 3.5 percent of the crop, but by the later 1990s, when insecticide use had increased massively, corn crop losses were estimated to be around 12 percent.[55] In 1999, pest resistance to pesticides was estimated to cost American agriculture about $1.5 billion in increased pesticide costs and decreased yields. The positive correlation between increased use of pesticides and increased crop losses is a worldwide phenomenon.

How is it possible for losses to pests to increase when the amount and toxicity of pesticides are increasing? There are several possible explanations:[56]

• The industrialization of agriculture and the reliance on agrochemicals have produced higher yields but also have led to increased vulnerability of crops to pests.
• There are now more monocultures, less genetic diversity in crops, and less crop rotation. Planting the same field with the same crop year after year allows pest populations to build up. Many pests are specific to a certain crop and appreciate having a stable food supply. The buildup of pests can be broken by following, say, a deep-rooted plant with a shallow-rooted one, a broadleaf with a grass, or a crop that is harvested early in the year with one that is harvested late. Corn followed by soybeans is a traditional and commonly used rotation. However, recent evidence suggests that some insects can adapt to accommodate a regular, predictable rotation, for example, alternating years, and can change their life cycle to match the rotation cycle.
• There has been a reduction in tillage with more crop residues left on the land surface. This means more food for the pests to eat year round.
• Crops are increasingly being produced in climatic regions where they are more susceptible to insect attack. To increase production, farmers are extending the borders of farms into less favorable areas.
• Herbicides are being used that alter the physiology of crop plants, making them more vulnerable to insect attacks.
• Pests are becoming more resistant to pesticides, a normal survival-of-the-fittest phenomenon. This effect has worsened over time. In 1938, there were 7 insect and mite species known to be resistant to pesticides. By 1984, this number had risen to

477. Resistant weeds were unknown in 1970 but numbered 48 in the late 1980s. By 1999, about 1,000 species of insects, pathogens, and weeds were resistant to commonly applied pesticides. The resistant species include 550 insect and mite species, 230 plant diseases, and 220 weeds.[57]

Considering the recommended dosages of pesticide used on corn in the United States, the impact on nontarget pests has led to a threefold increase in aphids, a 35 percent increase in corn borers, a fivefold increase in corn smut disease (a fungus), and a total loss of resistance to southern corn leaf blight.[58] Farmers and pesticide producers have locked themselves in a race with rapid pest adaptation and evolution, an evolutionary arms race. The pests may be winning. As one weed expert noted, there are weeds in Australia that are immune to seven different types of weed killers. The Australians joke that they have weeds immune to weed killers that have yet to be invented. After more than fifty years of this evolutionary rivalry, there is abundant evidence that pests of all sorts—insects, weeds, or pathogens—develop resistance to just about any chemical that humans throw at them. Successes in the battle against crop pests are only temporary; the war is endless. As Rachel Carson wrote decades ago in *Silent Spring*, "The chemical war cannot be won, and all life is caught in its violent crossfire."

The problem with pest resistance to pesticides was highlighted in 1998 by a study in England. The researchers examined grain from 279 storage silos and found that 81 percent of them contained mites and 27 percent contained beetles.[59] The investigators also studied 567 cereal-based food products in stores: bread, breakfast cereals, and cookies. Twenty-one percent of these foods contained mites. The technique used to sample the mites from food kills the creatures, so the researchers do not know which of the foods contained living mites.

Although mites are tiny and do not add much bulk to a diet, this cannot be said of other creatures eaten regularly in many parts of the world. For many of the world's peoples, contamination of the grain crop by tiny creatures simply adds another type of food to their diets. For example, edible insects such as caterpillars, grasshoppers, and grubs (insect larvae) are rich in protein, and according to the United Nations Food and Agriculture Organization, their use should be encouraged.[60] North America is one of the few places in the world that does not use insects as a significant food source. But they are cheap, easy to prepare, and abundant. Many environmentalists have suggested bugs could solve the world's hunger problems without damaging ecosystems. What is needed is an economical way to raise, process, and package them and, probably most difficult, make them accepted as foodstuffs.

The development of a new pesticide takes many years for all the testing and licensing, and switching to a new compound is a short-term solution in any case because resistance arises again much more rapidly. Development of a new pesticide takes, on average, $80 million, but the typical time before a pest develops resistance is only ten to twenty-five years, after which the pesticide's usefulness decreases.[61]

• With increasing international trade and transport, new pests sneak across borders and oceans and conduct crop terrorism. About 40 percent of U.S. insect pests are nonnative, as are 40 percent of our weeds and 70 percent of our plant pathogens.[62] Are these unintended imports a harbinger of something a human bioterrorist might do?

How do pests become immune to a pesticide? The ubiquitous corn earworm caterpillar, now largely resistant to most insecticides, serves as an example.[63] When the pest starts munching on a plant, it immediately activates the plant's natural defenses. First, the plant releases signaling chemicals, which stimulate the production of toxins within the plant that should kill the pest when it eats them. But the earworm counterattacks. The chemicals the plant uses for its toxic defense switch on a set of genes inside the pest. These genes lead to the production of enzymes in the caterpillar's gut that break down the plant's toxins, and many synthetic pesticides, into harmless by-products.

In the absence of pesticides, plants have developed their own protective systems. For example, when a caterpillar chomps on a cornstalk, the plant releases a chemical SOS that summons tiny parasitic wasps no longer than an eyelash.[64] The wasps land on the caterpillar and lay eggs. Within two days, the eggs hatch, and larval wasps crawl into the caterpillar and eat it. Nature has its own mechanisms for protecting plant offspring. Researchers have identified the gene in corn that generates the plant's SOS signal, have transplanted it into another plant, and found that a squadron of wasps appeared when needed.

• Pesticides kill not only the target organism but also their natural enemies, creatures that normally keep the unwanted insect under control.

For example, the brown planthopper, a quarter-inch-long sucking bug that attacks rice plants, was almost unknown in Asia before the introduction of rice varieties created artificially. Scientists discovered that the planthopper is normally controlled by parasites that destroy its eggs and by the wolf spider that preys on the planthopper itself. But the pesticides sprayed on the new rice variety killed these natural predators along with the planthoppers. In some areas the planthopper population rose in direct proportion to the amount of insecticide sprayed. The most successful defense against the planthopper was found to be an ecological one: farmers used the minimum amount of pesticide that also allowed natural predators to help in destroying the invaders. When this combined approach was used, overall yields rose by 15 percent as pesticide use declined by 60 percent. The lesson here is that working with nature is superior to working against it. Nature knows best how to maintain stability in natural systems.

• Quality controls in the marketplace have become more demanding, resulting in a higher proportion of crops being classed as pest damaged. Everyone today expects picture-perfect fruits and vegetables. And so comparison between the real-world appearance of these foods in the field, on the tree, or on the vine, and their perfect, unblemished appearance in the supermarket shows that culling has been significant between the farm and the market.

The cost-effectiveness of pesticides, like the cost-effectiveness of different forms of energy, can be calculated in two ways, with very different results. In 1992, it was estimated that each dollar invested in pesticide controls returned four dollars in saved crops. But the picture changes in estimating the dollar costs associated with pesticide effects on human health, water pollution, natural ecosystems, wildlife, soil fertility and structure, interference with natural pest controls and pollinators, and government inspection and controls associated with pesticides. This latter group of costs is termed "external costs," and is estimated to be $4 billion to $10 billion per year.[65]

Industrial agriculture has severely changed the natural ecology, and changes in one part of an ecosystem affect the rest of the system, often in unexpected ways.[66] As ecologists say, everything is connected to everything else. Thus, pesticides should be used sparingly and targeted accurately. They are useful but dangerous to non-target plants, animals, and humans, and it is important to determine the appropriate balance between the need for improved crop production and the unintended effects of pesticide use. This includes killing natural predators and parasites that may have been maintaining the population of a pest species at a reasonable level. These problems have stimulated the growing interest in chemical-free organic farming (chapter 4).

Recently a new and potentially serious effect of pesticides has been discovered.[67] One of the pesticides used on alfalfa interferes with nitrogen fixation by the bacteria that inhabit the roots of this legume, lowering crop yields by more than a third. If legumes are unable to restore the nitrogen content of the soil, additional nitrogenous fertilizer will have to be added. This will increase both the cost of growing crops and the number and extent of near-shore dead zones.

Crop Farming and Crude Oil

The bottom of the oil barrel is now visible.
—Christopher Flavin, environmentalist, 1985

U.S. agriculture, once fully dependent on energy from the sun, has become so reliant on fossil fuels that it now takes more than 1 calorie of fossil fuel to produce 1 calorie of food. Eighteen percent of the oil Americans use in a year is devoted to food production.[68] The manufacture of fertilizers and pesticides could not exist without petroleum and natural gas, which are used as both sources of energy for production and a base from which the fertilizers and pesticides are produced. A lot of these essentials are manufactured, and a lot of crude oil and natural gas are used. In addition, farmers use motorized equipment—tractors, combines, harvesters, irrigation pumps—which use gasoline or diesel fuel.

Growing food accounts for one-fifth of agriculture's oil consumption. The other four-fifths are used to move, process, package, sell, and store food after it leaves the farm. About 28 percent of the energy used in agriculture goes to manufacture fertilizer, 7 percent goes to irrigation, and 34 percent is consumed as diesel and gasoline by farm vehicles used to plant, till, and harvest crops. The rest goes to pesticide production, grain drying, and facility operations.[69]

From farm to plate, the modern food system relies heavily on cheap oil. Threats to the oil supply are also threats to the food supply. As food undergoes more processing and travels farther, the food system consumes ever more energy each year. As fossil fuels increase in price, the cost of food is likely to increase. And the cost of fossil fuels has been skyrocketing in recent years. A barrel (42 gallons) of crude oil sold for about $29 at the end of 2002, $32 a year later, and $44 in 2004. It then hovered around $60 at the end of 2006 and more than $80 at the end of 2007, and hit $100 early in 2008.

The commercial production of nitrogen fertilizers is energy intensive and costly, because gaseous nitrogen must be combined under high temperature and pressure with hydrogen from natural gas (methane) to produce the nitrogen-bearing fertilizers. Hence, the cost of inorganic nitrogen fertilizer varies considerably with the price of natural gas, which doubled between 2002 and 2006. It averaged about $3.50 per 1,000 cubic feet in 2002 and rose to $7 in 2006. The costs of both oil and natural gas seem able to keep rising indefinitely. The cost of nitrogen fertilizer in 2007 was between 130 percent and 150 percent higher than in 2000; phosphate increased 79 percent and potassium 70 percent.

Nevertheless, ways to reduce agricultural dependence on fossil fuels are readily available, given modern technology and traditional, effective farming practices. With regard to fertilizers, it is estimated that livestock manures contain five times the amount of fertilizer currently in use each year.[70] But practically all large farms these days contain only crops or animals, not both. Hydrocarbon-based pesticides can be replaced by time-honored practices such as alternating crop types each year, eliminating monocropping, and regularly letting land remain fallow. The gasoline- and diesel-powered wheeled farm equipment can be converted to solar power, for example, by installing panels on the roofs of the tractors. In other words, America's farmers need to convert to sustainable agriculture.

Soil Erosion: The Dirt Is Moving

When the soil is gone, men must go; and the process does not take long.
—President Theodore Roosevelt, 1908

Erosion is the process by which running water, wind, ice, or other geological agents wear away the land surface. The first thing to go is the uppermost layer, the topsoil. By removing topsoil, erosion reduces the capacity of the soil to support plant life and undermines the ability of farmers to increase food production. Because soil erosion happens so slowly, it is hardly noticed during one generation of say, thirty years. Soil appears to be here forever.

Industrial agriculture has greatly increased the rate of soil erosion (removal) and deterioration (loss of fertility), and many soil scientists and environmentalists are concerned. We see the effects of human interference in the mountains of surface sediment, much of it productive topsoil, washed from plowed agricultural areas bordering the muddy Mississippi River. Approximately 40 percent of the mud in U.S. streams comes directly from cultivated land; another 12 percent emanates from pasture and rangeland. A further 26 percent comes from the erosion of stream banks,[71] and much of this mud probably originated in farmland. Water and wind wash or blow away tens of billions of tons of topsoil every year. Some estimates say that 90 percent of U.S. cropland is losing soil faster than nature can replace it.[72]

Although all researchers agree that farmland in general is being eroded at a noticeable rate and that most of the erosion comes from agricultural land, there is considerable disagreement about the rate and amount of loss.[73] Some researchers believe the loss is small but needs monitoring. Others see the loss as rapid and a serious threat to the nation's agricultural productivity. Such conclusions are difficult to verify because nearly 75 percent of the topsoil washed from one farmer's land is not carried directly to the ocean but is deposited downstream on another farmer's land, and then perhaps on another's.[74] It may take hundreds of years or more for the soil that started upstream to reach the ocean. In addition, the loss of productive potential owing to erosion can be masked for a time by increased inputs of fertilizer, irrigation, and higher-yielding plant varieties. But all investigators agree that more measurements are needed. Existing data are not adequate to determine the seriousness of America's soil erosion problem.

The rate of soil loss on ground surfaces is greatly increased by farming because tilling the soil loosens it, making it more easily moved. Two-thirds of America's soil degradation results from crop farming and nearly all the rest from overgrazing.[75] Furthermore, tilling rips from the soil last season's plant roots, and it is roots that hold soil in place. In the Upper Mississippi valley, conversion of the region's natural landscape to primarily agricultural uses has boosted surface erosion rates to values three to eight times greater than those characteristic of presettlement conditions.[76] The amount of soil eroded from plowed ground is fifty to a hundred times more than is eroded from unplowed grassland. We have lost about one-third of our total national topsoil since farming started about 350 years ago.[77] It is estimated that because of erosion, it now requires 2 bushels of Iowa topsoil to grow 1 bushel of corn.[78] However, the rate of erosion on U.S. cropland declined 42 by percent between 1982 and 2004, in large part because of conservation tillage,[79] and the removal from production of the most highly erodible cropland under the federal government's Conservation Reserve Program, which pays farmers of highly erodible land to maintain permanent cover on the land.[80]

The old saying, "An ounce of prevention is worth a pound of cure," certainly applies to erosion problems. It is much easier to limit erosion to tolerable rates than it is to repair eroded land and to wait hundreds of years for new topsoil to form. One of the most important keys to a successful erosion control program is to predict when and where excessive erosion is likely to occur so that something can be done to prevent it.

Crop rotations are effective in controlling erosion and were used by American farmers for 300 years, until the advent of industrial monocultures in the mid-twentieth century. Yearly alternations of corn or beans with mixed grain or winter wheat reduces erosion by 40 percent. Alternating corn or beans with hay pasture generates a 70 to 90 percent reduction.[81] Clearly fallow land (uncultivated) suffers much less from erosion than plowed land. The biblical directive to let land lie fallow every seven years was sound advice. Crop rotations have an additional benefit because they remove the specifically required food supply of last year's insects. Also, most pathogens (disease-producing microorganisms or viruses) are crop specific and do not infect multiple crops.

Another effective way to reduce soil losses to erosion is no-till cultivation. In conventional farming, the results of the previous year's crop are plowed into the soil, destroying the roots and exposing a vulnerable soil surface to erosion. This traditional plowing method overturns up to 8 inches of soil. This surface is then smoothed, the seeds of the new crop planted, and the soil between the crop rows stirred to rip out weeds. The soil is left unprotected until the new seeds germinate and develop roots, which takes several months in the early spring. The soil is thus left bare and subject to significant erosion by spring rains. Conventional plowing and planting leaves only 1 to 5 percent of the soil protected by crop residue. A residue of just 20 percent reduces soil erosion by 50 percent; a cover of 50 percent reduces it by 80 percent.[82]

No-till farming is spreading rapidly. A principal objective of no-till farming is to reduce the amount of soil disturbance and leave the soil covered, and therefore less exposed to removal by wind and rain. Residue cover is one of the most effective and least expensive methods for reducing soil erosion. No-till planting systems, which leave the greatest amount of residue cover, can reduce soil erosion by 90 to 95 percent of that which occurs from cleanly tilled, residue-free fields.[83] No-till farming stirs less than 2 inches of topsoil in creating crop rows,[84] and there may be no plowing of the soil at all. Seeds are planted directly over the residue of last year's crop, regardless of whether the new crop is the same or different from the previous one. The soil is thus protected from erosion by spring rains; runoff is reduced by more than 95 percent. In addition, the lack of plowing leaves last year's plant roots intact, which reduces runoff and retains topsoil, and markedly decreases the production of wormburger meat.

As of 2002, reduced- or no-till farming was being used on nearly half of agricultural land in the central Great Plains, but adoption in other areas of the United States has been slow. No-till plots account for only 23 percent of our farmland.[85] One reason American farmers have been slow to adopt no-till farming, scientists say, is that government subsidies make them indifferent to the competitive advantage no-till gives foreign producers. Government subsidies decrease the farmers' incentive to make needed changes in agricultural practices.

Experience has shown that no-till farming results in less nutrient loss, increases in organic matter, preservation of porosity and permeability because of the reduced compaction of the topsoil and an increase in the worm population, higher crop yields, and lowered use of water, pesticides, and tractor fuel.[86] Total emissions of carbon dioxide from the decreased use of fossil fuel in a no-till system are 92 percent lower than in conventional agriculture.[87] The higher crop yields and lowered use of water occur because lessening the amount of tillage increases the amount of organic matter retained in the soil by 15 to 30 percent. And organic matter acts like a sponge and can absorb six times its weight in water.[88] Stirring the soil increases the oxygen supply in the pores, which speeds up the action of soil microbes that feed on organic matter.

No-till farming helped reduce water erosion by more than 40 percent between 1982 and 2004.[89] The amount of reduction varies with the type of crop residue, with corn residue being particularly effective. Midwestern farmers get consistent reductions of about 75 percent in soil loss when corn residue is left on the surface.[90] The implementation of appropriate soil and water conservation practices has the potential to essentially stop erosion completely and reduce water loss by at least 30 percent.[91] In 2000 no-till or reduced-till farming was practiced on 37 percent of the corn crop, 57 percent of the soybean crop, and 60 percent of the double-crop beans.[92]

Preservation of organic matter in the soil was less of a problem a hundred years ago when most farms raised cattle as well as crops, and abundant manure was available to be used as a soil supplement. In today's age of specialization, few farms raise both cattle and crops so that preservation of the soil's natural organic matter is a more pressing concern. Materials such as crop residues, commercially available animal manure, so-called green manure (leguminous plants such as alfalfa and red clover, and grasses), compost, peat, and wood chips can be added to maintain or immediately increase soil organic matter.

Clearly, the way crops are produced today differs greatly from the way they were produced a hundred years ago, particularly because of mono-cropping and the massive application of artificial fertilizers and pesticides. In the next chapter we will examine the effect of these changes on the production of humans' most basic crop—grain.

3

Grain Farming: The Basic Crop

Oh, beautiful for spacious skies, for amber waves of grain, for purple mountain's majesty, above the fruited plain.
—Katharine Bates, "America the Beautiful," 1913

Bringing in the sheaves, bringing in the sheaves [bundles of grain], we shall come rejoicing, bringing in the sheaves.
—George Minor, "Bringing in the Sheaves," 1874

America's grain production has been the envy of the rest of the world, particularly production of the major crops of soybeans, corn, and wheat. The huge annual crop has been produced with federal subsidies by industrial agriculture, in which the motto seems to be to obtain the largest yearly harvest possible without regard to the long-term effect on soil health or pesticide pollution. But the effects of this policy of industrial agriculture are becoming apparent: plants will not accept additional fertilizer, pesticides are present in the food we eat, the soil is being depleted of the organisms that produce soil organic matter, and the loss of topsoil to erosion is a serious problem on America's farms. The way farms are managed today is not sustainable, and we are all paying the price for current mismanagement. Our children and grandchildren will pay an even heavier price.

Most edible grains are members of the grass family, the largest in the plant kingdom (10,000 species), grown for their large edible seeds or kernels. The seeds are actually a fruit called a grain. Because they have kernels, the edible grains differ from their cousins, the lawn grass that covers 2 percent of the U.S. land surface.[1] In 2005, the harvested area of the major edible grains, soybeans, corn, wheat, and rice, covered 74, 71, 51, and 4 million acres, respectively. Harvested acreage may rise or fall a few percent each year depending on variations in rainfall, federal subsidies, international price variations, and other factors. Minor edible grains include sorghum, barley, oat, rye, triticale, and millet. Only soybeans are not considered a cereal crop.

Soybeans

Run, they're handing out tofu again!
—Panic scream of child trick-or-treating at Halloween

The legume we call soybean is often referred to as a miracle crop, and the United States is the world's largest producer and exporter. It contains much more protein than any other vegetable food (38 percent versus 10 to 15 percent) and is the world's chief provider of protein and oil. Unlike the cereal grains with their kernels, soybeans grow in pods similar to peas with two to four beans in each pod, contain an almost perfect balance of all eight of the essential amino acids, are an excellent source of many vitamins and minerals, inhibit the growth of cancerous tumors, and lower the bad cholesterol known as low-density lipoprotein (LDL) in human blood.[2] In addition, processed soybeans provide an environmentally friendly fuel for diesel engines, that is, biodiesel.

Soybeans as a Food Crop

Soybeans were virtually unknown as a food crop before 1920, when only 500,000 acres were planted. But by the early 1940s, the crop was planted on 10 million acres, and the area planted has increased steadily since then to 40 million acres in 1970 and 70 million in 2007.[3] In 2004, soybeans were planted on 28 percent of the total U.S. crop area, more than 75 million acres of soybeans. But in 2007, planted soybean acreage fell 11 percent as farmers planted more corn in response to the ethanol boom. Most soybean production is in the Upper Midwest. The United States produces 40 percent of the world's soybeans and is approximately tied with Brazil as the leading exporter of this nutritious bean; we export about one-third of our crop as either the beans or products made from them.[4] However, the USDA forecasts that America's soybean exports will fall by 23 percent by the 2009–2010 crop year, as acreage devoted to corn (and ethanol) increases

Europe, although resistant to most genetically modified (GM) crops, has approved U.S. soy products, perhaps because they use it to feed livestock rather than as food for humans, but their imports from American farms have fallen 50 percent since 1995 as their supermarkets switched to non-GM sources. Another factor in the reduced exports to Europe has been increased soybean production in other exporting countries: 38 percent since the 2000–2001 crop year.[5] Since 1990, the number of bushels exported from the United States has nearly doubled, while domestic consumption has remained the same. As was the case with the other major grain crops, exports are critical to the survival of American farmers.

Commercial fertilizer is applied to less than 40 percent of soybean acreage, and only 21 percent of the acreage receives nitrogen,[6] a much lower rate for both than for most other row crops because soybeans are legumes and can fix their own nitrogen. Soybeans are most commonly grown in a crop rotation with corn and, like corn, are heavily dowsed with pesticides. Pesticide use (mostly herbicides) for soybeans ranks second only to corn.

The exceptional nutritional value of soybeans was recognized in the 1930s by Clive McCay, a professor at Cornell University.[7] He was concerned about the poor quality of bread that was (and still is) being consumed in America, and his experiments with laboratory animals and dozens of dogs demonstrated that supplements of soybean products, dry milk, and wheat germ often doubled the life span of the smaller animals. Rats fed on a diet consisting exclusively of his "Cornell bread" and butter were healthier and lived longer than rats fed the same amount of store-bought commercial bread and butter. The rats fed the commercial bread were stunted and soon sickened and died. Because Cornell Bread, commercially named Triple Rich, contained his added healthful ingredients, that era's Food and Drug Administration (FDA) ruled that it could not be called "bread" and must be labeled "artificial." The product soon disappeared from stores. (The recipe and instructions for baking it are available to home bread bakers.)

Americans eat only 4 percent of the U.S. soybean crop despite its superior health benefits.[8] Animals eat nearly all of our soybean production as soybean meal. There are probably several reasons that American do not eat soybean products. Although abundant research documents the health benefits of soy, no new health claims have yet been formally approved, and although soy foods are gaining popularity among the young, they must compete with pizza and other pasta products. Perhaps the greatest obstacle soy products face is the reluctance of people to change their food habits. Changing from familiar beef hamburgers to the avant-garde soyburgers takes courage and leadership. And finding a restaurant that serves them can be a challenge. The relatively small amount of soy eaten by people is in the form of tofu, soy sauce, soybean oil, and soy flour used in some baked goods and high-fiber breads.

Many traditional dairy products have been imitated using soybeans, and products such as soy nuts, soy milk, soy yogurt, and soy cream cheese are available in some supermarkets. However, the main use of soybean products is as soy meal, a low-cost, high-protein food for farm animals.[9] Half goes to poultry, 25 percent to swine, 12 percent to cattle, and the rest to other uses. Soybean meal is the world's most important protein feed, accounting for two-thirds of world supplies. It grows in thirty-one states.

Nearly all soybeans are crushed to extract the oil from the resulting meal. One bushel of soybeans (60 pounds) yields 79 percent meal, 18 percent oil, and 3 percent waste.[10] The oil is the leading vegetable oil in the world and is found in margarine, salad dressing, and mayonnaise. Soybean oil accounts for 80 percent of all the vegetable oils and animal fats consumed in the United States.[11] In addition, the protein is used in such industrial products as plastics, soap, glue, ink, paint, cosmetics, and textile fibers.

Soybean Biodiesel

One acre of soybeans yields 60 gallons of biodiesel. A comprehensive analysis of the full life cycles of soybean biodiesel and corn ethanol, a competitive biofuel, reveals that biodiesel has much less of an impact on the environment and a much higher net energy benefit than corn ethanol.[12] Soybean biodiesel returns 93 percent more energy than is used to produce it, while ethanol made from corn provides only 25 percent more energy. Although diesel-powered passenger cars make up only 3 percent of sales in the United States (Americans use 3.4 times as much gasoline as diesel fuel), nearly all heavy-duty trucks, buses, ships, and farm equipment run on diesel fuel.[13]

The biodiesel industry is growing rapidly, from fewer than ten plants in 2000 to sixty-five in 2006 to one hundred seventy-two in January 2008. Biodiesel costs one dollar more per gallon to produce than diesel from fossil fuels and needs government subsidies to be economically competitive. However, the cost of the tax subsidy has to be balanced against the cost to the American economy of importing crude oil from foreign sources. The accounting should include not only the price of a barrel of crude oil, but also the cost of supply-line protection, military preparedness, and wars, such as the current one in Iraq, fought to protect U.S. oil supplies. The cost of this latest oil war (so far) is about half a trillion dollars. These costs are rarely considered in comparisons of the cost of biofuels versus the cost of imported petroleum.

Because of their molecular structure, soybeans are not suitable for making ethanol, and this is the reason corn is used. Similarly, corn is unsuitable for producing biodiesel. However, neither biofuel can solve America's gasoline crisis. Dedicating all current U.S. corn and soybean production to biofuels would meet only 12 percent of gasoline demand and 6 percent of diesel demand. Meanwhile, global population growth will increase demand for corn and soybeans for food.

The environmental impacts of the two biofuels also differ. Soybean biodiesel produces 41 percent less greenhouse gas emissions than diesel fuel, whereas corn grain ethanol reduces these emissions from gasoline by only 12 percent. And soybean growth requires much less nitrogen and pesticides than corn, which reduces soil and water pollution.

Corn

Sex is good, but not as good as fresh sweet corn.
—Garrison Keillor, prairie humorist

Table 3.1
Corn production in the United States, 2006

State	Bushels (billions)	Acres harvested (millions)	Yield (bushels/acre)
Iowa	2.1	12.4	166
Illinois	1.8	11.2	163
Nebraska	1.2	8.1	152
Minnesota	1.1	7.3	161
Indiana	.84	5.5	157
Ohio	0.47	3.2	159
Wisconsin	0.40	3.7	143
Missouri	0.36	2.7	138
Kansas	0.35	3.4	115
South Dakota	0.31	4.5	97
Total United States	10.5	70.6	149

Source: National Corn Growers Association, 2007.

Since World War II, the United States has been known as the breadbasket of the world, a title bestowed because of its enormous wheat production, the grain from which nearly all bread is made. But this has changed. America is now corn country. Even in Kansas, known for a century as the Wheat State, corn production pulled ahead of wheat in 2000, with Kansas producing 23 percent more corn than wheat in 2005. The growth rate in yield for corn production is four times that of wheat.[14] Today in the United States, two farmers out of every three, and 1 cultivated acre out of every 4, grow corn. Corn production is nearly four times that of any other crop and accounts for two-thirds of all the grain produced in the United States. Domestic use has increased every year since 1995. Iowa is the biggest producer, with Illinois close behind; four states produce more than 50 percent of the crop (table 3.1). In 2007, 90.5 million acres of corn were planted, 12 million more than in 2006 (and the most since 1944), with 30,000 to 35,000 plants per acre, or 2.7 to 3.2 trillion corn plants. The yield per acre has nearly quadrupled since 1950 from 40 bushels per acre to 149. The total crop harvested in 2006 was 10.5 billion bushels.[15]

Corn production dominates American grain farming for a number of reasons: federal farm policy, the genetics of corn growth, genetic engineering, technological advances in farm machinery, increased irrigation, global warming, and, most recently, the explosive growth of the biofuels industry. Unfortunately, this massive corn production has had negative environmental effects.

Federal Farm Policy

Almost all the subsidy money to farmers of edible crops is given to encourage the growth of corn, wheat, soybeans, and rice, and corn growers in 2004 received 62 percent of this money: $4.5 billion.[16] Wheat received only 17 percent, soybeans 13 percent, and rice 9 percent. In addition, the federal government rewards high corn production by guaranteeing growers the repayment of loans that become deficient when prices fall below the government loan rate of $1.95 per bushel. Corn prices that fell below $2.00 a bushel in recent years led to record payments to farmers. However, the recent ethanol boom has raised corn prices from a historically stable $2.00 per bushel to $4.00 a bushel, and it seems unlikely ever to drop below $1.95 again. Wheat prices have generally averaged $3.00 a bushel, above the $2.75 a bushel government floor.

Corn's higher yields and better subsidy support mean that it is much more economically attractive to grow corn than wheat. Wheat subsidies have not been adjusted much to account for the increased yields corn farmers obtain because of genetic engineering. Obviously the government wants farmers to grow corn, and farmers have responded. Corn production increased from 6.6 billion bushels in 1980 to 7.9 billion in 1990, 9.9 billion in 2000, and 11 to 12 billion in 2004 and 2005.

Federal farm policy has had important negative effects on the health of the American public.[17] The emphasis on corn and soybean subsidies has encouraged farmers to produce them in enormous quantities, to the detriment of the production of fruits and vegetables, which are not subsidized. As a result, the cost of fruits and vegetables to the supermarket shopper has increased nearly 40 percent in the past twenty years, while the cost of corn and soybeans has remained low. Corn and soybeans are the feedstocks for artificial sugar (high-fructose corn syrup, six times sweeter than sugar and one-third the price), fats, and oils, most of which are unhealthy. The consumption of high-fructose corn syrup has increased over 1,000 percent in the past thirty years, and consumption of added fats shot up more than 35 percent as the cost of high-fructose corn syrup, fats, and oils decreased by 10 percent over the same period. The unhealthy but cheap hamburgers and greasy french fries at fast food outlets are made possible by the feeding of most of America's cheap corn and soy meal to animals rather than humans. This misguided farm policy has made poor eating habits an economically sensible choice in the short term, a choice encouraged by food advertisers that spend twenty times more to promote their often unhealthy products than the USDA spends on nutrition education. (The resulting obesity epidemic is treated in some detail in chapter 11.)

Genetic Characteristics

On hot, still midsummer nights in the corn belt, farmers insist they can hear the corn growing. This facetious claim emphasizes the incredible speed at which corn grows, sometimes 4 inches a day. By late summer the corn may indeed be, as Richard Rogers and Oscar Hammerstein said, "as high as an elephant's eye, and it looks like it's climbin' clear up to the sky," perhaps 12 feet tall.

An acre of corn requires only 8 percent as much seed as an acre of wheat, yet the yield of grain is four times greater.[18] Also, wheat is subject to more diseases than other grains, and in wet seasons heavier losses are sustained from these diseases than is the case in other cereal crops. An added benefit of growing corn is that essentially everything in corn farming is now mechanized, from planting to harvesting to shelling the kernels from the cobs.

Another advantage corn has over wheat is its astonishing versatility. By-products of corn number almost 500.[19] Much corn is processed into oil, artificial sugar, and starches. Corn can also be distilled to produce alcohol fuel (gasohol) for cars and trucks, and high-fructose corn syrup is used as a sweetener in nearly all soft drinks. One bushel of corn sweetens more than 400 cans of soda.[20] As a result of the federal corn subsidy, the cost of soft drinks has decreased by 25 percent since 1985. The grain can also be processed to create a dust that absorbs 2,000 times its weight in water, and corn has a part in the production of toothpaste, paint, paper products, tires, plastics, textiles, drugs, explosives, wallboard, vitamins, and minerals, among other things.

Despite its astonishing versatility for human use, 70 to 80 percent of harvested corn is used to feed livestock. As is true for soybeans, animals eat most of America's corn crop. Cattle consume the most corn, followed by poultry and hogs.[21] One bushel of corn converts into 5.6 pounds of retail beef, 13 pounds of retail pork, 28 pounds of farmed catfish, or 32 pounds of chicken.[22] Americans see the results of the corn harvest primarily in the form of animal products such as milk, cream, cheese, butter, eggs, beef, pork, lamb, or poultry. An unhealthy bacon and egg breakfast, cheese sandwich, and glass of milk at lunch, or hamburger for supper were all produced with corn. About 25 percent of the items in a typical grocery store use corn in some form during production or processing.[23] Although the unprocessed corn is healthy, most of its processed edible products are not (chapter 9).

Genetic Modification

Major agribusiness companies like Monsanto have been spending large amounts of money developing improved forms of corn and soybeans, but not wheat, and their investments are paying off handsomely. Seeds have been engineered to resist drought, cold temperatures, and insects. The more resistant seeds have made it possible for farmers in colder climates with shorter growing seasons to produce successful corn harvests. For example, in North Dakota, corn acres have increased by 62 percent in recent years. GM corn has yielded huge gains in productivity and has helped produce record corn harvests each year since 2003. Corn yields rose by 30 percent between 1995 and 2005.

Corn exports, 20 percent of America's corn crop, have not been seriously harmed by the fact that 85 percent of the crop is now grown using GM seed. Europeans have refused to use or eat GM crops, but exports to countries outside Europe have taken up the slack.[24] In the 1994–1995 marketing year, 82 percent of corn imports by the European Union came from the United States. That was before GM corn was planted in our fields. In the 1997–1998 crop year, only 10 percent of EU corn imports came from American farmers, and since 2000–2001, the percentage has been 2 percent or less. In 2004, EU corn imports were up 17 percent from 1994–1995, but none came from the United States Recently there have been signs that the EU may be relaxing its opposition to GM products, and America's corn farmers may be the first beneficiaries. The fact that 70 to 80 percent of the corn produced here is used to feed animals rather than humans gives it an important advantage over some other GM grains as European importing restrictions are gradually lowered.

Corn Irrigation and Global Warming

Corn is a thirsty crop compared to wheat, its major food grain competitor, so plantings are centered in the wetter, eastern part of the Midwest (figure 3.1). However, thanks to genetic engineering and increased irrigation, low annual rainfall is less of a handicap to corn production than was the case decades ago. In addition, climatic changes during the past hundred years have been beneficial for corn growers. In the corn-growing region, annual rainfall has increased between 5 and 20 percent (figure 3.1), adding to the benefits from increased temperatures and carbon dioxide concentration (photosynthesis). Corn will not benefit from global warming as much as the other major food grains.

Plants photosynthesize by two chemical pathways, termed C-3 and C-4, that are affected to different degrees by increased carbon dioxide and increased temperatures. Those that use C-3 benefit more from higher levels of carbon dioxide than do C-4 plants. Among the major food grains, only corn uses the C-4 pathway.

Corn as Fuel for Cars: Ethanol

The Archer Daniels Midland company, the nation's top grain (corn, wheat, and soybeans) and corn syrup processor, spent nearly three decades lobbying Congress and the White House for the use of corn-based ethanol in gasoline. Now, with record-high gasoline prices, it has succeeded. When Congress passed the 2005 energy bill, it mandated an increase in the amount of renewable fuel that is blended into gasoline from the 4 billion gallons produced in 2004 to at least 5 billion gallons by 2007 and 7.5 billion by 2012 (figure 3.2). With oil prices at about $100 a barrel, sharply

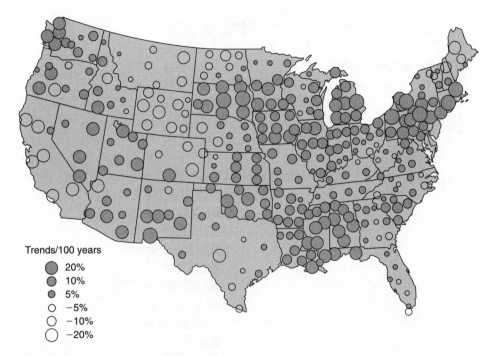

Figure 3.1
Precipitation trends over the past 100 years. *Source*: T. R. Carl, R. W. Knight, D. R. Easter-ling, and R. G. Quayle, Trends in U.S. Climate During the Twentieth Century, *Consequences* 1:1 [1995]: 5.

lifting the prices paid for ethanol, the average processing plant is earning a net profit of more than $2 a bushel on the corn it is buying for about $4 a bushel in early 2007, and that is before the tax credit of 51 cents a gallon given to refiners and blenders that incorporate ethanol into their gasoline. The rising price of a bushel of corn is occurring despite three years of record harvests, reflecting the increasing strength of ethanol.

The USDA predicts that world grain use will grow by 20 million tons in 2006. Of this, 14 million tons will be used to produce fuel for cars in the United States.[25] Cars, not people, will claim most of the increase in world grain consumption. In agricultural terms, the world appetite for automotive fuel is insatiable. The amount of America's corn crop used in U.S. ethanol distilleries was 5 percent in 2001 and rose to 14 percent by 2006. The USDA forecasts an increase to 31 percent by 2016.[26] Nevertheless, it supplies only 3 percent of America's automotive fuel.[27] Even if all the remaining corn were converted to ethanol, the total ethanol would offset only 12 percent of gasoline consumption.[28]

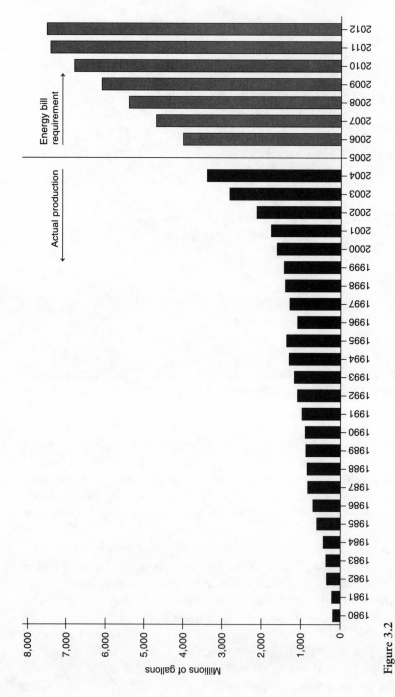

Figure 3.2
Trend in U.S. ethanol production, 1980–2012. *Source:* M. Wang, The Debate on Energy and Greenhouse Gas Emissions: Impacts of Fuel Ethanol, Argonne National Laboratory, 2005.

As of early 2008, there were 130 ethanol distilleries in operation (compared with 95 in 2006); 76 more are under construction, mostly in the Midwest; and hundreds more are in various stages of planning, according to the Renewable Fuels Association. When all of the plants under construction are completed, probably in early 2009, ethanol plants will need about 4.3 billion bushels of corn, 35 to 40 percent of America's corn production, to produce more than 12 billion gallons a year. (A bushel of corn yields about 3 gallons of ethanol.) In some U.S. corn belt states, ethanol distilleries are taking over the corn supply. In Iowa, the leading corn-producing state, 55 ethanol plants are operating or have been proposed. If all these plants are built, they will use almost all the corn grown in the state.

The amount of grain required to fill a 25-gallon gas tank with ethanol will feed one person for a year. The grain required to fill the tank every two weeks over a year will feed twenty-six people.[29] The stage is being set for a head-on collision between American's desire for fuel and need for corn-based food. With a U.S. ethanol subsidy of 51 cents per gallon in effect until 2010 (at least) and with oil consistently priced at more than $80 a barrel, distilling fuel oil from corn promises huge profits for distillers. At the end of 2007, it appears that the ethanol boom may be peaking because its effects on grain availability for people and animals are becoming more widely recognized.

As a result of federal regulations and subsidies, in the last quarter of 2005, Archer Daniels Midland's profit rose 18 percent compared to the same quarter in 2004. Profits in the corn processing division jumped 79 percent, from $132 million to $236.5 million. The company will earn an estimated $1.3 billion from ethanol alone in its 2007 fiscal year, up from $556 million in 2006, an increase in one year of 134 percent.[30] It controls 29 percent of the ethanol market, and ethanol profits rose 40 percent. And because of improved federal subsidies in 2005, the company's corn syrup profits leaped 150 percent. Archer Daniels Midland's stock price doubled from 2004 to 2005. As a report of the libertarian Cato Institute noted, "The Archer Daniels Midland Corporation has been the most prominent recipient of corporate welfare in recent U.S. history." No one questions the need for companies to make a profit, but whether tax money be used to finance it does require debate.

The sense in corn country in 2006 was that the "full speed ahead" signal had been received from Congress. As noted by Ken Cook, president of Environmental Working Group, "There is zero daylight" between Democrats and Republicans in the Midwest. "All incumbents and challengers in midwestern farm country are by definition ethanolics." Midwestern corn growers are ecstatic. Corn production, already America's chief grain product, increased from 10 billion bushels in 2000 to 12 billion in 2004. Between September 2006 and December 2006 the price of a bushel of corn rose 80 percent, and the price of food rose correspondingly. In

2007, dairy prices rose 14 percent (cattle are fed corn in feedlots) and other products whose production involves corn rose 5 percent. Government economists predict grocery prices will jump another 3 to 4 percent in 2008.

Energy Considerations

There is an acrimonious debate between most scientists and ethanol producers on one side, and a minority of scientists on the other side, concerning whether it takes more energy to produce 1 gallon of ethanol than the ethanol contains. Does making alcohol from corn create any new energy? Ethanol supporters calculate a 25 percent net gain, but calculations by two university professors indicate that making ethanol from corn requires 29 percent more fossil energy than the ethanol fuel actually contains.[31]

Regardless of which side is correct regarding the energy calculations concerned with the production of ethanol, all agree that corn-based ethanol contains only two-thirds as much energy as gasoline. So although 1 gallon of ethanol-blended gas may cost the same as regular gasoline, it will not take a car as far. Even if converting corn to ethanol yields a net energy gain, the fact that ethanol contains less energy than gasoline appears to make ethanol only marginally better or worse for the environment than using gasoline. But in the headlong rush to try to gain energy independence, which experts agree is impossible, scientific realities are taking a back seat to successful political lobbying. Using the government's enormous corn subsidies to underwrite increased research into wind power and solar power would be a more productive use of tax money.

Ethanol enthusiasts are aware of the looming conflict between using America's corn for automotive fuel and having enough corn to eat and use for other purposes. The solution proposed is to use stover rather than the corn kernels themselves. *Stover* is the name given to corn husks, corncobs, and other refuse obtained from corn plants. But the cellulose of which these materials are composed is difficult and expensive to turn into ethanol, and the technology is not yet economically profitable. The idea is to use enzymes made by GM organisms to dissolve the cellulose and convert it to ethanol. These enzymes could also potentially convert other materials to ethanol, for example, wood chips, switchgrass, algae, and other currently useless biomass.

But there is an enormous potential danger. Industrial-scale production is bound to have leaks and spills. Microbes with an enhanced ability to liquefy biomass might reduce the plant kingdom to green goo. Those who favor using the GM organisms say that they will not survive in the wild because they cannot compete with natural microbes. But living organisms are inherently unpredictable and mutate continuously. How can we guarantee that the GM proponents are correct? The possible consequences of an error are disastrous to contemplate.

Environmental Problems Caused by Corn Productivity

Because of subsidies, corn appears to be very cheap to supermarket shoppers. But everyone is paying part of the cost of the corn and corn-based products that most Americans buy. Forty-two percent of America's corn is used to make high-fructose corn syrup,[32] which is found in countless products and has completely replaced more expensive but arguably less harmful sugar in sodas and soft drinks. Federal sugar tariffs make sugar more expensive to import and use as a sweetener in processed food products. Archer Daniels Midland is a major lobbyist in Washington for high sugar import tariffs. We are subsidizing obesity (chapter 11).

Another downside to ethanol production is that Archer Daniels Midland burns coal to power its ethanol factories. Research suggests that ethanol made from coal-burning plants has no net benefit in terms of reducing air pollution.[33] The pollution caused by burning the coal nullifies the lesser pollution emitted by cars burning the ethanol. And ethanol plants are a major drain on surface and subsurface supplies of water because of their enormous water needs—as many as 2 million gallons per day. Brazilian ethanol made from sugarcane is cheaper to produce and more energy efficient than American corn ethanol and does not pollute the air or drain America's water. But corn ethanol production is protected by a prohibitive import tariff on imported ethanol of 54 cents per gallon, adopted in 1980 after lobbying pressure by Archer Daniels Midland on President Carter.

Farmers are paid an annual subsidy averaging $48 an acre not to raise crops on the 35 million acres of farmland set aside under a 1985 program for conservation. But the profit lure of ethanol may be great enough to push this acreage, much of it considered marginal, back into production. Marginal land requires extra fertilizer to make it economically productive.

Don Basse, president of AgResources, an economic forecasting firm, said in 2006 that "by the middle of 2007, there will be a food fight between the livestock industry and the ethanol industry. As the price of a bushel of corn reaches $3, the livestock industry will be forced to raise prices or reduce their herds. At that point the U.S. consumer will start to see rising food prices or food inflation."[34] Basse's prediction has come true: cattle prices increased more than 10 percent between 2006 and 2007, and the price of milk jumped as well, along with the cost of cold breakfast cereal and many other corn-based products.[35]

Corn and Cattle

Cattle are fed corn-rich feed in feedlots, which acidifies their digestive system so that *Escherichia coli* 0157 bacteria, a leading cause of food-borne illness, can survive. Tens of thousands of Americans are infected each year.[35] Antibiotics are routinely administered to cattle to control *E. coli* proliferation. Livestock farmers encourage the growth of harmful bacteria with the food they feed cattle in feedlots and then combat them with manufactured antibiotics; these substances filter down to consumers as they eat the remaining bacteria and antibiotics in the meat.

People who worry about public health do not have any control over agricultural subsidies, the food cattle eat, or the harm they generate in food products. A major concern of the USDA is to find ways of using the immense amounts of corn produced and helping agribusiness concerns make their products more cheaply. Agribusiness gives a large amount of money to elected representatives. Although cheap corn is touted as a way to help the United States compete internationally, the major beneficiaries are the food processors that are using corn domestically.

Fertilizer and Pesticides

Because of monocropping and the fact that corn is not a legume, an inadequate nitrogen supply in the soil is a persistent problem for corn farmers. Corn is the grain farmer's major user of nitrogen fertilizer. Corn acreage has expanded significantly, and over the past thirty years, a consistent 96 to 98 percent of corn acreage has received heavy applications of nitrogen compared to 21 percent for soybeans.[36] As noted in chapter 2, runoff of this nitrogen has had disastrous effects on waterways and the near-shore environment in the Gulf of Mexico.

Furthermore, corn yield has been greatly increased by the use of powerful and poisonous pesticides, which are sprayed on them about six times during a growing season. The effect on yield is made evident in the poetic doggerel sung by pioneers in the nineteenth century, who had to sow four corn kernels for every plant they hoped to reap: "1 for the maggot, 1 for the crow, 1 for the cutworm, and 1 to grow." Conventional corn production uses more pesticides than any other crop grown in the United States,[37] and runoff of pesticides from agricultural fields has had serious negative effects on waterways, the fish that live in them, and humans.[38]

Wheat

The people who buy wheat have both eyes.
Those who buy flour have only one.
Those who buy bread are blind.
—Ancient proverb, Ithaca, Greece

Wheat is grown on about 60 million acres in forty-two states, but planted acreage has been gradually declining since a high of 90 million acres in 1981. The chief cause of the decline in planted acreage has been increased production in other countries, but the recent boom in corn ethanol production is decreasing the acreage planted in wheat. In addition, Americans seem to be losing their taste for bread and cake products: per capita wheat consumption in the United States decreased by 9 percent between 1997 and 2005.

The main wheat belt extends south from the Dakotas to Texas and from Kansas northeastward into Ohio and Michigan. Kansas and North Dakota are the biggest producers. In contrast to soybeans and corn, most of the wheat crop is consumed by humans. We eat 36 percent of the crop, export 50 percent, use 10 percent to feed livestock, and save 4 percent of the kernels for seed.[39] The United States is the world's second leading producer of wheat, behind China, and is the world's largest exporter.

The average yield of wheat on American farms rose from about 17 bushels per acre in 1950 to 39 bushels per acre in 2006, largely reflecting technological progress with fertilizers and pest control. The 2006 wheat harvest was 1.8 billion bushels. Eighty-seven percent of wheat acreage receives nitrogen fertilizer.[40] One bushel of wheat that a combine takes only 9 seconds to harvest contains about 1 million kernels and produces seventy 1-pound loaves of bread.[41]

Wheat production varies annually because of changes in export needs. Half of the wheat crop is exported, so the planned production of wheat by farmers varies with anticipated productivity in other major wheat-exporting countries, such as those of the European Union, Canada, Argentina, Australia, Russia, and Ukraine. If their production expands, ours is likely to shrink. Deciding how much to plant each year is a judgment call for wheat farmers, with serious financial consequences for bad guesses.

Wheat is the only cereal grain that contains the group of proteins, collectively known as gluten, that possess the unique characteristic when mixed with water of forming an elastic mass that can be stretched or extended. It is the gluten in the central part of the wheat kernel that allows wheat flour to be formed into dough, which can be expanded and baked to produce bread-type products. The high gluten content of wheat is the reason most breads are made largely or entirely of wheat. Some people are allergic to gluten, a malady known as celiac disease. Those with this genetic abnormality (estimated to be about 1 percent of the population) must avoid gluten because it damages the lining of their small intestine so that normal digestion becomes impossible.

The main use of wheat is in the manufacture of flour for bread and cakes. A bushel of wheat yields approximately 42 pounds of white flour or 60 pounds of whole wheat flour. In general, wheat with hard kernels is used for bread flour and pasta and wheat with soft kernels for cake flour. Wheat is also used in the production of breakfast foods and, to a lesser extent, alcoholic drinks. Low grades of wheat and by-products of the milling, brewing, and distilling industries are used as food for livestock.

In 2006 crop researchers discovered a gene in wild wheat that, when bred into cultivated varieties, boosts the grain's nutritional value.[42] The strains now used commercially have a mutation that reduces their protein, zinc, and iron content.

Apparently in wheat, as in most other foods, eating what nature has provided is more healthful that what human ingenuity has created.

Genetically modified wheat is not commercially produced in the United States because the European Union and Japan are unwilling to buy it. Without these export markets, it is not economical for American farmers to grow it. Most corn and soybeans are fed to farm animals, and so as foreign importers gradually relax their prohibitions against GM products, these two grains will be the first to benefit. Wheat, however, is consumed largely by people, so GM wheat products will lag. Monsanto dropped an effort to produce the world's first genetically engineered wheat in 2004 because most of the nations that import America's wheat said they would not accept it. The wheat was engineered to be resistant to Monsanto's Roundup herbicide, which would have allowed farmers to spray their fields to kill weeds without damaging the crop.

4

Organic Food: As Nature Intended

Organic food is not just about a product; it is a philosophy in which the process of production is as important as the final result.
—Peter Hoffman, "Going Organic, Clumsily," *New York Times*, March 24, 1998

Organic agriculture arose in the 1970s as a reaction to industrial farms that confine animals, regularly feed them antibiotics, and use large amounts of poisonous artificial pesticides and chemical fertilizers on crops. Increasing numbers of Americans have become justifiably concerned, and even alarmed, about these practices of industrial agriculture, and for the past ten years, sales of organic food have been booming. Nearly two-thirds of American consumers purchased organic foods in 2005, an increase of 17 percent from 2004. All the major players in the retail food industry now have organic brands. Retail sales have grown at a rate of 20 percent per year or more since 1990,[1] while total food sales have increased only 2 to 4 percent yearly (population growth during the period was 1 percent).[2] Organic food sales in 2006 were about 3 percent of all food sales, up from 2.5 percent in 2005 and 1.9 percent in 2003.[3] The *Nutrition Business Journal* projects that increases will continue.[4] Some organic food products are sold in 73 percent of all conventional grocery stores.[5] The share of organic foods sold at discount outlets like Costco and Wal-Mart jumped from just 1 percent in 1998 to 13 percent in 1999 and is still rising. Their sales of organic foods increased 30 percent between 2002 and 2004.[6] Supermarkets and grocery stores in 2006 accounted for 53 percent of organic sales; natural food stores, 24 percent; and farmers' markets, 9 percent.[7] This is very different from the relative percentages in 1991, when only 7 percent of all organic products were sold in conventional supermarkets.[8] Major commercial food giants, such as Kraft, Gerber, Heinz, Nestlé, Coca-Cola, Unilever, and General Mills, have entered the organic market by buying small organic companies and producing organic baby food, flour, and other products. Organic food is now mainstream. Even snack foods, the bane of the weight-conscious and parents of small children, are organic. Forty-two percent of sales of organic food in 2003 were for

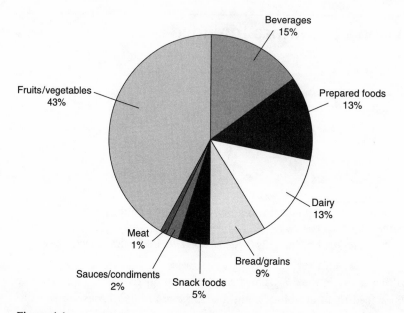

Figure 4.1
Share of organic food sales by category, 2003. *Source*: Organic Trade Association, 2004.

fruits and vegetables (figure 4.1), with vegetables edging out fruits.[9] The top eight were, in order, tomatoes, carrots, peaches, squash, leafy vegetables, apples, potatoes, and bananas. Ninety-three percent of the organic fruit and vegetable sales were for fresh produce rather than for frozen, canned, or dried varieties.

A growing number of medium-priced conventional restaurants now offer items they term organic on their menus. Many restaurants that serve only organic food have opened in recent years, and their locations can be found on the Internet.

Availability of Organic Foods

Business is a good game—lots of competition and a minimum of rules. You keep score with money.
—Nolan Bushnell, founder of Atari

Large organic food supermarkets exist throughout the United States, and more are opening almost daily. The biggest retailer of organic and natural foods and the largest in floor space is Whole Foods Market, with 263 stores in thirty-seven states. They average 32,000 square feet in size and carry more than 1,200 items. Its newest store in San Jose, California boasts 86,000 square feet, rivaling Costco or Wal-Mart in size. Coming in second is Wild Oats Market, bought in 2007 by Whole Foods, with 110 stores in twenty-four states. Its average store size is 25,000 to 32,000 square feet. Third in floor space is Trader Joe's with 294 stores, averaging 10,000 square feet in size and with more than 2,000 products, in twenty-three states. It has more stores than Whole Foods or Wild Oats, but the individual stores are smaller. In addition to these three major chains, many smaller organic markets exist throughout the country. Clearly the growth of organic food markets is phenomenal.

Fifty-four percent of Americans have tried organic foods and beverages.[10] According to the Organic Trade Association, 27 percent of Americans ate more organic products in 2004 than they did in 2003, and 11 percent now use organic products daily.[11] Consumers reporting weekly use grew from 9 percent in 2003 to 16 percent in 2004. The number of Americans who never eat organic dropped from 45 percent to 34 percent between 2000 and 2003.[12] Surveys forecast an average annual growth rate of 18 percent for organic foods from 2004 to 2008, with meat, fish, and poultry projected to increase by 31 percent and fruit and vegetables by 21 percent.[13]

More than 800 new organic products were introduced to consumers in the first half of 2000.[14] As an executive of the Whole Foods Market noted in 2004, "When I started in the natural foods industry more than 25 years ago, most organic items were in the produce aisle. But today, the selection stretches throughout the store from farm-fresh produce to handcrafted pastas, cereals, dairy products, wine, cheeses, chocolates, grains, vinegars, and almost every product imaginable."[15]

One of these products is pet food. According to the American Animal Hospital Association, 80 percent of American pet owners consider their pets to be their children rather than "companion animals,"[16] a point of view that provides the organic pet food industry with a large potential market. Every year Americans spend four times more on pet food than on baby food. In America, there are 70 million pet cats, 60 million pet dogs, 10 million pet birds, 5 million pleasure horses, and 17 million exotic pets such as rabbits, snakes, rats, hamsters, gerbils, guinea pigs, mice, and skunks.[17] In addition, although they are not yet sold in Whole Foods Markets, organic worms are available for environmentally focused fishers.[18] Currently available are organic red wigglers and European and African night crawlers. The African variety is touted as a "good trolling worm" but is not recommended for ice fishing. Organic worm castings (feces), exceptionally rich in nitrogen, phosphorous, and potassium, are also available for composting (vermicompost).

The main impediment to the rapid growth of organic foods is price: certified organic products are usually more expensive than conventional ones, and sometimes twice as expensive. One way to lower the cost of organic products is to purchase them at farmers' markets. Organic-only farmers' markets have been organized in many states. The renaissance in farmers' markets during the 1990s was fostered by state and local municipalities that wanted to revitalize neighborhoods and preserve regional farmland and open space. The number of these markets increased from 1,755 in 1994 to 3,706 in 2004, and 4,385 in 2007, for a compounded growth rate of 8 percent a year.[19] California leads the way with 443, and New York is second at 276. Farmers' markets have been a particularly attractive option to organic farmers, who use this marketing outlet much more heavily than conventional farmers do. About 80 percent of organic production comes from family farms, a far higher percentage than for conventional farming. Some states are producing directories of farm stands and pick-your-own farms, including organic directories. More than 40,000 farms engage in some form of direct marketing through farm stands, farmers' markets, or buyers' groups called community-supported agriculture.[20]

Organic farmers for the past twenty years have been pioneering community-supported agriculture (CSA), an innovative direct marketing arrangement in which consumers subscribe to the harvest of a CSA farmer for the entire upcoming growing season and pay for their produce in advance.[21] Under a CSA arrangement, consumers buy shares in the upcoming season's harvest. The share price goes toward production and distribution expenses, and consumers share the natural risks with the farmer. Typically CSA farms are small, independent, labor-intensive family farms. By providing a guaranteed market through prepaid annual sales, consumers essentially help finance farming operations. This allows farmers to focus on growing and somewhat levels the playing field in a food market that favors large-scale industrialized agriculture over smaller farms and locally grown food. Producers capture a much higher share of the consumer food dollar when they market their produce directly to consumers, eliminating the middlemen who get about 75 percent of the consumer food dollar in normal mass marketing such as supermarkets. As of early 2006, there was a growing list of 1,140 CSAs covering all fifty states in the database maintained by the USDA and the Robyn Van En Center at Wilson College in Pennsylvania. Most CSA farms use organic production systems.[22] Clearly, organic foods have entered the mainstream. (A listing and detailed location map of CSAs and farmer's markets can be found at localharvest.org/organic.jsp.)

Farmland managed under organic farming systems expanded rapidly in the United States throughout the 1990s, and that pace has continued as farmers strive to meet increasing consumer demand in local and national markets. In 2005, organic production systems were present in all states, and farmers dedicated 4 million acres of cropland and pasture to organic production systems. Nearly 2.3 million acres were used for growing organic crops.[23] Currently 0.5 percent of America's crop acreage is farmed organically on 8,400 certified organic farms; organic crop acreage increased from 638,000 in 1995 to 1.7 million in 2005. Edible grains dominate organic crop acreage; the percentage devoted to each crop is shown in table 4.1. Interestingly, although the most commonly purchased organic foods in

Table 4.1
Percentages of some certified organic crops planted on U.S. cropland, 2005

Crop	Percentage
Wheat	32.0
Corn	15.1
Soybean	14.1
Vegetables	11.4
Fruits	11.2
Oat	5.4
Barley	4.5
Rice	3.0
Millet	1.6
Rye	1.0
Sorghum	0.7
Total	50.3

Note: Most of the remaining 49.7 percent of the certified organic area was devoted to hay and silage or left fallow. Total organic area is 0.51 percent of all cropland.
Source: U.S. Department of Agriculture (2006).

retail outlets are vegetables and fruits, they form only 11 percent of organic crops.[24] But the percentage is growing rapidly. Salad vegetables are most abundant, with one-third of organic vegetable acreage planted with lettuce, tomatoes, or carrots. However, these plots form only a tiny percentage of the national lettuce, tomato, and carrot crop—4 percent, 1 percent, and 6 percent, respectively.[25] California is both the biggest conventional producer of vegetables in the United States and the biggest producer of organic vegetables, accounting for 41 percent of all certified acres.[26]

In the category of organic fruit, grapes are the dominant crop, accounting for 29 percent of organic fruit acreage (much of it for organic wines), followed by apples at 24 percent, citrus at 20 percent, and nuts at 10.5 percent. Berries and stone fruits (those with a single large pit such as peaches) are abundant in the remaining 27 percent.[27]

Maharishi Vedic City, a community in Iowa founded in 2001 and dedicated to "creating a national center for perfect health and world peace," has banned the sale of nonorganic food in the city.[28] A resolution passed by the city council in 2003 says, in part: "WHEREAS, throughout the U.S., only 2 percent of the food produced is organic, and 98 percent of the food is produced with poison, contributing to the deplorable state of health in the country,..." The city has established the five-member Organic Committee to decide on the acceptability of food products. The names of approved foods are published on the city's Web site. According to the mayor, nonorganic food can be bought outside the city and still be eaten legally in the city.

Organic Crops in Other Countries

Men and nations behave wisely once they have exhausted all the other alternatives.
—Abba Eban, Israeli politician (1915–2002)

Organic crops are thriving outside the United States. Sixty-five million acres of farmland is now organically certified worldwide, and about 10 percent of the organic food Americans eat is imported.[29] Far and away the world leaders in the adoption of organic agriculture are Australia, where organic plantings cover 27.9 million acres, and Cuba, where organic agriculture was adopted as the official government strategy after its highly successful introduction in 1990.[30] Part of the reason Cuba adopted organic methods was the collapse of the former Soviet Union, which had provided Cuba not only with fertilizers and pesticides, but also with the tractors and the oil necessary to run them. Cuba's Institute for Crop Protection (INISAV) has become a magnet for agronomists from other Latin American countries, who come to learn about growing food organically. INISAV has also developed organic herbicides and pesticides that are sold in Cuba and other countries and has 222 local centers devoted to producing natural agricultural weapons such as insects and larvae that eat crop pests and viruses that attack crop-damaging strains.

In Western Europe the amount of organic and in-conversion acreage increased an average of 22 percent each year between 1995 and 2001.[31] Retail sales are growing at 8 percent a year. In Denmark, 2.5 to 3.0 percent of food sales are organic products. Organic acreage is now 3.3 percent of total agricultural area, nine times the percentage in the United States. As of February 2005, Lichtenstein (a small nation 6 percent the size of Rhode Island) leads the way in the percentage of its agricultural land that is organic: 26 percent.[32] Nearly 80 percent of British households buy organic food. Europe now accounts for 46 percent of worldwide sales of organic food, with the United States at 37 percent.

The EU has supported organic agriculture through a variety of schemes, including paying subsidies, or "green payments," to farmers for converting to organic farming and for continuing organic farming.[33] The economic rationale for these payments is that organic production provides benefits, like health, that accrue to society and thus deserves support. In practice, green payments for organic production target new and existing organic farmers, partly to compensate new or transitioning farmers for the decline in yields when moving from conventional to organic production.

Support payments averaged $90 per acre in 2001 compared with $44 per acre for conventional farms.

As we will see in chapter 5, which deals with genetically modified crops, Europeans are much less inclined to take chances with their food supply than Americans. They are much less inclined to eat genetically modified foods.

What Is Organic Farming?

Fessing up to addictions and seeking treatment has become a fact of contemporary life, and farmers are not about to be left out. Maybe we'll end up naming the nineties the Detox Decade.
—Richard Nilsen, *Whole Earth Review*, 1990

Organic farming is an agricultural system that:[34]

1. Uses management practices that sustain soil health and fertility.
2. Uses natural methods of pest, disease, and weed control.
3. Demands the lowest possible levels of environmental pollution
4. Uses a minimum of off-farm imports.
5. Prohibits genetically engineered food and products. European countries have set a limit of 0.9 percent of genetically modified material permitted in food labeled "organic."
6. Has high standards of animal welfare.
7. Promotes enhancement of the landscape, wildlife, and wildlife habitat.

In the opinion of the environmentally focused prince of Wales, "Organic farming delivers the highest quality, best-tasting food, produced without artificial chemicals or genetic modification, and with respect for animal welfare and the environment, while helping to maintain the landscape and rural communities." Organic farming is roughly synonymous with sustainable agriculture. The focus is on the balance between what is taken out of the soil by growing crops and what is returned to it, without outside imports. It is the opposite of industrial agriculture. Sustainable agriculture integrates three main goals: environmental health, economic profitability, and social and economic equity. Sustainability rests on the principle of meeting the needs of the present without compromising the ability of future generations to meet their own needs, disturbing the local ecosystem as little as possible, and treading lightly on the earth because humans are but one strand in the web of life.

A comparison of the work requirements of sustainable and industrial farming found that sustainable farmers are more likely to own most of the land they farm, and they spend one-third more time on farm work than do industrial farmers.[35] Spouses of sustainable farmers spend three times more time on farm labor than do spouses on industrial farms, probably because the husband is already working more than 50 hours per week. The sustainable farmer's work is centered more on livestock and is therefore spread out more evenly during the year, while labor demands on industrial farms are more concentrated during crop-growing seasons. For sustainable farmers, the most difficult management problems are crop and soil practices; industrial farmers report marketing as a bigger management concern. This may reflect the complex agronomic and animal husbandry issues in sustainable agriculture, and the high-volume, low-profit-margin, cash-cropping strategies of industrial agriculture.

In a 2002 poll, 67 percent of Americans thought that organic food will become more common in the future.[36] They said they were willing to pay as much as 20 percent more for organic foods. Most of the consumers who do not eat organic foods said that price was a major barrier to purchasing them. However, owing to industry growth and consumer demand, the price of organic food is becoming more competitive with that of conventional food.

What Is an Organic Product?

For the first time in the history of the world, every human being is subjected to contact with dangerous chemicals, from the moment of conception until death.
—Rachel Carson, *Silent Spring*, 1962

After years of acrimonious public and governmental debate on the meaning of the term *organic*, the USDA in December 2001 defined four categories of organic foods.[37] The label "100% organic" certifies that the food was grown without pesticides, hormones, antibiotics, irradiation, or genetic modification, and "organic" means made with at least 95 percent by weight of such ingredients; both of these categories can display the USDA Organic seal. "Made with organic ingredients" certifies between 95 percent and 70 percent organic, and three organic ingredients can be listed on the label. Products containing less than 70 percent organic ingredients can identify organic ingredients on their ingredient list, but the word *organic* is not permitted on the front of the package. These standards are probably the most stringent in the world and apply to all food and fiber products (such as cotton towels) labeled as organic, including imported products, and require them to be certified by a USDA-accredited agency. Every phase of food production, from farm to retail shelf, is subject to inspection. The organic standards are statements about how food has been grown, raised, and processed; it is not an evaluation of the quality, healthfulness, or safety of the product, topics that are the subject of heated debate.

For organic livestock, the standards ban the use of antibiotics in feed or water and of hormones as growth promoters. Livestock would have to be fed organic feed and not given mammalian or poultry slaughter by-products or plastic pellets (as fiber).

The USDA has decided not to extend the organic certification program beyond crops and livestock, so rules will not be developed to cover farm-raised fish.

There is some cost and a great deal of cumbersome paperwork that must be completed for a farm to receive organic certification, and some small farms have decided not to apply because of this.[38] Larger corporate farms are better equipped to go through the certification process, and this may help them capture the growing market for organic food, which until now has been the domain of small-scale family farms.

Benefits of Eating Organic Food

We have now reached a state in which we are more particular about the purity of the gasoline for our cars than about the food we eat.
—Aharoni Cohen, manager of an organic food store, 2005

Most people who buy organic foods say they do so because they believe they are a good way to support small and local farmers (57 percent), are more nutritious (54 percent), are better quality (42 percent), taste better (32 percent), and are less likely to be contaminated with harmful pesticides.[39] Is there evidence to support these beliefs?

Support for Small and Local Farmers

Traditionally, organic farms have been smaller than nonorganic farms. A typical size is 50 to 100 acres, in contrast to an average of 449 acres in 2007 for a conventional farm. One of the largest organic farms occupies more than 4,000 acres in California.[40] The reason for the difference in average size is that organic farming is both labor intensive and information intensive. Studies indicate that 11 percent more labor is required of organic growers per unit of grain production.[41] The difference can be much greater where the crop is vegetables or fruit, although technological improvements are narrowing this gap. Organic systems also require additional management time in planning, pest scouting, and related activities. For this reason, organic management can be done better if a farm is not too large.

Nutritional Benefits

Although no one has claimed that foods grown by industrial agriculture are more nutritious than organically grown foods, some data indicate that organic foods are superior.[42] One study found that organic vegetable soups contained almost six times as much salicylic acid (the active ingredient in aspirin) as nonorganic vegetable soups.[43] Salicylic acid helps combat hardening of the arteries and is recommended as a daily health supplement by many doctors to help prevent heart attacks. A study in 1998 found that protein quality was superior in organic products, and research in 2001 found that organic crops had higher average levels of twenty-one nutrients, including vitamin C and iron.[44] Several studies have found that organically grown foods and those with lower levels of pesticides contained substantially higher concentrations of antioxidants and other health-promoting compounds than crops grown with higher doses of pesticides.[45] Organic corn and raspberries contained 56 percent more of the antioxidant compounds than corn grown using pesticides.

The pesticides and chemical fertilizers in modern agriculture apparently disrupt the ability of crops to synthesize antioxidants, compounds associated with reduced risk for cancer, stroke, heart disease, and other illnesses. In other words, nature is trying to help us remain healthy, but we are refusing the help. Certainly heart attacks, strokes, and cancer are more prevalent today than 100 years ago, but how much of the increase is due to the practices of industrial farming is uncertain. Nevertheless, according to the scientist who conducted the study of antioxidant variations, "Lots of these compounds are synthesized to protect the plant from insects and disease. So if we're protecting the plant with pesticides, the plants are not going to waste the energy to produce them."[46]

Better Quality

The feeling of consumers that organic produce in stores is of better quality than conventional produce probably results from their observation that it appears fresher: fewer brown spots in cauliflower tops, crisper lettuce, sturdier cucumbers, and so on. The explanation for these observations lies in the local source of the products. Organic crops are much more likely to be locally grown and are therefore fresher in the store than conventionally grown crops, which normally travel hundreds or thousands of miles to get from farm to store.

Better Taste

Flavor results from a mixture of many different and complex molecules in the plant. A healthy living soil provides a constant and more complex mixture of these molecules, and the result is more flavor. It is no surprise that chefs working in the highest-caliber restaurants prefer organic ingredients to conventionally grown ones.[47]

Freshness has a major effect on taste, as is easily detected by comparing organic home-grown backyard tomatoes with the conventionally grown ones from the industrial megafarms that supply most supermarkets. The difference in taste is striking. The standard tomato tastes like cardboard. However, keep in mind that "organic" does not necessarily mean locally produced. The tomatoes labeled organic may have originated in distant states or foreign countries and were picked when green and allowed to ripen off the vine during transit, like most tomatoes. Tomatoes that ripened on the vine would be too soft and semirotten by the time they were displayed in the store.

Industrial crop breeders have developed varieties of fruit able to withstand the rigors of shipping and mechanical harvesting, and in the process, many of the tastiest and juiciest fruits have been abandoned as unsuitable. Distance and durability have become central goals of industrial farming, although each step away from the source adds more cost, more use of fossil fuels, and probably more chemicals. Some supermarkets see local food as the next major development in food retailing; the word *locavore* has been coined to describe people who restrict their food purchases to food grown within a specified radius of their homes. Products sold in farmers' markets, organic or not, taste better than what are ostensibly the same products sold in supermarkets. They taste better because they are fresher and they are nutritionally superior as well.

What should consumers do when faced with choosing between an organic product that was shipped from afar and a nonorganic product that was produced locally? Almost all the organic food in the United States comes from California, much of it from five or six big farms that dominate the industry. It takes a lot of fossil fuel to get this produce to the East Coast. The term *organic* does not mean "locally grown by a small farmer." From the standpoint of taste, local is always better, and studies have revealed that buying local is also much greener than buying organic. We should always seek to limit the size of our foodshed—the area of land, people, and businesses that provides a community or region with its food. This is not easy to do because we have gotten used to having all produce available year-round. Eating seasonally brings with it a reduction in choice, and few Americans choose to vote for reductions in choice of anything. Strawberries are a summer fruit, but we are used to having them available all year. In addition, much food is imported from other nations, a worldwide food shed. Eating organically, locally, and seasonally is ecologically and environmentally desirable, but it is not easy to do and requires knowledge not readily available to consumers. Consumers should push supermarket managers to label their produce with point of origin and method of production.

Lower Pesticide Contamination

Conventional crops are grown with the aid of massive amounts of pesticides, and common sense suggests that some of these chemicals remain on or in the crop. Almost by definition, organically grown grains, vegetables, and fruit contain fewer pesticides than conventionally grown crops. But how much lower, and what is a "safe" level? Numerous studies have tackled these questions.[48] Each year the USDA checks the types and amounts of pesticides in thousands of samples of vegetables, fruits, and a few other products, domestic and imported, fresh and processed.

In 1998, Consumers Union analyzed the results of the testing done between 1994 and 1997 on twenty-seven food categories, covering 27,000 samples.[49] Of the twenty-seven foods, seven stood out as being hundreds of times more toxic than the rest: apples, grapes, green beans, peaches, pears, spinach, and winter squash. Lowest in toxicity were apple juice, bananas, broccoli, canned peaches, milk, orange juice, and canned or frozen peas and sweet corn. The Environmental Working Group has published a similar list.[50]

Consumers Union published an independent investigation in 1998, testing conventionally grown and organic peppers, apples, peaches, and tomatoes, and found the lowest amounts of pesticide in the organically grown products.[51] However, one-fourth of the organic samples had traces of pesticides, compared to 77 percent of the conventional samples. The amount of pesticide residue on all but one of the conventional samples was within federal limits.

In 2002, Consumers Union analyzed three extensive data sets of twenty major crops and found that conventionally grown samples consistently had pesticide residues far more often than organic samples (table 4.2). And the amounts of pesticide were higher in the conventional crops two-thirds of the time. These crops also contained multiple pesticides eight times more often than organically grown produce. (More extensive lists of vegetables and fruits that tend to be high or low in pesticide residues have been generated by the Environmental Working Group: foodnews.org/highpesticidefoods.php; foodnews.org/lowpesticidefoods.php.)

Pesticide poisons pose the greatest danger to children, born and unborn.[52] Pesticides cross the placenta during pregnancy, and research on children in New York

Table 4.2
Percentages of produce found with pesticide residue

	Conventional	Organic
Pears	95	25
Peaches	93	50
Strawberries	91	25
Spinach	84	47
All fruit	82	23
Grapes	78	25
Bell peppers	69	9
All vegetables	65	23
Lettuce	50	33

Source: consumer reports.org, 2002 consumerreports.org/main/content/aboutus.jsp?FOLDER %3C%3Efolder apartments has demonstrated a link between pesticide use and impaired fetal growth. Another study found that preschoolers fed conventional diets had six times the level of certain pesticides in their urine as those who ate organic foods. Another report found twice the level of some pesticides in the urine of children as in that of adults.[53]

Summary
Organic food products are not 100 percent free of pesticide residues, but they are much cleaner, and therefore presumably safer, than conventionally grown crops. No one knows for certain why organically grown products contain any pesticides at all. It is perhaps meaningful that when the pesticide analyses excluded now-banned substances such as DDT, the percentage of organic produce with pesticides dropped by nearly 50 percent. Perhaps some of the pesticides were obtained from soil previously used for industrial agriculture that had yet been completely cleaned of pesticides. DDT, for example, is persistent in the environment. Or perhaps the residues resulted from drift from a neighboring field or contamination in trucks and warehouses. Or perhaps there was mislabeling at some point. But the data from many studies leave no doubt that crops grown using conventional methods are much more likely to contain pesticides than organically grown crops.

It is worth emphasizing that the amounts of pesticide residues on all produce are almost always within current federal allowances. And the health risks associated with dietary pesticide residues are still uncertain and subject to debate. But risk is relative, and lower exposure undoubtedly translates into lower risk. Buying organic minimizes dietary pesticide exposure. Those who choose to buy conventional foods should wash the produce purchases thoroughly in a highly diluted solution of liquid dish soap before eating them. This usually removes about half of the pesticide. However, some pesticide gets into the internal tissue of some types of produce and cannot be washed off.

Farming Techniques: Conventional Versus Organic

In simplest terms, agriculture is an effort by man to move beyond the limits set by nature.
—Lester R. Brown and Gail W. Finsterbusch, *Man and His Environment: Food*, 1972

To damage the earth is to damage your children.
—Wendell Berry, *The Unsettling of America*, 1997

Organic farming is essentially traditional farming, based on knowledge and techniques honed over thousands of years of agriculture, before the chemical farming revolution. Although the number of farmers in the United States has been decreasing for decades, an increasing number of farmers are choosing to farm organically. The average age of organic farmers is 47.5, nearly seven years younger than conventional farmers.[54] There were 5,000 organic farmers in 1997, 6,600 in 1999, 7,800 at the start of 2000, and about 10,000 in 2006. This is about 1/2 of 1 percent of all U.S. farmers.[55] It is noteworthy that this small percentage of farmers produces, on farms averaging about one-fifth the size of conventional farms, 2 percent of food sales.

There was a 15 to 20 percent increase in organic acreage each year during the 1990s. Organic farming involves natural processes, often taking place over extended periods of time, and a holistic approach. (Holism is the philosophy that things can have properties as a whole that are not explainable from the properties of their parts, that is, interactions among simple things can produce unanticipated results. It is the opposite of reductionism.) Most modern farming focuses on immediate, isolated effects and reductionist strategies. (Reductionism is a philosophy that believes the nature of complex things can always be explained by simpler or more fundamental things. It is the opposite of holism.) The differences are most clearly seen in the way each system deals with crop fertilization, pest control, and crop diversity.[56]

Fertilization

Organic farming relies on natural products to fertilize crops, replacing the many nutrients in the soil that have been depleted by previous crops. The methods include spreading natural manures collected from livestock on the farm, "green manures" (field or forage crops applied while green or soon after flowering), or cover crops, preferably nitrogen-rich legumes. These natural products decay and add organic matter to the soil and release a wide variety of macro- and micronutrients to enrich the soil. This process is driven by microorganisms and allows the natural production of nutrients throughout the growing season. Three-quarters of the nutrient value of all feed consumed by animals is returned in manure.[57] There is a saying among organic gardeners that chemical fertilizers may feed the plant, but organic fertilizers feed the soil.

In conventional agriculture, purified bags of easily dissolvable compounds of nitrogen, phosphorous, and potassium are applied to the soil on a schedule. All other macro- and micronutrients are addressed separately. If excess nitrogen, phosphorous, or potassium has been applied, other chemicals are applied to neutralize them. Alternatively, the excess may be flushed away with water.

Synthetic chemicals such as pesticides and fast-acting manufactured fertilizers interrupt or destroy the microbiotic activity in the soil, reducing the soil to being merely an anchor for plants. In conventional agriculture, plants receive only air, water, and sunlight from their environment. Everything else must be distributed to the plant by the farmer, often from inputs manufactured and transported thousands of miles to reach the farm. Plants are fed only the most basic elements of plant life and so must depend on the farmer to fight all of nature's challenges: pests, disease, and drought.

Pest Control

Although plants over the millennia have evolved many types of natural defenses against the organisms that want to kill them or eat them, only about 5 to 10 percent of the crops grown today have significant built-in insect resistance and only about 1 percent have significant weed resistance.[58] To get the amount of crops we have come to expect, human intervention is needed. In organic farming, some pest populations are tolerated while the farmer concentrates on the long-term health of the soil and its surroundings. The farmer may encourage predatory beneficial insects to flourish and eat pests, or plant companion crops that discourage pests. In the EU a certain species of naturally occurring bacteria is sprayed onto wheat, barley, and oat seeds to combat fungal diseases.[59] It is nontoxic and has proven 98 to 100 percent effective. Soap rather than pesticides may be sprayed on plants to protect them. Crops may be rotated from year to year to interrupt pest reproduction cycles. This system of using several natural mechanisms to control pests is termed *integrated pest management* (IPM) and can also include the release of hordes of sterilized male pests to decrease the reproductive capacity of the pest species.[60] Studies of IPM in Britain showed increased harvests, and farmers' profits increased by 20 percent. Experiments are being conducted in several countries to identify bacteria, fungi, insects, or artificial pesticide-laced lures that will prey on and kill crop pests and diseases.[61]

In conventional farming, specific insecticides are sprayed on the crops to kill a particular insect pest. Chemical controls can dramatically reduce pest populations for a while, but they kill natural predatory insects and cause an ultimate increase in the pest population. As noted earlier, we lose a higher percentage of our crops to insects today than before modern insecticides were developed. One farmer who switched from conventional to organic agriculture described the devastation caused by insecticides as appalling. He examined his field three weeks after planting alfalfa and applying a potent insecticide and found "there was nothing but dead bugs, dead birds, dead worms—and little alfalfa shoots. Hardly any of the dead things were crop pests. We stood in amazement. It was shocking. It was the last time I ever did anything like that."[62] The observation of this farmer is not surprising because only 1 percent of pesticides applied actually hit the target organisms. Most reach non-target sectors of the agroecosystem or spread to surrounding ecosystems.[63]

It is worth noting that minor crops—defined as vegetables, fruits, and nuts—comprise only 2 percent of the acreage planted in the United States but use about 17 percent of the pesticides.[64] The development of pest control substitutes for these crops is slow when compared with their development for the major grain crops. The minor crops are grown on far fewer and more diverse acres than the major crops, making private investments for safer pesticides for minor crops less profitable and therefore less likely.[65]

Some of the 99 percent of pesticides that do not reach the target pests reach the farmers who apply them. Studies show that farmers who work with pesticides develop Parkinson's disease, immune system malfunctions, mental illness, and several kinds of cancer more often than the general public does. Recently researchers have found that men with higher levels of three common pesticides in their urine have dramatically lower sperm counts and a higher incidence of irregular sperm.[66]

The effects of organic farming on the immediate environment have been studied by the Soil Association, Britain's organic certification body, which examined studies conducted in Europe between 1987 and 2000.[67] It found substantially greater levels of both abundance and diversity of species on organic farms, including:

• Five times as many wild plants, including 57 percent more species. Several rare and declining wild plants were found only on the organic farms.
• Twenty-five percent more birds at the field edge and 44 percent more in the fall and winter.
• Sixty percent more insects that birds eat.
• Three times as many nonpest butterflies.
• One to five times as many spiders and one or two times as many spider species.
• Dramatic increases in soil biota, including earthworms.

A similar study by the British Trust for Ornithology in the UK found 109 percent more wild plants and 85 percent more plant species on organic farms than on non-organic farms.[68] Furthermore, organic farms supported 32 percent more birds and 35 percent more bats.

Organic farmers occasionally use naturally occurring pesticides on their crops, for example, botanical insecticides such as rotenone (an insecticide derived from the roots of tropical plants that is highly toxic to fish) and pyrethrum (an insecticide made of dried African chrysanthemums), sulfur and copper compounds, and a variety of other traditional pesticides permitted in organic agriculture. Some commentators have suggested that residues of these natural pesticides are present in organic foods and are as harmful as the pesticide residues of conventional, artificial crop chemicals.[69] There is no evidence to support these claims. The botanical insecticides tend to break down rapidly in the environment and are used by a relatively small fraction of growers, ordinarily only as a last resort. Consequently these substances are not expected to leave residues in foods. They are therefore exempt from tolerances as set by the EPA, and no agency routinely tests for them.

Crop Diversity
Organic farming is characterized by crop rotations and the use of cover and forage crops.[70] For example, corn yields are generally 5 to 20 percent higher when grown

with a two-year rotation with soybeans than under continuous corn cultivation. Researchers are not sure what causes this rotation effect (that portion of the enhanced yields not explained by the added nitrogen from the legume), but it may result from the negative effect on soil-borne pathogens. The rotation may have its beneficial effect by "starving" the pest in years when a nonhost crop is planted. For example, wireworms feed on potatoes but not on alfalfa; hence a rotation of potatoes with alfalfa can be beneficial in reducing wireworm populations. The worms basically starve when the field is planted with alfalfa. However, insects, like humans, are clever when they are hungry, and there is some evidence that insect pests can adapt their life cycle to a regular rotation of crops.[71]

Another natural method of pest control is altering the time of planting. If the pest is one that emerges early, the farmer can plant a crop that goes in later. The opportunity to use this method is limited by the length of the growing season or type of crop, and so may not be practical for many crops.

Teaching the Young

Speak to the earth, and it shall teach thee.
—Job 12:8

In the past few years, a movement called Edible Schoolyards has arisen. The object of this creative idea is to give school children firsthand experience with soil, seeds, water, plants, and bugs so they will appreciate how crops grow and learn to appreciate healthy and tasty food. The edible plants are grown organically, without pesticides, artificial fertilizers, genetic modifications, or other undesirable aspects of factory agriculture. Urban public school students are provided with a 1-acre garden and a kitchen classroom, and students learn how to grow and harvest seasonal produce and prepare nutritious food. Children learn about the connection between what they eat and where it comes from, with the goal of fostering environmental stewardship.

Food-related activities are woven into the entire curriculum. Math classes measure garden beds. Science classes study drainage and soil erosion. History classes learn about pre-Columbian civilizations while grinding corn. The students learn about the botany and history of the vegetables they are eating. The concept of the program is to engage the students in interactive education in which they develop a new relationship with food.

Biodiversity

This concentration on a few species at the expense of crop diversity makes us extremely vulnerable to catastrophic interruptions in the food supply, through natural or engineered disaster.
—Jean Mayer, *Business Week*, 1990

The problem of pest control is tied to the problem of maintaining genetic diversity, because plant responses to stresses such as pathogens or pests are partly controlled by genetics.[72] Flexibility in response to these stresses is greater when there is more genetic diversity present at the population or landscape levels. The green revolution of the 1950s and 1960s resulted in farmers' planting fewer varieties of crops so that they could focus on a few high-yielding varieties; in addition, the varieties that are now planted have been bred to a high degree of genetic uniformity within each variety. These approaches are a change from past practices, in which farmers planted a large number of different, often locally adapted varieties (cultivars). Natural pest barriers formed by genetically resistant crop varieties have been dismantled. How many of us are aware that the people of the Andes developed 5,000 varieties of potato? Today only about 100 are regularly grown. As few as five varieties probably account for 90 percent of U.S. acreage.[73]

The restriction of the potato crop to so few varieties leaves the tuber open to decimation by the various fungi and other creatures that plague these tubers. The classic case is the potato famine in Ireland in 1845. Potatoes were introduced into Ireland in the mid-1700s, and by the 1800s, Irish peasants were eating a daily average of ten potatoes per person.[74] Potatoes supplied about 80 percent of the calories in their diet. Potato fodder was used to feed their animals, making the people dependent on potatoes not only for vegetable nutrition but also for milk, meat, and eggs.

In the 1840s, disaster struck. Three successive years of a microscopic fungus and heavy rains rotted the potato crops in the ground. Without potatoes, the peasants and the animals went hungry. More than 1 million of Ireland's 8 million inhabitants died of starvation, and almost 2 million emigrated. The population of Ireland was reduced by 23 percent. Genetic diversity in the potato crop might have mitigated this disaster. Of course, Americans eat a much more varied diet than the people in Ireland did 170 years ago, so that decimation of one type of crop such as potatoes would not cause a famine such as the Irish faced. But the principle is clear: it is not a good idea to concentrate on only a few varieties of a crop. A new strain of this potato blight hit potato crops in Europe, Asia, and Latin America in the 1980s. By 1994 it had spread to North America and today threatens millions of dollars in crops. It is resistant to all fungicides yet developed.

In 1966, the fungal disease known as Karnal bunt swept through the U.S. wheat belt, ruining over half of that year's crop and forcing the quarantine of more than 290,000 acres. The popularity of only a few varieties of wheat with low resistance to the disease facilitated the crop failure.[75]

In 1954, a virulent strain of black stem wheat fungus named Ug99 wiped out 40 percent of the North American wheat crop. Since then, farmers everywhere have grown varieties that resist stem rust, but in 1999 the fungus reappeared, this time in Uganda. It has evolved to attack these resistant varieties; almost no wheat crops anywhere are immune.[76] Since 1999, stem rust spores carried by the wind have spread widely and are about to enter North Africa, Turkey, the Middle East, and India. Scientists estimate it will take five to eight years to breed new varieties of wheat that can resist the evolved Ug99 fungus, but the spores may reach the United States before then.

About twenty pampered plant species make up the bulk of modern agricultural production; eight are grass species. Americans have no concept of the wide variety of colors and flavors of each grain, vegetable, and fruit that used to be available to their palates. The National Academy of Sciences concluded in 1972 that "U.S. agriculture is impressively uniform genetically and impressively vulnerable." Only one genus of rice, composed of two species, is cultivated today, a far cry from the large number of the black, purple, and red varieties that grew in the wild. Most Americans probably believe all rice is white, a reasonable assumption based on a walk through the supermarket.[77]

Today's corn on the cob is unrecognizable from its wild relatives. Corn was once a spikelet of erratic seeds that sprouted at the top of the plant like a roadside grass, with no tidy row of plump kernels to sink your teeth into. This was the natural state of corn. New varieties developed by cross-breeding during the past hundred years by farmers became as domesticated as house pets, changing characteristics to suit their human trainers or cultivators. These new plants became dependent on a package of added fertilizer and pesticides, washed down with irrigation water. None of them would survive on their own if tossed out into the real world. In 1970, the southern corn leaf blight destroyed 60 percent of the corn crop in one summer.[78] Seventy-one percent of U.S. corn acreage in 1991 was planted with just six varieties. Crops that are very similar to each other in yield and appearance are also similar in their susceptibility to disease.

In the early 1900s thousands of varieties of apples were grown. Eighty-eight percent of them are extinct today, and just two varieties account for more than 50 percent of the current apple market.[79] Pear varieties have been similarly decimated. In 2000, 73 percent of all the lettuce grown in the United States was iceberg.[80] We have lost hundreds of varieties of lettuce with flavors ranging from bitter to sweet and colors from dark purple to light green. Do you think that nearly all tomatoes are shaped like the earth: smooth, round, and wider at its equator than at its poles? If so, you are wrong. There are hundreds of varieties that are not cultivated commercially, with colors ranging from near-black to near-white. They come in an astonishing variety of shapes: long and cylindrical, pointed, lobed, and ridged. Some are striped, and some have fuzzy skins like peaches. Some are meant for canning, some for slicing, and some to be eaten while you are standing in the garden looking for tomato worms. The monoculture of conventional industrial agriculture has reduced the natural diversity of nearly every major food crop in terms of varieties grown, color, size, and flavor. The UN Food and Agriculture Organization (FAO) estimated that more than three-quarters of agricultural genetic diversity has been lost in the past hundred years.[81]

The Rural Advancement Foundation International conducted a study of the seed stock readily available in 1903 compared to the inventory of the U.S. National Seed Storage Laboratory in 1983.[82] They found an astounding decline in diversity. The United States has lost nearly 93 percent of its varieties of lettuce, 95 percent of its cabbage, 93 percent of its carrots, 94 percent of its cauliflower, 81 percent of its tomatoes, 94 percent of its cucumbers, 94 percent of its peas, over 96 percent of its sweet corn, about 91 percent of its field corn, more than 95 percent of its tomato, and almost 98 percent of its asparagus varieties. Ninety-seven percent of the crop varieties once listed by the USDA have been lost in the past eighty years. This represents not only an environmental disaster but also a staggering reduction in food choices available to us and future generations.

Nine varieties of wheat and six varieties of corn occupy half of all the land devoted to these crops.[83] In 1903, seed catalogues in the United States listed 408 edible pea varieties. Only 25 remain today and, by 1970, just 2 varieties comprise 96 percent of America's commercial pea crop. On average, perhaps 90 percent of the crop varieties grown 100 years ago are no longer in production and are not maintained in major seed storage facilities.[84] The concentration on a few crop cultivars at the expense of crop diversity makes us extremely vulnerable to catastrophic interruptions in the food supply through natural or engineered disaster.

In an attempt to stem these losses, forty-eight countries in 2004 signed the International Treaty on Plant Genetic Resources for Food and Agriculture, which has three main objectives: encouraging the conservation of genetic diversity, promoting the cultivation of a wider variety of crops, and sharing the benefits that come from exploiting these plants. The United States has not yet ratified the treaty.

As daunting as the loss of crop diversity is for humankind, there is the possibility of an even more disastrous calamity. Suppose an Armageddon-like catastrophe such as plague, nuclear war, or a large asteroid struck the earth and wiped out most crops. How might the survivors redevelop agriculture? In March 2007, work began on a project to protect edible crops against such an event.[85] The project, located on a remote Arctic island, is called the Svalbard International Seed Vault. It will take about a year to build and will contain the biological foundation for all of agriculture. When completed, the vault will house the seeds of up to 3 million different crops, to serve as a last-resort seed bank to permit the regeneration of the world's food supply. Eventually the facility will house samples of every known crop variety that can be grown from seed, from the tropics to the highest latitude. It will be an insurance policy for human civilization.

Unfortunately, there are questions about the long-term viability of stored seeds.[86] Experience at the Leibnitz Institute of Plant Genetics and Crop Plant Research in

Germany indicates that the vast majority of plant seeds cannot be stored for more than forty years without losing some of their ability to regenerate, however low the temperature at which they are stored. The seeds must be used every so often to grow plants in greenhouses, field plots, or laboratories, or they will die. This requires money, manpower, knowledge, and rigorous quality management, things often in short supply.

The genetic diversity of livestock has been similarly diminished in recent decades.[87] The FAO has warned that we are witnessing a potentially catastrophic loss of domestic animal breeds throughout the world, with 1,350 of the 6,300 breeds registered by it at risk of extinction or already extinct. The reasons include war, diseases, and urbanization. But the biggest threat is farmers' preferences for breeds best suited for intensive agriculture. In 1962, in her seminal book *Silent Spring*, Rachel Carson argued that humankind's best chance for long-term survival depends on having a minimal impact on planetary ecosystems and that the biodiversity on which we are ultimately interdependent must be maintained. We ignore her warning at our peril.

Productivity

I hope, some day or another, we shall become a storehouse and granary for the world.
—George Washington, 1788

Intensive agriculture and factory farms have increased their crop yields but clearly have generated serious environmental problems and cannot be sustained indefinitely. Can sustainable (organic) agriculture match the productivity of conventional agriculture? The answer from most studies is yes. Nearly all scientific surveys have found that yields from organic fields are comparable to those of conventional systems, especially over the long term.[88] One study, however, found yields to be 20 percent lower.[89] Organic farming is not a return to farming like our grandfathers did. It is a sophisticated combination of old wisdom and modern ecological understanding and innovations that help harness the yield-boosting effects of nutrient cycles, beneficial insects, and crop synergies. It is heavily dependent on technology, and not only the technology that comes from a chemical plant.

The longest-running study comparing organic with conventional farming practices is the Rodale Institute Farming Systems Trial, a review of which was published in the journal *BioScience* in July 2005.[90] The results to date of this twenty-two-year study revealed that organic farming used 30 percent less fossil energy, conserved more water in the soil, caused less erosion, maintained better soil quality, and conserved more biological resources than conventional farming. And in drought years, organic corn yields were 22 percent higher than conventional ones, and nitrogen levels in soils farmed organically increased by up to 15 percent.[91]

When a farmer is converting from conventional farming to organic methods, yield will be significantly less because soil fertility has been decreased by chemical applications and repeated monocultures and artificial pesticides and fertilizers are no longer being used to compensate. But within a few years of sustainable agriculture, the soil will recover its lost potential and give yields that are about equal to those obtained using conventional farming methods. No costly artificial fertilizer or pesticides will be needed, half the fuel energy per unit yield will be obtained, and the number of nutrient-cycling microbes, worms, and helpful fungi will be increased. An added benefit of farming organically is that organic systems are more resilient in maintaining productivity in drought years that lead to disastrous failure in conventional agriculture.

Nevertheless, the transition from conventional to organic farming involves increased managerial and production costs. Organic farming is labor intensive and requires specialized equipment and other substitutes for synthetic chemicals, and the farmer generally must pay higher prices for organic seeds and other specialized inputs. And because of the need for longer crop rotations to control pests and crop diseases and the need to plant other than high-value crops for maximum pest control, net returns can be reduced.

Because organic systems use crop rotations rather than continuous monoculture, the acreage devoted to any given crop will be lower, and there will be a different mix of production over the long term. That is, an organic farm can yield as much corn as a conventional farm in any given year, but over a four-year period, the conventional farm will produce more total corn. But the conventional farm will produce no soybeans or other companion crop used in rotation by the organic farmer. For the organic farmer, the key concept is diversity and the health of the soil. For the conventional farmer, the key concept is monoculture and maximum short-term profit. To some extent, organic farming is a calling similar to a religious ministry or deciding to teach chemistry in high school rather than accepting a higher paying job in industry.

Costs, Prices, and Profits

Money may not be everything, but it's way ahead of whatever's in second place.
—Classic American saying

In the U.S. Midwest, farmers who produce corn and soybeans organically find that because of fewer outside inputs, their net profits equal or surpass those from conventional production.[92] In Britain, it was found that "friendly farming techniques" can be up to 40 percent more profitable than conventional methods.[93]

This is powerful news for America's beleaguered farmers but may not translate into lower prices for consumers because the larger conventional farms are heavily subsidized by the federal government and many of the much smaller organic farms are not. Nonorganic produce does not carry its full cost. In addition, the external costs generated by conventional farming practices are not paid by the individual consumer, who does not realize that these production costs are passed on to the federal government and are paid for by tax revenues, costs such as polluted waterways, more dirty air because of additional fuel use, and the storage and disposal of surpluses. In the United States, the total environmental and public health costs of pesticide use alone are equivalent to almost $1 in externality costs for every $1 of pesticide sold in the country.[94]

The rapid growth of the organic food sector of American agriculture (figure 4.2) has recently resulted in the entry of Wal-Mart, the nation's largest grocery retailer, into the organic food market. It has announced that its organic produce will be only 10 percent more expensive than conventional produce, a major saving for the organic food consumer. Some organic food advocates emphasize the beneficial aspect

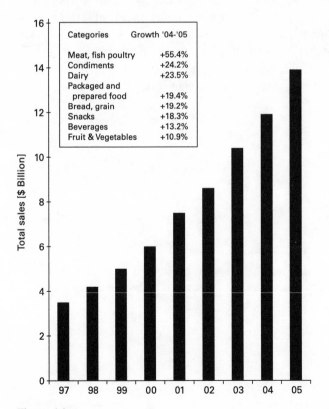

Categories	Growth '04-'05
Meat, fish poultry	+55.4%
Condiments	+24.2%
Dairy	+23.5%
Packaged and prepared food	+19.4%
Bread, grain	+19.2%
Snacks	+18.3%
Beverages	+13.2%
Fruit & Vegetables	+10.9%

Figure 4.2
Growth of various sectors of organic food, 2004–2005. *Source*: Organic Food Association, 2005.

of having giant corporations in organic food retailing. They point out that it will expand the amount of land that is farmed organically and the amount of organic food available to the public. Others concerned with organic food emphasize the dangers that behemoths like Wal-Mart pose. The company has a reputation for driving down prices by squeezing its suppliers, which will include farmers, whose share of the food dollar has already been declining for decades.

There is also the concern that large companies will use their muscle in Washington to weaken organic standards. This has already happened. In 2005 the Organic Trade Association, which represents Kraft, Dole, and other large corporations, successfully lobbied in Washington for a measure that allows certain synthetic food substances in the preparation, processing, and packaging of organic foods.

Another danger is the almost certain increase in the amount of organic food imported from other countries. Today only 10 percent of organic food sold in the United States is imported, but Wal-Mart's purchasing power almost guarantees the globalization of organic food. Like every other commodity global corporations touch, organic food will henceforth come from wherever in the world it can be produced most cheaply, which often means Third World countries whose organic standards would not pass muster in the United States. It is difficult enough to certify the safety of conventional products entering the country (see chapter 9). How can Americans expect verification that an imported foreign product adhered to American organic standards?

5

Genetically Modified Food: Food Fights Among Adults

If we do not change the direction we are going, we are likely to end up where we are headed.
—Chinese proverb

What is it that 92 percent of Americans say they do not want to eat but eat anyway? Answer: Genetically modified food.[1] What is it that 90 percent of Americans want labeled but is not? Answer: Genetically modified food.[2] What products, unwanted by most Americans, are present in two-thirds of processed foods in your supermarket? Answer: Genetically modified soy, corn, canola, and ingredients derived from them.[3] What is going on in the United States with regard to our food supply? Why are the clear and overwhelming desires of Americans about the food they eat being ignored by the government? Why doesn't the Food and Drug Administration, the Department of Agriculture, or the Environmental Protection Agency do something about this? Can anything be done about this situation?

How serious is the danger posed by genetically modified (GM) food? Growing GM crops is prohibited by the European Union (although seven member countries grow small amounts), where 70 percent of the people oppose them.[4] They are considered to be so dangerous by some African countries (Mozambique, Zambia, Zimbabwe) that even though they are facing starvation, they have refused to import these foods unless they are first ground into powder so they cannot be planted.[5]

The United States dominates the growing of GM crops (table 5.1). Twenty-one other countries grow smaller amounts of GM crops. An additional 29 countries have granted regulatory approvals for GM crops for import and feed use and for release into the environment. In 2005, GM crops represented almost 7 percent of world farm acreage,[6] and the percentage has been increasing at an average of 13 percent a year between 1996 and 2006 (figure 5.1). GM acreage has risen from about 4 million acres in 1996 to 282 million acres in 2007.[7] Most of the GM acreage in the United States (table 5.2) is grown on industrial farms. In 2007, 91 percent of America's soybean acreage and 73 percent of corn acreage was planted with GM seed.[8] However, many governments and people around the world believe there are

Table 5.1
Percentages of genetically modified crops in the countries that plant them. Total GM acreage is 282 million

Country	Percentages
United States	53.5
Argentina	17.6
Brazil	11.3
Canada	6.0
India	3.7
China	3.4
Paraguay	2.0
South Africa	1.4
Uruguay	0.4
Australia	0.2
Philippines	0.2
Romania	0.1
Mexico	0.1
Spain	0.1

Note: Eight other countries have trace percentages of these crops.
Source: ISAAA (2007).

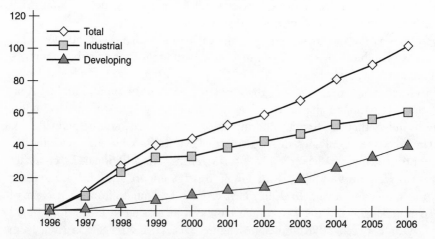

Figure 5.1
Global area of genetically modified crops in millions of hectares, 1996–2006. *Note*: 1 hectare = 2.471 acres. *Source*: Clive James 2006. (isaaa.org/resources/publications/briefs/35/executivesummary/default.html

Table 5.2
Acreage and percentage of land area planted with genetically modified crops

Rank	Country	Acres (in millions)	Percentage
1	United States	134.9	53.6
2	Argentina	44.5	17.7
3	Brazil	28.4	11.3
4	Canada	15.1	6.0
5	India	9.4	3.7
6	China	8.6	3.4
7	Paraguay	4.9	1.9
8	South Africa	3.5	1.4
9	Uruguay	1.0	0.4
10	Philippines	0.5	0.2
11	Australia	0.5	0.2
12	Romania	0.2	0.1
13	Mexico	0.2	0.1
14	Spain	0.2	0.1

Source: International Service for the Acquisition of Agri-biotech Applications (2006).

serious concerns about eating foods that contain GM ingredients. In this chapter I consider the reason GM crops were developed, see who is for and against growing and eating such products, and try to reach some conclusions about their safety and effects on the environment.

What Is a Genetically Modified Organism?

It would be a bleak world indeed that treated living things as no more than separable sequences of information available for disarticulation and recombination in any order that pleased human whim.
—Stephen J. Gould, *Discover*, 1985

What's happened more and more ... is that the gene for common sense and judgment has been eroded all to hell and it doesn't function anymore.
—Norman Borlaug, Nobel laureate and father of the green revolution

Those who approve of genetic engineering of plants view them as a refinement of traditional crop improvement.[9] Traditionally crop improvement was accomplished by selecting and crossing the best-looking, largest, or best-tasting samples of a certain plant and saving the seeds to plant for next year's crop. The genetic composition of the plant was modified by making crosses and selecting superior genotype combinations ("better" genes). The technical term for this process is *hybridization*. For example, a tomato plant that bears sweeter fruit might be crossed with one that has better disease resistance. To do this, it takes many years of crossing and backcrossing generations of plants to obtain the desired trait. Along the way, undesirable traits may appear in the tomatoes because there is no way to select for one trait without affecting others.

Examples of recent successes in conventional hybridization have occurred in China, where the country's most popular strain of hybrid rice has been cross-bred with a wild cousin, boosting yields by 30 percent; and in West Africa, where a new hybrid variety of dryland rice grows so fast it outcompetes the weeds while resisting drought and disease and delivering more protein. Philippine farmers are selectively breeding rice for tolerance to salinity. Corn breeders in Africa have developed more than fifty new varieties that are tolerant to drought. In Australia a researcher has developed a form of "sentinel" corn that turns red when it needs watering, signaling growers to water the rest of the crop. The investigator is currently working on developing self-cloning crops that spread like dandelions.[10]

Thousands of years of selective breeding have boosted the yields of crops, the milk production of cows, the quantity of meat on cattle, and the colors and sizes of our flowers and dogs. Corn was bred by ancient Native Americans probably from the scrawny wild grass known in Mexico as teosinte with an "ear" barely an inch long; the modern plant that dominates American crops is nothing but grass with monstrously hypertrophied seeds. The wild tomato has been transformed from a fruit the size of a marble to today's giant, juicy beefsteak tomato. Virtually everything humankind exploits has been genetically modified in a major way. Hybridization has included chemical and radiation mutagenesis to create new varieties, involving major and quite unknown genetic modifications followed by a selection of useful outcomes. There has never been a public outcry against these techniques, which are certainly not natural, possibly because they were not aware of them. But we should be aware that the crops grown for food are radically different from those that existed in the natural state. The crops we eat could not survive in the wild. They are, in some sense, artificial. The same is true of the farmed animals we eat. Domestic animals are not just wild species that have become domesticated; they are wild species that have been completely transformed by humans. This was the necessary precondition of our ability to use them.

Hybridization, fertilizers, irrigation, and pesticides gave rise to the green revolution of the 1950s and 1960s, an international movement that believed malnutrition in developing countries was caused mainly by protein and food energy deficiencies. It resulted in substantial increases in the production of the major staple grains: rice, wheat, and corn. However, cereals provide few nutrients other than calories and protein, and milling (in the case of rice and wheat) removes most of their vitamins and minerals.[11] Because the cereals have displaced the local fruits, vegetables, and legumes that traditionally supplied these essentials, the diet of many people in the developing world is now dangerously low in iron, zinc, and vitamin A and other micronutrients.

Hybridization has been, and still is, an important tool for plant improvement, but it has limitations. First, it can be done only between two plants that can sexually mate with each other. The two plants being mated must be of the same or closely related species; the mixing or recombination of genetic material can occur only between two plants that share a recent evolutionary history. Hybridization employs processes that occur in nature. The new traits being emphasized are limited to those that already exist within the species. Because of this, traditional plant breeders know where the gene or genes being transferred are going to end up on the host plant's genome, not necessarily in a biochemical sense but in a practical sense. Second, when plants are mated in this way, many traits are transferred along with the trait of interest, including traits with undesirable effects on yield potential. And it takes a long time, normally many years of cultivation, to determine such things. Moreover, there seems to be a limit to the yields of crops. Their physiological potential may have been reached.[12]

The genetic engineers who work for the major chemical companies believe that GM is simply a technological extension or refinement of traditional hybridization. GM has the advantage that we can change individual genetic characteristics rather than mixing all those in one plant variety with all those in a different variety. It works like this.[13] All organisms are constructed of cells, and each cell contains an assortment of things needed for the organism to function. One of these is elongate structures called chromosomes, and each cell of the organism contains an identical set of chromosomes. Each chromosome consists of a single molecule of DNA, the helical assemblage of linked atoms that contains the genetic code for the organism. Each species has a specific number of chromosomes.

A gene is a segment of a chromosome, and each chromosome harbors many genes. A chromosome can be said to be a group of linked genes and is the instruction manual of the plant.[14] The genes are the "words" that describe the plant and tell it how to reproduce itself, and the words are written using four letters, A, C, G, and T, that are abbreviations for the four chemicals of which the chromosome or DNA is composed. The same four letters (chemicals) make up the genes in all living organisms but are arranged differently in different organisms, just as the words in a novel are arranged differently from the words in a comic book. That is the reason a dog is not a cat, a fly is not a worm, a wheat plant is not an oat plant, and *The DaVinci Code* is not a Superman comic. A gene, an assemblage of A, C, G, and T, in some order, can function in many organisms or books like words can. A word such as *color*, *chew*, *shoe*, or *car* has the same meaning regardless of which type of literature it appears in. Genes are the instruction manual the plant uses to copy itself.

The number and arrangement of genes in a cell is called its *genome*, the size of which varies with the complexity of the organism. The human genome has about 25,000 genes. Genes are very small; for example, a half-pound steak contains 750 trillion genes.[15]

A GM organism is one that has been modified by the insertion of a new gene by humans—a gene that is foreign to the plant. The gene holds information that will give a new trait or characteristic to the organism. The gene may mean "smell," or "vitamin A," "kill grasshoppers," or some other desired characteristic. A biochemical process is used to cut up strings of DNA and select the required genes, which are then inserted into the DNA of the plant that is to be engineered. The DNA segment, the gene, can be from the same or a different organism, or it may be a sequence synthesized in the laboratory. Producing a genetically modified organism (GMO) alters the cell biology of a living organism in ways that are not possible by natural processes. We have overcome the sexual barrier. For example, a gene from a fish can be inserted into a tomato plant, bacterial genes can be put in corn, or a daffodil gene can be put into a rice plant. A strawberry can be given a flounder gene that makes it frost resistant, a bacterial gene that confers antibiotic resistance, and a virus gene that "turns on" the other added genes. A strawberry is no longer restricted to acquiring genetic material from other strawberries. Potatoes have been given genes from bees and moths to protect them from potato blight fungus.[16] They were also given a gene from a chicken to confer bacterial resistance.[17] Grapevines have been injected with silkworm genes to make the vines resistant to a disease spread by insects.[18] Another gene-splicing novelty is a goat that carries a gene from spiders, allowing it to produce silk in its milk.[19] Jellyfish genes have been added to wheat to make the plants glow whenever they need water.[20]

And this is only the beginning. Between 2005 and 2007, a group of scientists on a specially equipped sailboat trolled the world's oceans for bacteria and viruses and examined their DNA, resulting in a doubling of the number of genes in earth's biological dictionary.[21] And the rate of discovery of new genes was as great at the end of the voyage as it was at the start, suggesting that the biological dictionary is far greater than we had imagined. The possibilities for interspecies gene transfer seem endless.

The ability to genetically modify an organism is possibly the most technologically powerful development the world has ever seen. And with exceptional power comes exceptional risk. The technology has the awesome power to break down fundamental genetic barriers not only between species, but between humans, other animals,

Figure 5.2
Cornfish, a possible new creation. *Source*: J.-F. Podevin.

and plants (figure 5.2). In the extreme of a science-fiction film, all the world's genes could be tossed in a dish to see what kinds of plants and animals might appear as the genes combine in previously unknown ways. Figure 5.2 shows a possible new creation of a cornfish. Do you debone it or butter-and-salt it? Do you sprinkle it with dill and coriander before cooking, or coat it with butter and salt after heating? Does it have a fishy taste? When alive, can it conduct photosynthesis?

Gene splicing has dangers as well as benefits. When the very nature of living things is being changed, it is never certain what characteristics may show up in the future. Perhaps the gene insertion affects not only the one characteristic of the organism that is being changed or added to, but has ramifications in other areas of the organism's functioning that do not show up until many years or reproductive cycles later. This possibility is certainly a matter for concern. However, GM foods have now been eaten by hundreds of millions of people for more than a decade, perhaps some billions of person-years of consumption, and there is no evidence that anyone's health has been harmed. There has not been a single verified allergic reaction or GM-caused stomachache or even a case of hiccups.[22]

But it is worth keeping in mind that the function of most genes in an organism is not known, and interactions between and among genes are known to occur. Our understanding of gene functioning is still in its infancy, and tinkering with the unknown is always risky business, particularly when the unknown is the very basis of life. And an organism, such as a plant, once created, reproduces and mutates forever. Its pollen will drift on the wind and contaminate everything it touches in neighboring areas. It cannot be put back in a bottle like spilled salt on the kitchen floor or cleaned up like a gasoline spill on the driveway. A new type of organism has been created by humans, something new in the biological world. We are reengineering life. Is the laboratory creation of GMOs another example of human technology outstripping human wisdom? The jury is still out on what Europeans have dubbed "frankenfoods."

Why Were Genetically Modified Crops Developed?

I have the feeling that science has transgressed a barrier that should have remained inviolate.
—Erwin Chargaff, father of molecular biology

Anything humans can figure out how to do, they will do sooner or later, if not for a specific reason, then just for the fun of it. We want to be the masters of creation. All of human history demonstrates this. If we are clever enough to do it, we will. It may be cloning humans and other creatures, constructing superbombs, erecting buildings that reach into the stratosphere, synthesizing chemicals that will kill a million people with one drop in a water supply, growing babies in test tubes and choosing their gender, or capping volcanoes so they do not explode. Arguments for and against each invention or technological development can always be made by proponents and opponents. Discussions among ethicists or futurists in the society about such proposed new developments are uncommon, and even when they occur, their conclusions are generally ignored or unknown by the public and politicians in the rush to show how clever humans are. Commonly a new way to make money is involved, and money is the mother's milk of politics, a field of human endeavor in which preparing for the next election often takes precedence over everything else.

In the case of genetic engineering, money has been the prime motivator. The reason gene splicing was developed is so that agrobiochemical companies could own their own seed supply and control the means and methods of food production, and profit, at each link in the food chain. The farmer is put in an even more serflike position than would otherwise be possible. The major chemical companies have invested billions of dollars in the technology, have political clout in Washington, and have successfully lobbied government agencies to allow them to be essentially self-regulating and allow GM foods to enter the food chain and the environment without announcing it to the public. The FDA has ruled that GM products are "substantially equivalent" to conventional products and therefore need not be labeled, regardless of what the public wants.[23]

With this rationale, one could argue that frozen peas are substantially equivalent to fresh peas and need not be labeled as frozen, or that additives such as preservatives do not make a product significantly different and need not be mentioned on a label. But both of these differences from the natural condition of consumable products are required to be identified. Apparently the U.S. government has decided that altering the genetic makeup of a plant is a less drastic intervention than adding a preservative to a product or freezing it. Sticking a new gene in a plant is apparently not an "additive," a definition that can be described only as Orwellian. Although only four countries are significant growers of GM crops, at least thirty-seven have labeling requirements for crops with transgenic content above certain thresholds.[24]

Consumers have a wide variety of religious, ethical, and environmental preferences in their food choices, and they cannot exercise them without comprehensive
labeling. For example, it is possible to put pig genes into the DNA of a tomato. Although many Jews and Muslims will say, "If it looks like a tomato, then it is a tomato," others may not want to eat a tomato with pork genes. Vegetarians may not
want a tomato with pig characteristics, and Hindus may not want a fruit laced with
cow genes.

The National Academy of Sciences, America's most prestigious scientific organization, and the International Council for Science, an umbrella organization for
more than a hundred national academies of science, agrees with the FDA on the issue of substantial equivalence.[25] The Union of Concerned Scientists does not agree,
and has raised alarms about possible dangers. So have nearly 700 scientists from
seventy-four countries in an open letter submitted to the World Trade Organization
in 1999.[26] The commercial companies that produce GM foods support the FDA and
use its conclusion as the basis for strongly opposing the labeling of products that
contain GM ingredients. They are aware that, as stated in 1994 by Norman Brak-
sick, an executive in one of Monsanto's affiliates, "If you put a label on genetically
engineered food you might as well put a skull and crossbones on it." The American
public is skeptical about eating GM foods, as are the majority of Europeans and
Japanese. As of 2008, Europe's supermarket shelves are free of almost all biotech
produce. Top retailers shun GM foods.[27] Many wheat-importing countries have
said they will not buy any U.S. wheat if biotech varieties are grown. Because of
this, Monsanto in 2004 put on hold its plans to market genetically engineered
wheat.[28] Wheat would be the first GM crop used primarily for human food to be
commercialized. Corn, soybeans, and cotton are used primarily for animal feed or
fiber.

The Precautionary Principle

Reports that say something hasn't happened are always interesting to me, because as we
know, there are known knowns; there are things we know we know. We also know there
are known unknowns; that is to say we know there are some things we do not know. But
there are also unknown unknowns—the ones we don't know we don't know.

—Donald Rumsfeld, secretary of defense, 2003

Much of the dispute about the use of GM crops revolves around the application
of the precautionary principle, which says that when an activity raises threats of
harm to human health or the environment, precautionary measures should be taken
even if some cause-and-effect relationships are not fully established scientifically.[29]
The proponent of an activity, rather than the public, should bear the burden of
proof.

Current decision-making approaches ask, "How safe is safe?" "What level of risk is acceptable?" "How much contamination can a human or ecosystem assimilate without showing any obvious adverse effects?" These policies give the benefit of the doubt to new products and technologies, which may later prove harmful. The approach stemming from the precautionary principle asks a different set of questions: "How much contamination can be avoided while still maintaining necessary values?" "What are the alternatives to this product or activity that achieve the desired goal?" "Does society need this activity in the first place?" The precautionary principle does not require industry to provide absolute proof that something is safe. It does not deal with absolute certainty. On the contrary, it is specifically intended for circumstances where there is no absolute certainty. The perpetrator or company must demonstrate beyond a reasonable doubt that the activity or product is safe; it is not up to the rest of society to prove that it is not.

In the case of GM foods, industry believes it has satisfied the precautionary principle by years of experimental testing before their products were introduced into the food supply. Their point of view is supported by the FDA, the National Academy of Sciences, a UK government investigative panel, and the fact that there is no evidence of harm to humans after many years of consumption of GM foods. The public and many scientists are not so sure. They cite the dangerous loss of biodiversity, loss of nutrient value in crops, contamination of traditional and organic crops by fugitive GM seed, increasing control by a very few large chemical companies over America's food supply by the patenting of GM seed, the increased control of farmers by major corporations, and the fact that farmers' bottom line is not improved by using GM seeds. Weed management costs may be lower, but seed costs are higher, or yields may be slightly higher, but fertilizer costs are also higher.

Farmers say they like the convenience of GM crops. They used to have to spend much time and effort spraying only the weeds around the plant, not the plant itself, but modern technology has produced Roundup Ready GM plants. These are plants that are resistant to glyphosate, a chemical produced by Monsanto that kills all green plants. Farmers can blanket-spray the whole field with glyphosate. They can cover more acres more quickly when spraying and do not need to worry about weed management. But the development of weed resistance to herbicide-tolerant (HT) sprays is increasing and is spreading around the world. About one-quarter of midwestern farmers report they have weeds resistant to glyphosate.[30] As of 2005, 181 species of weeds are known to have developed resistance globally, and the number is growing rapidly. As the number increases, it will defeat the purpose of Roundup Ready crops on a farm. Anti-GM scientists are fearful that "superweeds," resistant to all available herbicides, may develop.[31]

Recently, research has shown that Monsanto's Roundup can kill human cells at very weak doses, disrupts the human endocrine system, and disrupts sex hormones at nontoxic levels. The investigators concluded that the herbicide is far more toxic than glyphosate alone.[32]

What Have Been the Chief GM Objectives So Far?

Genetically engineered organisms are the latest element of a continuing process by petrochemical and pharmaceutical companies to control the world economy. It is a system based on agro-tyranny.
—Tom Evans, 2000, U.S. Congressman from Iowa, 1981–1987

Three GM food crops dominate the commercial market: soybeans, corn, and canola. (GM cotton is also widely grown, and some of it is consumed in the form of cottonseed oil.) The agricultural traits that have been added to these crops include herbicide resistance on 48 percent of American agricultural acres; pesticide resistance on 21 percent; changes in ripening behavior or resistance to bacteria and viruses or sterility promoters, 9 percent each; and changes in fat content, 4 percent.[33] Eighty-nine percent of America's soybean crop in 2006 was genetically modified, as was 75 percent of canola and cotton and 61 percent of corn. All of these percentages continue the steady upward climb of previous years in GM acreage. Plants are made herbicide tolerant by introducing bacterial genes that either allow plants to tolerate the effects of a weed killer (Monsanto's Roundup, which contains glyphosate) or chemically break the weed killer down into a nontoxic form. Herbicide tolerance is intended to make weed control easier for farmers. Spraying a broad-spectrum herbicide will kill the majority of unwanted plants without harming the GM crop. GM varieties of corn, soybeans, and canola are immune to Roundup pesticide and termed "Roundup Ready." Aventis produces herbicide-resistant Liberty Link soybeans, resistant to its herbicide Liberty, the trade name for the chemical glyphosate.

Syngenta tried for twelve years to use conventional breeding (hybridization) to develop corn that was resistant to the corn borer, a dominant insect pest in corn, but ended up with a variety that reduced the pest damage only about 10 percent.[34] Gene splicing has almost completely eliminated damage from the corn borer.

Plants are made insect resistant by using a set of insecticidal toxins produced by the soil microorganism *Bacillus thuringiensis* (Bt) that act by binding to the gut of the insect that ingests them, killing them.[35] Bt corn, Bt soybeans, and Bt canola are big moneymakers for the companies that produce such seeds, and the courts have ruled that they can be patented. Courts have ruled that gene transfer in effect creates a new organism despite the fact that the FDA has ruled that the "new organism" is "substantially equivalent" to the old one. Because it is legally a new organism, the creator can control its use and distribution. A farmer cannot save seed from this year's Bt corn, soybean, or canola plants and plant it next year. This is what was done previously with the conventionally hybridized crop, but it is illegal with Bt crops. The farmer must buy seed each year from Dow, DuPont, Monsanto, Syngenta, or one of the other chemical giants. Seed is no longer free; it must be purchased anew each year. Twenty years ago, there were thousands of seed companies. Today ten companies supply one-third of the global seed market.[36] Ninety-one percent of all GM crops grown worldwide in 2001 were from Monsanto seeds.[37]

There is, of course, the expectation that widespread resistance by insects to the Bt toxin will develop over time, and all the agrobiotech companies are working on insecticidal toxins as potential successors to Bt crops. The goal of the research is to engineer not just for toxins that could replace Bt but for other toxins that kill pests unaffected by Bt, such as the corn rootworm. The scenario does not differ essentially from the current one; in place of a pesticide treadmill, we would substitute a gene treadmill. The perpetual arms race between farmers and pests would continue but would include an additional biological dimension. Transgenic plants, designed to secrete increasingly potent combinations of pesticides, would vie with a host of increasingly resistant pests.[38]

The agrobiotech companies are currently working on GM products that for the first time will have added value for consumers. This may increase the acceptance of GMOs by the public in the United States and elsewhere. Monsanto has developed GM soybeans that have reduced or no transfats, fats implicated in heart disease.[39] In its pipeline are soybeans with more unsaturated fats, which it hopes will lower a persons "bad" cholesterol, and soy enhanced with omega-3 fatty acids, which also provide a cardiovascular benefit.[40] Researchers at the Scottish Crop Research Institute have taken two genes from algae and one from a fungus and added them to a weed to produce essential fatty acids normally obtainable only from fish.[41] Other researchers in Europe have added bacterial genes to potatoes to enable them to make extra beta-carotene, the pigment in carrots that on consumption is converted into vitamin A.[42] Private and governmental interests have developed an unpatented GM variety of rice that is resistant to drought, pests, and disease.[43] It seems clear that future GM products will carry traits that are beneficial for nutrition and health. Healthful products that are profitable for the agrobiotech giants will arrive in the near future.

Golden Rice

Perhaps the most publicized healthful GM product so far has been yellow or golden rice. Rice is the world's number one food staple, eaten every day by 3 billion people, nearly half the world's population. Over 90 percent of the world's rice is produced and consumed in Asia, and rice provides 50 to 80 percent of their daily calorie intake.[44] But rice contains far less iron than any other cereal grain and contains large amounts of a compound that binds iron from other sources in the digestive system, preventing the physiological absorption of as much as 98 percent of all iron digested.[45] In addition, rice is the cause of vitamin A deficiency in about 4 million children in the world. This deficiency is thought to kill up to 2 million people a year and leads to blindness in 500,000 others in developing countries.[46] GM golden rice contains two genes from a daffodil and one from a bacterium to generate rice that contains high levels of beta-carotene, the precursor to vitamin A, which gives the rice its golden hue.

Although golden rice has promise, it is several years from approval and is still far from a solution to the vitamin A deficiency problem. The vitamin, like many other nutrients, is most soluble in fat or oil and cannot be absorbed by the bodies of people with fat-deficient diets.[47] And poor Asians eat little fat or oil. Another problem is that for beta-carotene to survive cooking, it should be microwaved or steamed, not boiled. Not many folks in rural Asia have microwaves. In summary, there has been a lot of hype but few proven benefits. If a GM cynic were looking for an example of a much-hyped, very profitable, but useless GM creation, golden rice might be it.

In 2005, scientists completed a genetic map (genome) of the rice plant, the first crop plant to be genetically deciphered.[48] This promises to be a great aid to researchers struggling to improve the healthful qualities of the most important food crop in much of the world. In the short term, completion of the rice genome is expected to speed conventional breeding programs, allowing researchers to produce rice strains that resist drought and disease and grow in colder climates and at higher elevations. These are critical needs as Asia's rapid urbanization reduces the land available for rice cultivation.

Biopharming

One type of GM crop that has been open-air field-tested at several hundred sites in fourteen states is the production of pharmaceuticals from plants. Called *biopharming* by proponents and *pharmageddon* by opponents, the objective of these genetic modifications is to produce "edible vaccines." Plants such as corn and soybeans and animals such as cows (milk) and chickens (eggs) are turned into factories for producing drugs. A modern drug biofactory can take seven years to build and cost $600 million. Growing the same drug in a field could cut costs in half.[49] Why build expensive factories when you can simply grow chemicals? Why not turn a farmer into a pharma? You could be eating your vitamin supplements, cough remedies, and antibiotics in your waffles instead of swallowing them as tablets with orange juice. The breakfast cereals that now occupy one hundred linear feet of supermarket shelf space could expand to perhaps five hundred with specialty items such as sugared penicillin oat flakes or dietetic streptomycin wheat squares. The biotechnology industry already makes dozens of drugs by moving human or other genes into bacteria, yeast, or hamster ovary cells, encouraging these genes to make proteins.[50] So why not move these genes into corn, wheat, and soybeans?

The danger is that crops experimentally transformed to produce pharmaceutical or other industrial compounds planted in a restricted area for testing will mate with crops meant for human consumption, with the unanticipated result of novel chemicals in the human food supply. Seventy-three percent of engineered biopharmaceuticals are incorporated into corn, a prolific pollinator (soybeans are 12 percent; tobacco, 10 percent; and rice, 5 percent).[51] As we are well aware, pollen can drift on the wind or be carried by insects or birds for many miles. Effective segregation is impossible, and experimental plots of "pharmed" crops contaminated fields of traditional soybeans in Nebraska and Iowa in 2002.[52] Because soybeans do not look like corn, they could be easily identified and removed from the contaminated fields. But if the corn had come up inside a corn field, it could have cross-pollinated, and there would be no way to tell where it was. One hundred fifty-five acres of corn surrounding the test site were harvested and destroyed. Without realizing it, we may eat other people's prescription drugs in our cornflakes: blood-thinning or clotting agents, insulin, growth hormones, diarrhea medicines, AIDS drugs, contraceptives. And some of these drugs, such as plant-derived birth control hormones, would not be destroyed in the human gut before they are absorbed into the bloodstream.[53] A researcher at Arizona State University genetically modified potatoes to produce a hepatitis-B protein, and more than half the people who ate the potato produced large numbers of antibodies to hepatitis-B.[54]

Because of these dangers, the major biotech companies in North America agreed in 2002 not to plant GM crops that produce drugs or chemicals in important food-producing regions.[55] Research continues elsewhere, however. The ban was voluntary and was not mandated by the federal FDA. Environmental and consumer groups and the Grocery Manufacturers Association would like to see only nonfood crops such as tobacco used for the production of pharmaceuticals. This would make it easier to trace and control pharm crops from the field to the factory and ensure that residues do not end up in the food supply.

The View from Europe

We come inevitably to the fundamental question: What are people for? What is living for?
—Marya Mannes, *Life*, June 12, 1964

Europeans have been adamant in their refusal to adopt GM crops, although the resistance appeared to be weakening in 2004.[56] The resistance can be interpreted as cultural rather than as based on scientific concerns.[57] Americans tend to view food merely as fuel that keeps bodies operating; mealtimes are viewed as necessary interruptions in the daily pursuit of fame and fortune. It is like our view of stops necessary to add gasoline to the depleted tank in our car during a drive to visit a treasured relative. Evidence for this can be seen in the way we eat, often fast food eaten in cars, at desks, or standing next to the kitchen counter. In Europe, as in some ethnic American families, dining is the highlight of the day. Food is savored at length and is an integral part of culture. Europeans and ethnic Americans often linger over supper and even lunch for hours, turning meals into social occasions. From this perspective, tampering with food by genetic engineering is abhorrent to most Europeans. As expressed by a member of the French Parliament's committee on environmental safety in 2000, "The general sense here is that Americans eat garbage food, that they're fat, and that they don't know how to eat properly."[58]

The way Europeans view food is connected to the reason they have refused to reduce farm subsidies during the seemingly interminable rounds of discussions about them in recent years. While much of American agriculture is devoted to large-scale, single-crop agribusiness, European farms are smaller and often family run, with greater crop variety on farms. The average size of farms in the fifteen-member European Union in 2003 was 47 acres[59] compared to 441 acres in the United States. Farms in Europe are more like farms in the United States before World War II. So while the switch to GM crops is seen simply as a change in business operations in the United States, it is perceived in Europe as a change in the way families earn their livelihoods.

Also, a greater percentage of Europeans than Americans live in rural areas. Changes in their mode of farming by conversion to an agribusiness type of organization are seen by many Europeans as threatening to their very way of life. Such changes are also seen as threatening the traditional variation in the distinctive, locally produced food products that are a source of regional pride.

The View from the Third World

It seems clear that without meeting the basic needs of human beings, concern for the environment has to be secondary. Man has to survive, answer, and attend first to his basic survival needs—food, housing, sanitation—and then to the environment.
—Walter Pinto Costa, 1985, Brazilian member of the World Health Organization

Although big companies spend billions on GM research, little of it goes into looking for specific ways in which GM crops might benefit poor farmers in the less developed parts of the world. The green revolution of the 1960s in the United States was for the most part publicly funded and targeted to help poor farmers. The GM revolution of the 1990s and early 2000s has been largely privately funded. The rise of GM crops and proprietary control over the important aspects of GM processes has certainly been financially rewarding for the large agrobiotech corporations in the United States, but there has been little benefit for the world's 852 million people who live on the edge of starvation; 40,000 die every day from hunger-related causes.[60] It is generally agreed by both sides of the GM debate that GM crops commercialized so far are meant to help farmers in industrialized countries boost their yields and profits, not to help the needy.

Few of the GM seeds now being sold by big biotech companies will make much difference to poor Third World farmers. With their poverty, tiny plots, and desperate need to raise yields of indigenous crops, they do not need the sort of high-tech soybeans carpeting the American Midwest. Some of the crops and traits that have been gene-spliced into canola, corn, cotton, and soybeans are indeed useful to subsistence farmers in sub-Saharan Africa and rural Asia, but they are in need of help with crops grown in their countries, most of which are of much lesser importance in the United States. They need protein-enriched potatoes, fungus-resistant bananas, virus-resistant sweet potatoes, drought-tolerant barley, insect- and disease-resistant cassava, cowpea, rice, wheat, millet, and sorghum. To Westerners, these are orphan crops, much like the orphan diseases we sometimes hear about, the search for whose cures is not seen as commercially profitable by private businesses. In the case of these orphan diseases, research languishes because too few people contract the disease. In the case of agricultural research, the cause is not lack of clients but lack of concern. In addition to help combating insects, weeds, and pathogens, there is a need for crops that can tolerate droughts and excess salinity, as well as the poor soils and toxic levels of aluminum that are common in underdeveloped humid tropical countries. Some GM research is occurring in the poor countries, with help from private foundations in the Western world, but without additional financial help from the wealthy nations, progress is slow.

One obvious reason for the neglect or back-burner position of these needs is that improvements to staples that feed the world's poor offer little potential profit to biotechnology companies that answer first to their shareholders. In arguments about the safety of GM crops, neither the proponents nor the opponents are people of the world who go to bed hungry. They are mostly North Americans and Europeans who conduct their disputes after a meal of overly rich Western food.

Contamination by GM Crops

Pollution doesn't carry a passport.
—Thomas McMillan, Canadian environment minister, 1987

GM pollen carried by the wind or by flying creatures can contaminate neighboring traditional or organic crop fields despite the wishes of the neighboring farmer. Non-GM crops can become "polluted" with genes from the genetically engineered (GE) crops, and biological diversity will suffer. Since their introduction in 1996, GE seeds have contaminated food crops around the world. More than fifty incidents of contamination have been documented in twenty-five countries on five continents, and those are only recorded incidents.[61] For example, researchers in the southern United States have found that more than half of the wild strawberries growing within 150 feet of a strawberry field contained marker genes from the cultivated strawberries. In the central United States it was found that after ten years, more than a quarter of the wild sunflowers growing near fields of cultivated sunflowers had a marker gene from the cultivated sunflowers.[62] In Canada in 2002, following trials of herbicide-resistant canola capable of withstanding a broad-spectrum weed killer, canola in a nearby field acquired similar resistance to three different herbicides.[63] In a UK study, pollen from a canola plant fertilized plants up to 15 miles away, perhaps by bees.[64] In 2006 GM rice was found in U.S. commercial rice supplies, leading Japan to ban imports of American long-grain rice. The United Kingdom demanded that a test be developed to detect GM grains in future American rice shipments.

Everyone who suffers from hay fever knows how far pollen can travel. The record for rapid propulsion is held by the bunchberry dogwood, a ground-covering plant.[65] It accelerates its pollen with a g-force 800 times greater than a rocket taking off. But pollen is moved by more than the wind. It also hitches a ride on passing insects such as bees. Pollen has evolved over millions of years to become highly effective at transporting its genes. Pollen from GM plants can be expected to land on traditional plants. It cannot be stopped. Legislatively required distances between GM crop fields and traditional crop fields are very short compared to the distances pollen can travel and, at best, can only slow the contamination process.

In a pioneering study released early in 2004,[66] the Union of Concerned Scientists asked two independent labs to examine samples of traditional corn, soybean, and canola seeds. The labs found contamination in half the corn, half the soybean, and more than 80 percent of the canola varieties. Contaminating the traditional varieties of crops is to contaminate the genetic reservoir of plants on which humanity has depended for most of its history.

Nearly a hundred farmers in the United States have been sued by Monsanto because Roundup Ready crops were found on their land,[67] with some farmers saving the GM seed and infringing on the company's intellectual property rights. The farmers believe they are being unfairly persecuted by Monsanto because they are downwind from neighboring farmers' GM crops.[68] A member of Parliament in the UK has compared Monsanto's lawsuit to having a burglar break into your house, smash up your belongings, and then demand payment. The farmers are fighting back. In Canada, there is a class action lawsuit by canola growers against Monsanto because of widespread contamination of their traditional crop by GM pollen.[69] Clearly both farmers and Monsanto believe they are being harmed by fugitive GM pollen. Farmers in the United States have lost overseas contracts because their conventional soybean crop was contaminated by cross-pollination from a neighbor's Roundup Ready soybeans.[70]

Perhaps most concerned about fugitive GM crops are America's 12,000 organic farmers.[71] Many of them have been unable to sell their produce because of GM contamination.[72] As one farmer put it, "I no longer see Greenpeace as a radical or nut group."[73] Genetic contamination can come through the sharing of expensive equipment like combines, elevators, and trucks. Grain elevators can be contaminated. And contamination can also come through seeds, whose purity can no longer be guaranteed by seed companies. The costs associated with trying to keep organic seed separated from GM seed are mounting. For farmers, it includes buffer zones whose needed size is uncertain, cleaning equipment, inspections of crops and processing facilities, and frequent testing. Seed testing costs on average about $10 a bag. After-harvest testing can cost $400 per sample.[74] Some organic farmers have given up farming altogether because of financial losses caused by fugitive GM seeds on their crops. Crops in British Columbia have tested positive for genetic modification although grown organically for more than ten years.[75] In 2002 a survey by the California-based Organic Farming Research Foundation found that 8 percent of organic farmers had already lost certification because of GM contamination.[76] Virtually all of the seed corn, canola, and soy in America is contaminated.[77] Over time, there is a possibility that organic farms, perhaps our best attempt to maintain sustainability in agriculture, will be destroyed.[78] In an attempt to slow the progress of this possibility, Denmark, which has no GM farmers as yet, passed a law in 2005 to tax farmers who grow GM crops about $6 an acre. The money will be used to compensate organic or conventional farmers who cannot sell produce at its usual price because of contamination from a GM farm nearby.[79]

Will Genetic Engineering Produce More Food?

Given the present capacity of the earth for food production, and the potential for additional food production if modern technology were more fully employed, the human race clearly has within its grasp the capacity to chase hunger from the earth—within a matter of a decade or two.

—Donald J. Bogue, *Principles of Demography*, 1969

GM crops were marketed on the promise of significant yield increases "because they're driven by progress." However, no commercial GM variety has yet been engineered specifically to have a higher yield; the focus of genetic engineers so far has been on weed and pest management. For this reason, all cases where farmers have had increased yields have been because of decreased damage from pests or reduced weed competition. A cultivar grown in the traditional way should have the same yield as the same cultivar grown with GM seed if insects and weeds are not involved.

What is the experience of farmers using Monsanto's Roundup Ready herbicide? There are conflicting results. Research in 2001 on the effects of GM crops in India, South Africa, China, and Mexico revealed increased yields and reduced inputs of chemical pesticides, as predicted by the agrobiotech companies.[80] And a study by the National Center for Food and Agricultural Policy found the same results in the United States: yields up, farmers' incomes up, and pesticide use down.[81]

However, a study in 2003 by the Northwest Science and Environmental Policy Center found that the planting of GM crops since 1996 has increased pesticide use by about 50 million pounds. Between 1996 and 1998, the cultivation of GM crops reduced pesticide use by 25 million pounds, but GM crops caused an increase in pesticide use between 2001 and 2003 by 73 million pounds.[82] According to the study's author, the reason for the change from less pesticide to more is that farmers must increasingly spray more herbicide on the GM crops to keep up with evolving tougher-to-control weed species. This interpretation is supported by results in both the United States and Canada where "superweeds" have resulted from accidental crosses between neighboring GM crops that were modified to resist different herbicides.[83] A principal reason for converting from traditional crops to those with genetic modification was the claim of better weed and insect control with lower pesticide use. It now appears that the validity of this rationale has lasted for only a few years.

In an exhaustive study of results in North America by the Soil Association in the UK,[84] investigators in Canada, the USDA, and Charles Benbrook, former chairman of the Agriculture Committee of the National Research Council in the United States and now an independent agronomy consultant, have shown that farmers growing Monsanto's Roundup Ready soybeans, the most popular GM crop, experienced yield decreases, generally between 5 and 11 percent, in comparison to farmers who planted conventional soybeans. Decreases in yield reached 19 percent in Iowa. A farmer in Mississippi was awarded $165,742 in damages from a Monsanto subsidiary by a state court for the reduced soybean yield. The Monsanto hype about improved yield was thus costly. Monsanto's GM oilseed rape (canola) caused a yield decrease in Canada of 7.5 percent compared with conventional seed. Of the GM crops studied so far, only GM corn has shown a slight increase in yield, 2.6 percent, not enough to compensate the farmer for the 25 to 40 percent increase in cost of Monsanto's GM seed over traditional seed. And many farmers formerly saved seed from the previous year's traditional crops and did not buy seed at all. In 2001 the USDA concluded that biotechnology so far would "most likely not increase maximum yields."[85] The poor overall performance of most GM varieties may be due to a general problem with GM crops. The task that the new gene performs requires additional energy, which will detract from the plant's capacity to grow normally. So the benefit of decreased labor time in spraying for bugs and weeds and ease of planting and harvesting may not be a good trade-off.

Do We Need Increased Food Production?

At this moment we produce twice as much food as needed to give every person on Earth a physiologically adequate diet. The real problem is poverty. Excess population is a symptom of poverty, not the other way around.
—Barry Commoner, 1990

One of the reasons given for the introduction of GM crops is the 852 million people in the world, 13 percent of the world's population, who live on the edge of starvation. Heart-rending photos of these desperate people can be seen regularly in the print press and on TV. Every five seconds someone dies from lack of food; 25,000 people will die of hunger today. In contrast, in the 365 days of 2006, a record high 10,000 people died from terrorism.[86] Half of sub-Saharan Africans are malnourished, a figure that is expected to increase over the next few decades. GM proponents say that food scarcity will surely diminish with the increased yields that are bound to come as gene "therapy" in crops becomes more widespread and our understanding of gene function increases. How valid is this point of view?

Is the volume of food produced in the world each year adequate to feed the hungry and malnourished? Experts who have studied this question over the past twenty-five years unanimously agree that the answer is yes.[87] The world today produces enough grain alone to provide every human being on earth with 2790 calories a day. Humans need only about 1600 to keep healthy; 2790 is therefore enough to make most people obese![88] And this estimate does not include vegetables, fruits, beans, nuts, root crops, grass-fed meats, and fish. Increases in food production between 1961 and 2000 outstripped the world's unprecedented population growth by 23 percent,[89] and world grain yield per acre increased steadily from 1.1 tons per acre in 1950 to more than 3 tons per acre in 2004.[90] A smaller percentage of earth's inhabitants are in food trouble today than thirty years ago, despite a 62 percent increase in population (figure 5.3).[91] According to the FAO, the number of undernourished people in developing countries dropped from 920 million in 1980 to 799 million in 2000, even though the world's population grew by 1.6 billion over that period.[92] As noted in chapter 3, sunlight, photosynthesis, and soil nutrients are doing well enough on the earth as a whole to adequately feed everyone. Hoped-for increases in crop yield resulting from genetic engineering are not required to feed the world. The problems lie elsewhere.[93]

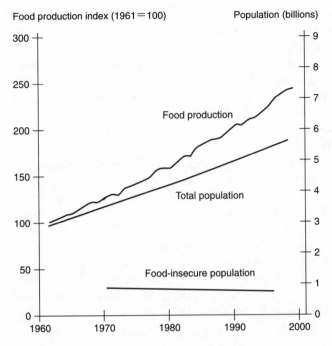

Food production index (1961 = 100)   Population (billions)

Figure 5.3
World food production and population, 1960–2000. *Source*: K. Wiebe, *Linking Land Quality, Agricultural Productivity, and Food Security*, USDA Agricultural Economic Report 823, 2003, p. 1. *Note*: Food insecurity is defined as chronic undernourishment. 1961 = 100.

The reasons for hunger are:

• Tyrannical, undemocratic governments, as in North Korea and Zimbabwe. As Nobel laureate A. K. Sen pointed out more than two decades ago, there are no famines in democratic, representative governments with a free press. The corrupt governments in power in many of the underdeveloped countries control all resources and means of transportation and communication and care little about the masses of hungry citizens. Hunger is caused by a shortage of democracy.[94]

• Inequitable distribution of land. In most Third World countries, the trend is that fewer and fewer people control more and more farm and pastureland. The poor lack access to land, having been driven off so the land could be used for growing high-priced export products instead of diverse crops for local populations. Small farmers are forced to use less fertile marginal lands. As a result, during industrial agriculture's prime years (1970–1990), the number of hungry people in every country except China increased by more than 11 percent.[95] In Africa, for example, while severe famines have occurred in the last decade, industrialized agriculture has achieved record yields for its cash crops, such as coffee, bananas, and flowers, exported to wealthy countries. Nearly 80 percent of all malnourished children in the developing world in the early 1990s lived in countries that boasted food surpluses.[96] Most of the "hungry countries" have enough food for all their people right now.

• Poverty. Those without land on which to grow food or the money to buy it go hungry no matter how dramatically technology pushes up food production. Most Third World people cannot afford to buy food from Western sources, so increased yields in Western countries will not alleviate world hunger. Genetically modified food does not offer a solution to poverty. Hunger can be solved only by a system that promotes food independence.

• Armed conflicts. During the seemingly interminable civil wars in sub-Saharan Africa, food is often used as a weapon. It saves bullets to starve opponents to death. Recent examples of this technique include Sudan, Kosovo, and Angola. Areas planted with land mines also reduce the amount of land available for planting crops.

• Excessive consumption of meat. In 1970 one-third of the world's grain was fed to livestock; today it is estimated to be 40 percent because of increased meat consumption in rapidly developing areas such as China.[97] Using grain to feed people has a lower priority in the current food system than does the profit to be made from feeding it to livestock to enhance the global spread of luxury diets, most of which have deleterious effects on the health of both humans and ecosystems.

While considering world hunger, we should not overlook the United States, which provides more than half the food aid that feeds hungry people around the world. In America, more than 30 million people cannot afford a healthy diet, and 8.5 percent of U.S. children are hungry or, as the USDA now describes it in politically correct terms, have "very low food security,"[98] despite the huge storehouses of cheese, milk, and butter maintained by the government. The USDA estimates that 3.9 percent of households in the United States experienced hunger at some time during 2004. The average daily prevalence of hunger was probably between 0.5 percent and 0.8 percent.[99] Hunger is as unnecessary here as in the impoverished nations. Hunger is real; food scarcity is not.

Hence, even if genetic modifications of cereal crops eventually result in increased yields, hunger among the world's masses will persist. Helping them requires political changes in African and Asian countries that are beyond U.S. control. It also requires financially unrewarding efforts by the international agrobiotech companies to improve the crops important in water-scarce and soil-poor sub-Saharan Africa and parts of Asia. More production in Western countries will not be of much help to the poor and dispossessed in the Third World. The causes of poverty are political and social, not biological or technological. As the FAO concluded, "Bluntly stated, the problem is not so much a lack of food as a lack of political will."[100]

Chicken, Eggs, Turkey, and Duck: Fowl Weather

Man is the only animal that can remain on friendly terms with the victims he intends to eat until he eats them.
—Samuel Butler, nineteenth century

Heart attacks are God's revenge for eating His animal friends.
—Bumper sticker

The consumption of massive amounts of meat in the United States is a relatively new development in our food consumption. Meals based on grains and vegetable protein such as beans did not constitute a fringe diet; that was the way most people ate. Meat and even eggs were considered luxuries, eaten on special occasions or to enhance the flavor of other foods. Recipes in cookbooks from the 1800s and well into the twentieth century focus on stretching a small amount of meat over many meals. Instead of having bacon for breakfast, a hamburger for lunch, and steak for dinner, people reserved meat for Sundays or to celebrate holidays. Today most Americans gorge on meat and dairy products, a habit supported by industrial agriculture and factory farming.

Industrial farming has transformed the raising of farm animals. In the early twentieth century, a typical farm grew a few tens of acres of grain, perhaps a few dozen cattle and hogs, and some chickens to provide eggs for the family's breakfast. Today's animal farm is more like a factory for processing edible animals than an idyllic pastureland. With the change has come incredible cruelty to the animals and animal diseases transmitted from them to humans.

The world's appetite for meat and other animal products, and the ability to obtain them, has quadrupled during the past forty-five years[1] and now forms about half of our daily calorie intake (table 6.1). In 2003, humans slaughtered nearly 46 billion animals, about seven times the human population.[2] That's 126 million a day, more than 5 million every hour, 83,000 thousand each minute, or 1,400 each second. Ninety-three percent of those killed were chickens. We now have 22 billion

Table 6.1
Animal products in the human diet in percentage of calories per person per day

Product	Percentage of calories
Poultry	17
Milk	15
Pork	14
Cheese	13
Beef	12
Eggs	5
Fish	3
Others	21

Source: *AAAS Atlas of Population and Environment* (2000, p. 60).

farm animals, including 16 billion chickens (20 billion according to some estimates), 1.34 billion cattle, 800 million hogs, 256 million turkeys, and 22 million ducks.[3] And the industry expects a 50 percent increase in livestock by 2025, mostly because of increasing affluence in developing countries, particularly China, where half the world's pigs are raised and eaten.[4]

Throughout the world, eating meat is seen as a sign of wealth and prosperity, and consumption is expected to grow by 2 percent each year.[5] World meat production has more than doubled in the past fifty years, from 38 pounds per person per year in 1950 to 86 pounds today.[6] And there are a lot more people alive today than in 1950. As a result, the overall demand for meat has increased fivefold. Despite the urgent pleas of animal rights groups and vegetarians, there is no sign that this trend will reverse in the foreseeable future. In 2002, hogs accounted for 38 percent of consumption; poultry, 30 percent; beef, 25 percent; and other animals (sheep, goat, deer, rabbit) 7 percent.[7] Clearly, humans like to eat the flesh of other animals. Twenty percent of this consumption was in the United States.[8] Americans alone eat more than 1 million animals per hour.[9]

America and Meat

For I was hungered, and ye gave me meat.
—Apocrypha

America is a nation of meat eaters. According to an animal rights organization, in 2003 we killed an estimated 9 billion chickens, 425 million laying hens, 291 million turkeys, 133 million pigs, 41 million cattle, 25 million ducks, and 6 million sheep,

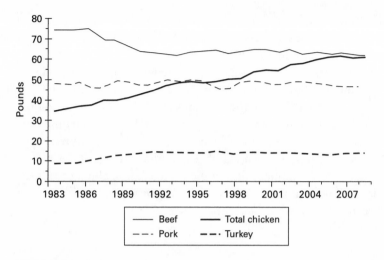

Figure 6.1
Per capita consumption of beef, pork, chicken, and turkey in the United States, 1983–2008. *Source*: American Meat Institute.

a total of about 10 billion animals.[10] In 2003, meat consumption (beef, pork, and poultry) reached 218 pounds per person (and 16 pounds of fish), 77 pounds above the average consumption in the 1950s.[11] But chicken has overtaken beef in popularity. The number of fowl in the United States has increased consistently since 1940, when it was only 500 million, to the present 9 billion (97 percent chickens, 3 percent turkeys).[12] Chicken for dinner became more popular than pork in 1985 and beef in 1992 (figure 6.1).[13] Most of the chicken Americans eat (58 percent) is bought in supermarkets and groceries, 25 percent in fast food restaurants, and 17 percent in other restaurants. How much of this is due to the Colonel and his "finger-lickin' good" chicken is uncertain, as is the contribution of lightly breaded Chicken McNuggets, introduced to the hungry world in 1983.

Chicken McNuggets, although descended from chicken, are chicken only through the miracle of advertising.[14] As the Canadian humorist Stephen Leacock put it, "Advertising may be described as the science of arresting the human intelligence long enough to get money from it." The chicken nuggets are made by pushing chicken carcasses through a giant teabaglike screen to produce a slurry of protein. To this slurry is often added large amounts of water and soya proteins to restore the texture of meat and flavorings and sugars to make up for the lack of meat. The mixture is then made solid by the addition of polyphosphates and gums.

The number of cattle and calves raised in the United States has been declining for thirty years, from 132 million head in 1975 to 97 million in 2006.[15] But beef consumption has remained steady at about 93 pounds per person per year because of the increased weight of each animal. Ten percent of the cattle are dairy cows that produced 20.7 billion gallons of milk in 2005.[16]

Pork consumption has been fairly constant despite the larger-than-average families of Orthodox Jews and Muslims, who do not eat pork. Hogs in the United States number 61 million head, and pork consumption stands at 63 pounds per person per year, numbers that have changed little for decades.[17] The number of sheep on American farms has decreased consistently from a high of 51 million in 1940 to only 6 million in 2004.[18] It seems that mutton and lamb are on the way out of the American menu. Goat has never been a popular dish in the United States, although it is the world's most popular red meat and milk source and is considered a holiday treat in several ethnic American cultures. Deer and rabbit are delicacies not even present on most American restaurant menus.

Chicken

Both the jayhawk and the man eat chickens, but the more jayhawks, the fewer chickens, while the more men, the more chickens.

—Henry George, *Progress and Poverty*, 1879

Chickens are the only animals that you can eat before they are born and also after they are dead.

—Author unknown

Although the chicken arrived in the Americas from Polynesia no later than 1304, the United States is now home to about half the world's chickens. The 9 billion chickens raised for consumption in the United States, called broilers (the word derives from a combination of the two traditional methods of cooking chicken: boiling and roasting), tend to have moister and more tender meat and larger breasts than chickens intended for egg production.[19] Most Americans prefer the larger breast meat of chicken. All chickens pass through similar paths of development. Breeders make certain that chicks hatch at close to the same time by artificially inseminating the mother hens. After fertilization, chicks hatch from eggs in 21 days at an average weight of 2 ounces and are raised for 42 days so they reach a weight of about 5 pounds, at which time they are sent to a slaughterhouse. The life span of a chicken can be up to fifteen years, so a chicken's life may be shortened by more than 99 percent.

The chicken has a beak (or bill) and does not have teeth. Any mastication occurs in the gizzard, accomplished by grit eaten by the small percentage of uncaged or free-range chickens. Many commercial poultry producers do not provide grit to their chickens; instead they are fed a ground feed of fine meal consistency that can be digested by the bird's digestive juices.[20] Farmers watch the birds carefully as they grow, giving them special mixes of food until they reach the desired weight. Improvements that accelerate the process can save farmers large sums of money, and one such money-saving method has recently been developed in Israel.[21] Eggs are injected with a nutrient solution three days before hatching so that the chick is born weighing 5 percent more and requires only forty-one days to be big enough for the slaughterhouse. What sounds like a small change will in fact save hundreds of millions of dollars for the poultry industry.

Whole broilers form only 13 percent of the chicken bought in the United States, having been replaced by more convenient products such as cut-up parts and specialty chicken items. American consumers prefer the white meat of chicken breasts and are willing to pay much higher prices for them than for whole birds or dark meat (thighs, legs, back, and neck; wings contain both light and dark fibers). Chicken farmers have responded to this preference by breeding chickens with larger breasts. In the 1980s, 10 percent of a chicken's weight was breast meat; now it is 21 percent.[22] An exception to the American preference for pure white meat is buffalo wings, spicy chicken wings that originated in Buffalo, New York, that contain both white and dark meat and have become a popular snack. Exports of chicken parts, particularly legs and thighs, have increased every year since 1984.[23]

History of Chicken Farming

Chicken farming in the first two decades of the twentieth century was pursued in the idyllic manner of the small family farm. The farmer might have a few dozen birds wandering around the farm, and the chicken meat was a by-product of egg-laying flocks. Chicken meat was not yet an industry. This started to change in the 1920s, when Cecile Steele of Maryland received by mistake a shipment of 500 chicks rather than the 50 she had ordered. Instead of returning the chicks, she built a small shed to house them and raised them indoors. When they weighed 2 pounds, she sold them for more money than their eggs were worth. Others heard of her success, and soon the Delmarva Peninsula was the center of a broiler industry. It remained so for about twenty years, until just after World War II.[24]

Raising chickens in large numbers and indoors meant they could not hunt and peck for insects, and so researchers developed specialized feeds for them. To get around the health problems the birds developed from the lack of sunshine in the sheds, vitamin D and cod liver oil were added to chicken feed. Then it was discovered that adding antibiotics to the feed caused the birds to gain weight more quickly. Producers quickly controlled the chicken's environment, from lighting, to temperature, to the amount and composition of food. The chicken industry was born.

A major development in the chicken industry came in the mid-1930s when an entrepreneurial truck driver from Arkansas named John Tyson started buying feed plants, building hatcheries, contracting with producers, building processing plants, and acquiring a fleet of trucks for shipping chickens around the country, a program designed to establish a vertically integrated chicken broiler industry.[25] Tyson foods came to own each of its millions of chickens from before they hatched to the day they were slaughtered. This standardized chicken production and propelled the shift toward factory farming. Tyson's techniques were applied to pigs and cattle in the 1960s.

In order to increase profits beyond those obtainable from raising garden-variety chickens, producers became interested in breeding designer chickens. Strains were created that developed a meaty, large-breasted carcass on a minimum of food. Before 1946 it took an average of 112 days and 60 to 100 pounds of feed to produce a 4-pound broiler. Today broilers eat less than half as much feed and reach 4 pounds in about 35 days.[26] The feed is mostly corn, which is inexpensive because it is heavily subsidized in American farm policy. Chickens are machines for converting 2 pounds of corn into 1 pound of chicken.

The vertical integration of the chicken industry in the United States had the same effect on the farmer as it did in the grain industry: the farmer became a small appendage in the body of enormous companies, with little control over the chickens. Large companies own most of the stages of production: as slaughterhouses, feed mills, hatcheries, and distribution. The chickens are raised on contract farms, but the corporations retain ownership during the entire life of the chicken until it is brought to the retail market. The slaughterhouses do not buy or sell chickens, and feed mills do not sell feed. These operations are parts of an integrated firm that purchases inputs such as feed, breeding stock, labor, and energy and turns these raw materials into chickens sold by the marketing department of the company. Most of these companies skip wholesalers and deal directly with the retail outlets, so the chicken is sold only twice: once to the supermarket and once to the customer. Tyson Foods controls 30 percent of the poultry industry, and 98 percent of the world's broilers are descended from birds supplied by just three companies, a rather restricted gene pool.[27]

Certified organic chicken numbers have soared in recent years, with California leading the way. Annual growth rates are estimated to be 23 to 38 percent through 2010. The number of broilers jumped from 38,000 in 1997 to 10.4 million in 2005. Organic laying hens increased from 538,000 in 1997 to 2.4 million during the same period.[28]

Chicken Liberation Groups

The days when the chicken was considered a sacred animal symbolizing the sun are long gone, but there is still much concern by certain parts of the environmentally conscious and vegetarian communities about the restricted and brief lives of chickens and the inhumane conditions in which they are raised. The lobbying organizations of these animal rights activists include Compassion over Killing, Compassion in World Farming, and the group that is perhaps best known, the Animal Liberation Front. These organizations regard with equal horror the needless suffering of all animals, man and beast, and point to the suffering of chickens as an example of human bestiality to animal colleagues. The animal rights groups believe that quality-of-life issues are involved in raising chickens. They point out that farm animals on America's factory farms literally never "have a nice day." From birth, they are caged, crowded, deprived, drugged, mutilated, and smothered on factory farms. A 2003 Gallup poll found that 62 percent of Americans favor passing strict laws concerning the treatment of farm animals.[29]

Chickens that reach America's tables today are raised on huge plantations termed factory farms. Farmers found they could increase productivity and reduce operating costs by mechanization and assembly-line techniques, much as Henry Ford had done decades earlier with automobiles. Animals are dealt with as food-producing machines rather than as living creatures that suffer and feel pain. A factory farm may contain hundreds of thousands or even millions of chickens.

The greatest problem with raising so many animals indoors is the spread of disease, which was largely solved with the development of antibiotics in the 1940s. In the United States, an estimated 70 percent of all antibiotics are administered to farm animals to promote growth and compensate for the unsanitary and confined conditions on factory farms.[30] The routine, medically unnecessary use of antibiotics to promote growth is making disease-causing bacteria more resistant to the drugs, which diminishes their power to treat life-threatening diseases in humans.[31] *Consumer Reports* found that 74 to 89 percent of supermarket chickens were infected with Campylobacter and 3 to 27 percent with Salmonella. Tests revealed that 84 percent of the chickens infected with Salmonella bacteria were resistant to one or more common antibiotics such as tetracycline, streptomycin, ampicillin, and seven others. Sixty-seven percent of chickens infected with Campylobacter were resistant to one or more of the drugs.[32] The antibiotic drugs from animal operations are being found in public waterways as well.[33]

Antibiotics are ineffective against viruses, as the epidemic of bird influenza that began in 2003 illustrates: the mortality rate from this disease is almost 100 percent. In the past, when chickens were dispersed on the farm, diseases were more easily controlled, and the worldwide death of millions of birds would not have been possible. The spread of the deadly H5N1 bird flu virus in recent years is believed by disease experts to be mostly a result of crowded and unsanitary conditions on densely populated factory chicken farms. Air thick with viruses from infected farms is carried for many miles, and integrated trade networks spread the disease through many carriers: live birds, day-old chicks, meat, feathers, hatching eggs, eggs, chicken manure, and animal feed. Chicken feces and bedding from poultry factory floors are common ingredients in animal feed.

Many of the antibiotics used in animals, including penicillin, tetracycline, and erythromycin, are similar to those used for people.[34] Because of the importance of antimicrobials in human medicine, the European Union began phasing out all growth-promoting uses of antibiotics in animals in 1998. In January 2006, it banned all nonmedicinal uses of antibiotics in animals.

A chicken's life on a factory farm begins at a hatchery, without any contact with its mother. Newly hatched chicks are moved to a broiler house, where they are thrown from a crate onto the floor of a building averaging 40 feet wide and 500 feet long. Twenty to thirty thousand birds or more are crowded together in each of these buildings, allowing only 0.5 to 1.0 square foot of space per bird. The smell of ammonia from the decomposing urine and feces of thousands of birds can be over-powering, and because nearly 90 percent of American chickens are raised in the hot and humid southeastern states (table 6.2), adequate ventilation for the crowded chickens is impossible. During heat waves, million of birds suffocate before reaching slaughter weight at 6 weeks of age.[35]

Most chicken droppings, known euphemistically as chicken litter, are used as fertilizer on crops. The droppings are richer in nitrogen, phosphorous, and potassium than cattle manure,[36] but as a fertilizer specialist said, "It's popular with farmers, but even guys who use it will tell you it's a pain and it stinks. The rule of thumb is that you don't spread it next to a church on Sunday." However, a chemical has been discovered that reduces the nose's ability to detect the odor of farm manure.[37] The fragrant organic compound exploits a process known as "olfactory cross adaptation"—the loss of sensitivity to one smell caused by constant exposure to another. The chemical could be sprayed on farm waste to help farmers and their neighbors coexist more peacefully.

Chickens excrete more waste than is needed for fertilizer, with the result that large amounts of toxic runoff occur from this unwanted by-product of chicken life. Expensive lawsuits have resulted. Arkansas, the home of many of America's biggest poultry farms, regularly sloshes excess chicken droppings into the Illinois River watershed, which supplies drinking water for twenty-two public water utilities in eastern Oklahoma.[38] The phosphorous from the poultry litter equals the amount generated by 10.7 million people, 1.7 times the combined populations of Arkansas and Oklahoma.

Recently increasing attention has been paid to the arsenic content of chicken litter.[39] Arsenic in the waste results from arsenic compounds added to poultry feed to promote growth and prevent parasitic infections. The U.S. Geological Survey has calculated that about 700,000 pounds of arsenic are added to farmland in the United States each year, and it is well known that crops grown in soils contaminated

Table 6.2
Broiler chicken production in the United States, 2006

Rank	State	Number produced (in thousands)	Pounds produced
1	Georgia	1,382,100	7,186,900
2	Arkansas	1,185,400	6,282,600
3	Alabama	1,053,400	5,688,400
4	North Carolina	749,000	5,093,200
5	Mississippi	803,800	4,662,000
6	Texas	628,300	3,330,000
7	Delaware	269,100	1,803,000
8	Kentucky	289,000	1,589,500
9	South Carolina	227,100	1,408,000
10	Oklahoma	249,400	1,346,800
11	Virginia	256,200	1,332,200
12	Maryland	271,800	1,304,600
13	Tennessee	213,500	1,088,900
14	Pennsylvania	144,900	782,500
15	Florida	75,000	442,500
16	West Virginia	89,700	358,800
17	Ohio	45,600	241,700
18	Minnesota	45,900	229,500
19	Wisconsin	38,300	168,500
20	Nebraska	5,100	30,600
	Other states	859,400	4,424,700
Total		8,872,000	48,794,900

Source: Department of Agriculture.

with arsenic can accumulate arsenic, a recognized human carcinogen. The European Union stopped using arsenic compounds for poultry in 1998.

As a result of selective breeding, chickens today reach a slaughter weight of about 4 pounds rapidly (increasing their weight twice as fast as they did thirty years ago) so quickly that in about a quarter of them, the weight causes their legs to collapse, and they become lame. Their legs cannot support their extremely overdeveloped bodies. Experiments have shown that chickens in pain selectively choose feed dosed with painkillers, an ability they share with sheep. Apparently the birds are aware they are hurting and recognize that some feeds relieve the pain. The rapid growth also puts excessive stress on the chicken's body, which results in congestive heart failure.[40] Obesity in a chicken has the same ill effects as obesity in a human (chapter 11). An animal activist organization reported in 2003 that 857 million animals of various kinds (probably mostly poultry) did not make it to the slaughterhouse, having suffered lingering deaths from disease, malnutrition, injury, suffocation, stress, extermination, or other deadly factory farming practices.[41] Currently there are no federal welfare laws regulating the raising, transporting, or slaughtering of poultry. As an animal activist publication says, "Every day in the United States, over 20 million birds are abused, then cruelly killed, without ever getting the legal recognition as animals."[42]

The Austrian legislature, in recognition of the many atrocities committed against poultry, in 2004 unanimously passed an anticruelty law forcing farmers to uncage their chickens, making them "free range," in American terminology. Violators are subject to a fine of $2,500 (at the present exchange rate), and in cases of extreme cruelty they could be fined $20,000 and their animals seized by the authorities.[43]

In the United States, Whole Foods Market, the largest retailer of organic foods, is now carrying meat with labels saying "animal compassionate," indicating the animals were raised in a humane manner. Other retailers are making similar animal welfare claims on meat and egg packaging, with terms such as "free farmed," "certified humane," "cage free," and "free range." Many of these terms have no legal definition, however. The shopper may pay twice the normal price for products carrying these labels.

Perhaps most intensely concerned about chicken welfare are the Japanese, who hold an annual memorial service for the poultry they have killed during the previous year.[44] The service held in a Tokyo hotel on April 28, 2004, was particularly intense because of the forced destruction of 300,000 birds to rid the country of the bird flu epidemic that had ravaged flocks. Across Asia about 100 million chickens were sent to early graves. The leader of the service, dressed in black, stood mournfully in front of an altar decked with a pyramid of eggs in clear plastic cartons surrounded by white daisies and lilies. Among the dignitaries at the service was the Japanese minister of agriculture, fishery, and forestry. A director of the Japan Poultry Association spoke for all the mourners when he said, "We wish to express our regret to chickens for having to kill them, while also giving thanks to them for providing us with food. I don't know how chickens feel about it, but humans should show appreciation." He was joined in his sad reminiscences by the head of the Oita Prefecture Poultry Association, who added, "There were so many chickens that had to be sacrificed because of the bird flu. So a memorial service is extremely important." Over two hundred officials and poultry producers observed a minute of silence and then bowed to the stack of eggs. Not wanting to miss a promotional opportunity, poultry officials passed out 6,000 eggs, including some from the altar, at the entrance to the hotel.

At about the same time, Buddhist monks from all over Asia gathered in Hong Kong to pray for the souls of the millions of chickens and ducks that were slaughtered in 2003 to prevent the spread of bird flu. Meanwhile, in California, a couple was fined for allowing their chicken to cross the road.[45]

In Britain, a small-scale pet craze for chickens is brewing, and there are runaway sales for the compact, colorful Omlet Eglu coop. For about $600, buyers receive a coop complete with feed, egg boxes, and two chickens. According to the coop's creator, "People give the chickens names, buy them treats such as strawberries, and get very upset if anything happens to them."

Chicken Feathers

The 9 billion chickens killed every year in the United States leave a lot of feathers behind. With billions of birds massacred every year, the volume of feathers adds up to an estimated 2 to 3 billion pounds per year.[46]

In slaughterhouses, feathers are rubbed off the birds, taking blood and bits of skin with them. They then fall into a water flume that flows around the abattoir, breeding bacteria all the time. By the end, they are so contaminated that they cannot be put into a landfill until they have been sterilized.[47] Imagine a product so biologically repulsive that it must be sterilized before burial. Most of the feathers are burned, buried, or ground up into feather meal and fed to livestock, disposal methods that are costly and controversial. But solutions to the feather dilemma are being pursued by chemists in chemical and electronic laboratories across the United States and in professional symposia with titles such as "Nonfood Applications of Proteinaceous Renewable Materials."[48] Chicken feathers are made mostly of a tough, strong, lightweight protein also found in hair, hoofs, horns, and wool. Because of their good insulating properties, prototype fabrics made from chicken and turkey feathers might be put into skiwear or building insulation. Could a vegetarian in good conscience work in such a building? They are highly absorbent and could be used to filter heavy metals from wastewater or to soak up oil spills. The feathers are also being converted into biodegradable plastic containers, hurricane-proof housing, and electronic circuit boards for computers.

The raw materials the scientists need are being supplied by Tyson Foods, the chicken-producing giant. Because chicken feathers are relatively inexpensive, they could be cost-competitive with materials such as fiberglass and, consequently, ideal for use in automobiles. Although still in an early phase of development, products made of chicken feathers appear ready to fly.

Chicken Fat Biodiesel

Because the cost of soybean oil has been rising in response to the increasing use of soybean biodiesel as a fuel for cars and trucks, some entrepreneurs are turning to chicken fat as a fuelstock for biodiesel.[50] Animal fats are cheap and plentiful, costing 19 cents a pound, while soybean oil costs 33 cents. The chicken fat is supplied by Tyson Foods, the nation's biggest producer of leftover fat from chicken, cattle, and hogs. In 2006 the company established a renewable-energy division to coordinate its chicken fat biodiesel activities.

Rooster Combs

The red combs of roosters and hens are one of the world's richest sources for the sugar molecule hyaluronan, a compound some doctors are calling the next big thing after Botox for removing wrinkles.[51] Hyaluronan production is a response to the presence of testosterone, so roosters have more of it than hens do. In response to this possible financial opportunity, the drug company Pfizer has entered the poultry business and has selectively bred roosters to have huge combs. However, there is a physiological limit to comb size, Pfizer found. When the combs get too big, the rooster's head sags from the weight. Combs have generally been thrown away after the bearer is slaughtered to be eaten, so combs are readily available.

Eggs

It has, I believe, been often remarked, that a hen is only an egg's way of making another egg.
—Samuel Butler, *Life and Habit*, 1878

The Lives of Chickens

The average number of egg-laying hens in the United States is about 342 million, and an average hen on a factory farm lays 260 eggs per year. The average American eats 257 eggs per year, so each of us eats the total yearly production of one hen. Egg consumption has been increasing at a rate of two eggs per year for the past ten years.[52] Eggs are classified by weight, which reflects their size, and despite their apparent simplicity are fairly complex objects. All forty-eight contiguous states produce eggs, but Iowa is the leading producer (table 6.3). There are approximately 260 egg-producing companies with flocks of at least 75,000 hens each, and these companies produce about 95 percent of all the layers in the country.[53] In 1987 there were about 2,500 companies, so it is apparent that the trend in egg farming is the same as for grain: fewer and larger operations. Ninety percent of the companies that existed twenty years ago have either gone out of business or been absorbed by larger competitors.[54] Among the 260 egg-producing companies still in business,

Table 6.3
Egg production in the United States, 2006

Rank	State	Number produced (in millions)
1	Iowa	13,811
2	Ohio	7,507
3	Pennsylvania	6,687
4	Indiana	6,593
5	Texas	5,039
6	California	4,962
7	Georgia	4,811
8	Arkansas	3,267
9	Nebraska	3,129
10	Minnesota	2,940
11	Florida	2,938
12	North Carolina	2,636
13	Michigan	2,391
14	Alabama	2,002
15	Missouri	1,903
16	Mississippi	1,546
17	Illinois	1,307
18	Washington	1,298
19	Wisconsin	1,284
20	South Carolina	1,280
	Other states	77,331 (85% of total)
Total		90,900

Source: USDA.

flocks of 100,000 hens are not unusual, and 64 farms have more than 1 million hens; 11 companies have more than 5 million.[55]

Of the 74 billion chicken eggs America produced in 2003, we exported 350 million, less than 1 percent.[56] Transportation of these large numbers of eggs can be tricky. Canada and Hong Kong are our biggest customers.

The single comb white leghorn hen dominates today's egg industry. This breed reaches maturity early, uses its feed efficiently, has a relatively small body size, adapts well to different climates, and produces a relatively large number of white-shelled eggs, the color preferred by most consumers outside New England. It takes close to six months for a female chicken to mature sexually and start laying eggs. Newly hatched leghorn chicks are sorted by gender when they are 1 day old; 280 million newborn males (equal to the number of female egg-layers) from egg-laying strains are separated and discarded.[57] They will never lay eggs, and they do not grow fast enough to be raised profitably for meat, so they are of no commercial value. They are commonly crushed or ground up, sometimes while they are still alive, suffocated in trash bags, or killed by gassing.

The females that survive become some of the 280 million layer hens that supply eggs. Layer hens live crammed into stacks up to two stories high of "battery cages" 18 by 20 inches in size, five or six to a cage, so that each hen has space equal to three-quarters of a sheet of notebook paper. They spend their brief and almost immobile lives generating multitudes of eggs. Bright lights are used to stimulate egg production. A hen requires 24 to 26 hours to produce a single egg, but one layer produced seven eggs in one day. Occasionally a hen produces double-yolked eggs throughout her egg-laying career. One chicken laid an egg with nine yolks. It is rare, but not unusual, for a young hen to produce an egg with no yolk at all. White-shelled eggs are produced by hens with white feathers and ear lobes.[58] Brown-shelled eggs are produced by hens with red feathers and ear lobes. The shade of color in a yolk depends on the hen's diet. Natural yellow-orange substances such as marigold petals may be added to light-colored feeds to enhance colors, but artificial color additives are not permitted.

Because of the excessive numbers of eggs they are forced to produce, today's laying hens commonly suffer from a condition known as cage layer fatigue, caused in part by insufficient calcium in the hen's body.[59] Egg shells contain 35 to 36 percent calcium. The amount of calcium used for yearly egg production is thirty times greater than the amount in her entire skeleton and, despite calcium supplements administered by the farmer, hens' bones still become brittle. Thirty percent experience broken bones, a percentage considerably greater than players experience in the National Football League.

The decrease in calcium in the hen's bones may be reflected in her eggs.[60] About 10 percent of eggs are lost through shell breakage, but a new technique may help poultry farmers spot which eggs will later crack before they are sent to the packing plant. The key to spotting weak-shelled eggs is a machine that uses a pulse of sound energy to "tap" each egg and make the eggshell resonate. By analyzing the frequency of the resonance, the sensor can determine the condition of the eggshell.

After one year in commercial egg production, hens' laying rates drop off, and they are considered spent. Although most spent hens are killed after one year, about 20 percent are force-molted.[61] Molting is a loss of feathers, a natural occurrence common to all birds. Force molting involves keeping the birds in darkness and withdrawing food and water for 4 to 8 weeks, which shocks their systems into another egg-laying cycle. In recent years, with fewer and fewer slaughterhouses taking spent hens, the number of force-molted hens has been increasing. Some are force-molted more than once. When these hens are finally killed, about half are slaughtered and used for human consumption in soups, pot pies, and other processed foods, where their bruised and battered bodies can go unnoticed. The other half are ground up alive, manually decapitated, crushed, composted, or otherwise discarded, because with an abundant supply of meat birds on the market, slaughterhouses have no use for them.[62]

Hens bred to be superlayers remain economically productive for no more than two years, at which time they are sent to the slaughterhouse. Their natural laying span in a free-range setting is up to twenty years.[63] Many birds become depleted in minerals (thin egg shells) and get osteoporosis because of excessive egg production and may die from fatigue. As a result the hens lay an average of 266 eggs per year (the world record is 371), about ten times more than their wild counterparts.[64] One side effect is that factory-farmed hens are commonly plagued by reproductive problems such as clogged oviducts and prolapsed uteruses.

The battery cages are stacked in long rows (figure 6.2) in huge warehouses. Beneath the rows of cages are computer-controlled conveyor belts that bring in food and water at appropriate intervals and remove eggs and excrement to different storage areas. The overcrowded hens experience severe feather loss as they rub

Figure 6.2
Caged white leghorn hens on a factory farm. Photo courtesy of the Sierra Club.

constantly against the wire cages. Every normal chicken behavior—including nest building, dust bathing, even walking and stretching their wings—is thwarted. The frustrated hens are driven to excessive pecking, and in order to reduce the resulting injuries, part of their beaks are cut off, which makes eating and drinking painful.[65] The beaks are cut off using a specially designed guillotine-like device with hot blades. An infant chicken's beak is inserted into the instrument, and the hot blade cuts off the end of it. The procedure is rapid, taking about 4 seconds; about fifteen birds per minute can be mutilated. Debeaking is a painful procedure for a chicken, and the pain may last for the rest of the bird's life.[66]

Prospects for a Better Life

In response to an increasing chorus of concern about the welfare of America's chicken population, the egg industry in mid-2004 unveiled a new animal care certification logo to appear on egg cartons nationwide. Only farms that adhere to new animal care guidelines may use the logo. So far egg producers representing 200 million layers, 80 percent of the industry, have agreed to participate in the program.[67] Participating producers will be audited yearly through an independent certification program to ensure the new standards are being met.

The guidelines are based on recommendations from an independent scientific advisory committee commissioned in 1999 to review the treatment of egg-producing hens. The guidelines place top priority on the comfort, health, and safety of the chickens and include:[68]

- Increased cage space per hen
- Standards for molting procedures
- Standards for trimming of chicks' beaks, when necessary, to avoid pecking and cannibalism
- Maintaining a constant supply of fresh feed, water, and air ventilation throughout the chicken house and monitoring for ammonia
- Standards for daily inspection of each bird as well as proper handling and transportation
- Availability of a new training video to instruct producer staffs on the proper handling of chickens to avoid injury to the animals

In a poll conducted in 2003, two-thirds of consumers familiar with the logo believed that it signified that the animals were cared for and treated humanely.[69] Twenty-six percent said they would be more likely to buy eggs that carried the logo. In August 2004, the Federal Trade Commission determined that the logo is deceptive because tight confinement, debeaking, and forced molting are still permitted. Experts disagree over what constitutes humane treatment of animals.

The McDonald's hamburger chain, home of the Egg McMuffin, bowed to pressure from animal rights and public health groups and announced in 2000 that it would require its egg suppliers to provide 72 square inches per hen, a 50 percent improvement for most hens.[70] Other McDonald's guidelines will prohibit starving hens to induce them to molt. McDonald's also now requires its suppliers to stop giving birds certain classes of antibiotics to promote growth and gives preference to indirect suppliers that do not use these drugs over those that do. McDonald's is one of the largest chicken buyers in the United States, so the decision to change its standards will likely have a domino effect on the entire meat industry. Burger King and Wendy's have indicated that they will match McDonald's new space allowance and have hired specialists to research and devise new standards to improve animal welfare. McDonald's is considering ways to eliminate the painful removal of much of the hen's beak to keep them from attacking each other. In early 2007, Burger King announced that it would start getting 2 percent of its eggs from hens that are not confined to small cages and that the percentage would more than double by the end of the year. The company said that this seemingly small percentage "is a huge portion of the cage-free eggs available for processing as most cage-free eggs go into the retail grocery business." The percentage of chickens raised in a cage-free environment has risen from 2 percent a few years ago to nearly 5 percent today.

The way most eggs are produced is designed to minimize the cost to farmers and consumers. Departures from the battery cage system increase costs. Egg prices vary considerably across the United States, but comparisons reveal the range of prices shown in table 6.4. Although egg consumption decreased by 8.6 percent between 2002 and 2006, sales of organic eggs increased by 32 percent.[71]

Europeans have been more progressive than Americans with regard to the welfare of chickens. In Switzerland the battery cage system of producing eggs became illegal in 1981.[72] Swiss egg producers now allow their hens the opportunity to scratch on a floor covered with straw or other plant material and to lay their eggs in a sheltered, soft-floor nesting box. The EU has agreed to phase out the standard wire cage

Table 6.4
Cost of chicken eggs produced to different specifications

Conventional (battery cages)	$0.79–1.50
Low cholesterol/high omega-3 (special diet)	$1.99–3.29
Cage free (live on barn floor)	$2.09–3.49
Free range (roam at will)	$2.79–4.49
Organic (vegetarian diet; no biocides)	$2.79–4.99

Source: *USA Today*, April 11, 2006, p. 7D.

altogether by 2009 and to provide a nesting box.[73] In 2004, Austria mandated that farmers must not cage their chickens.[74]

In 1997, the EU ratified the Treaty of Amsterdam, agreeing on a common aim of "improving protection and respect for the welfare of animals as sentient beings."[75] The legal meaning of this phrase remains to be seen, but it is a major step away from the original EU framework treaty in 1957, which classified animals as "agricultural products." The Swiss egg producers and the EU banning of wire cages noted above are examples of progress.

Turkey

The basis of any real morality must be the sense of kinship between all living beings.
—Henry Salt, *The Creed of Kinship*, 1935

The turkey has come a long way, mostly down, from the days when Benjamin Franklin thought it would have been a better national symbol than the bald eagle. In his day, all turkeys were native game birds that could fly up to 55 mph, run up to 20 mph, and had a life span of ten years. Even after they were domesticated, turkeys traditionally spent the bulk of their lives outdoors, climbing trees, socializing, and breeding. But they were generally small by today's standards, a situation that was not favored by commercial turkey breeders. They can grow large, but they grow slowly. To grow to 20 or more pounds, a Heritage bird may be twice as old when slaughtered as the modern Broad Breasted White.[76] The average industrially farmed turkey today weighs 28 pounds.[77] Small size and slow growth are not good for profits. Also, the seventeenth-century turkeys had a balance between white and dark meat that does not suit the palates of modern consumers, who prefer white breast meat. In 2004 the average American ate 14 pounds of turkey annually and paid an average of $1.05 per pound.[78] Israelis eat the most turkey: 28 pounds per person annually.

The 256 million commercial turkeys produced in 2005 are barely recognizable from the ancestor that greeted the pilgrims almost 400 years ago.[79] The turkey in the supermarket freezer compartment at Thanksgiving, a time when Americans eat 15 percent of their yearly turkey consumption (8 percent at Christmas),[80] was born and raised on a factory farm.[81] In their natural state, the 7 million wild turkeys[82] are discriminating eaters and search out the exact food they want: insects, grass, and seeds. This is not a diet to produce the gigantic birds that consumers and commercial breeders prefer, so when the bird is about five days old, the farmer snips off about one-third of the bird's upper beak without anesthesia with a red-hot blade. Clipping the upper beak transforms the lower beak into a kind of shovel so the turkey can no longer pick and choose what to eat. It must gorge on the highly fortified, protein-rich corn-based mash provided by the farmer. Fortified food and selective breeding have produced today's enormous Broad-Breasted White turkey, which can grow so large so fast that the weight of their huge breasts causes hip disorders and bowlegs, and many of them fall under their own weight. Like broiler chickens, factory-farmed turkeys are prone to heart disease and leg injuries as a result of their grossly overweight bodies. Studies have shown that over 90 percent of male breeding turkeys have hip disease. Because of their deformed shape, turkeys cannot fly or mount and reproduce naturally, so their sole means of reproduction is artificial insemination. The tens of thousands of sexually mature turkey hens on an average turkey farm are inseminated weekly, and the toms are milked two or three times a week for semen collection.

The turkey's toenails are also clipped when the beak is mutilated. The clipped toenails, together with the shovel beaks, prevent the huge birds from picking at the feathers of their neighbors, and even cannibalizing them, in the crowded conditions of industrial turkey production.

Mass-produced turkeys hatch from their eggs twenty-eight days after fertilization and live to be twelve to twenty-six weeks before being slaughtered, when they weigh about 30 pounds. The world record turkey, a male gobbler (hens do not gobble), was 86 pounds. Full-grown tom turkeys are too big to fit in a standard American oven, so oven turkeys are probably females. Farmed turkeys spend the first three weeks of their lives confined with hundreds or thousands of other birds in a brooder, a heated room where they are kept warm, dry, and safe from disease and predators. Reaching puberty in the fourth week, the turkeys grow feathers. At this point, as many as 50,000 turkeys that hatched at the same time are herded from brooders into a giant barn whose floor is covered with wood shavings to absorb the enormous amounts of waste that thousands of birds produce each day. As with chicken warehouses, the stench of ammonia burns the eyes of the turkeys and the humans who enter the warehouse to care for them. The barns generally lack windows and are illuminated by bright lights 24 hours a day, keeping the turkeys awake and eating. Every bit of natural instinct and intelligence has been bred out of these industrial turkeys, and farmers joke that they will run into a burning building or drown themselves by looking up at the rain.

When large enough to be commercial, the birds, weighing 12 to 60 pounds, are transported to a slaughterhouse where they are hung upside down in shackles on a moving line. After perhaps 6 minutes hanging by their legs on the line, they are stunned by having their head and neck dragged through an electrically charged water bath. Poultry slaughterhouses commonly set the electrical current lower than what is required to render the birds unconscious because of concern that too much electricity would damage the carcasses and diminish their value. Many birds emerge from the stunning tank still conscious. Following stunning, the birds have their throats cut by a mechanical blade before entering a scalding tank that loosens the feathers for plucking. Inevitably the blade misses some birds, who pass into the tank of boiling hot water, that is, they are boiled alive. This occurs so commonly to millions of birds every year that the industry has a term for them: "redskins."

According to the U.S. Census Bureau, there were 8,436 turkey farms in the United States in 2002, each containing an average of 32,600 turkeys. In 2006, Minnesota (with 45 million birds) and North Carolina (with 37 million) raised the most birds, but thirty-three states have turkey farms. Annual production rose regularly from 35 million birds in 1948 to 301 million in 1997 before dropping to 265 million in 2006.[83]

The Salmonella and Campylobacter bacteria rampant in factory-farmed chickens are common in turkeys as well. *Consumer Reports* found Campylobacter in 14 to 35 percent of the turkeys in supermarkets.[84]

In recent years there has been increasing interest in organic turkeys, with the number increasing from 750 in 1997 to 144,000 in 2005. Michigan and Pennsylvania are the major producers.[85]

Duck

Then all of us prepare to rise and hold our bibs before our eyes, and be prepared for some surprise when father carves the duck.
—Ernest Vincent Wright, "When Father Carves the Duck," 1891

About 24 million ducks are raised annually in the United States to provide a fattier and more expensive fowl alternative to chicken and turkey.[86] Nearly two-thirds of the birds are produced by Maple Leaf Farms in California, Wisconsin, and Indiana. The rest are produced in specialized duck farms in a few commercially important duck production areas. Ninety-five percent of commercial ducks are of the Pekin breed, brought to the United States from China in 1873.[87] They have been developed as a meat and egg animal and are favored by factory duck farmers because they reach slaughtering weight early. Poultry scientists have used selective breeding to engineer fast growth rates for more profitability, as was the case with chickens. Under commercial conditions, Pekin ducklings are ready for market when seven to nine weeks old. If ducklings are kept longer than eleven to twelve weeks, new pinfeathers begin to come out, making it difficult to pick them clean for another several weeks. A duck's normal life span is fifteen to twenty years.

Pekins are artificially inseminated and produce eight to fifteen eggs at a time, but broodiness was bred out of them so they would lay more eggs. Pekin ducks rarely sit on an egg clutch and raise a brood. Hatcheries produce large numbers of ducks and brood and rear them in about the same way as baby chicks: in large sheds with thousands of others packed into 1 to 2 square feet of space per duck. Almost all ducks are raised indoors and so have no use for the webbing on their feet. Because they grow rapidly and feather early, ducks do not require as long a brooding period as chickens, and ducks are easier to raise because they are hardy and are not susceptible to many of the common poultry diseases. Ducks are raised primarily for meat; few Americans eat duck eggs.

The holiday retail duck market is greatest from Thanksgiving through New Year's Day. The Peking duck served in Chinese restaurants is very likely a Pekin duck.

7

Cattle, Milk, Swine, and Sheep: Raising Cholesterol

Red meat is not bad for you. Now blue-green meat, that's bad for you.
—Tommy Smothers

Beef, it's what's for dinner! Pork: the other white meat! Real men eat beef! Life is just a bowl of pork chops! Great cheese comes from happy California cows! We love our lamb! Ahh, the power of cheese! Got milk? Meat lovers know! The ad campaigns for meat and milk products never end. But who is winning, and what does it mean for animal farming and human health and longevity? Should we all become vegetarians or even vegans? Having lost the meat consumption race to the chickens, cattle and hogs are battling it out for second place. What does this mean for farmers and the animals involved, the cattle and hogs most Americans enjoy eating? Are these four-legged eatables as unhappy and persecuted as the chickens?

Cattle

We kill these weary creatures, sore and worn, and eat them—with our friends.
—Charlotte Gilman, *The Cattle Train*, 1911

My favorite animal is steak.
—Comedienne Fran Lebowitz

The United States has the largest domestic cattle industry in the world and is the world's largest producer of beef, primarily high-quality grain-fed beef for domestic and export use.[1] We have come a long way from the late nineteenth and early twentieth centuries when a 475-acre area in Chicago was the world's largest meat-packing district.[2] Sandwiched among the slaughterhouses were associated producers of glue, leather, violin strings, and other derivative industries. The cows, hogs, and sheep ready to be slaughtered were kept in vast pens, and after their throats were slit, the carcass parts of the animals were loaded on railroad cars and distributed around the country. Chicago's meatpacking industry evaporated in the mid-twentieth century with the decline of the railroads and the development of refrigerated trucks. Meat plants no longer needed to be near the railroads, so companies moved their slaughterhouses into rural areas, which offered cheaper land and labor.

Most beef found in the grocery store comes from steers, castrated males raised especially for the table. Four firms control over 80 percent of the beef market, led by Tyson Foods, which also dominates the chicken market.[3] American farms raised 97 million cattle in 2006 (of a world total of 1.4 billion), continuing the decline from the 1975 peak of 132 million.[4] Texas has the most cattle, with 13.9 million head. Nebraska, Kansas, Oklahoma, and California follow in rank order. Beef cattle dominate; about 40 percent of Texas's cattle are raised for beef. In 2004, the United States produced 25 billion pounds of beef, and each American ate 67 pounds of it.[5] Forty-two percent of it was hamburger and 20 percent steak.[6] Only 2 percent of our production was exported.[7] The United States both imports and exports cattle, but the types imported are different from those exported. Imports dominate exports, five to one.[8] Both the exports and imports involve mostly Canada and Mexico. Most exports go to Mexico; imports mostly come from Canada.[9]

A heifer (a cow that has not yet produced a calf), like a human, is pregnant for about nine months and may have her first calf when she is two years old. And like humans, cows occasionally have twins. A cow can live for twenty-two years.

The Cost of Beef

The price of animal feed rose sharply in 2007, largely owing to an ever-increasing proportion of the world's grain crops being diverted for the production of biofuels. Wheat rose 70 percent and soybeans by more than 90 percent. Corn prices have also soared, and the increase in beef prices reflects this trend.

Many consumers complain about the cost of beef at the supermarket, but unsympathetic vegetarians have pointed out that there is a more important high cost, and all beef eaters ignore it: the cost to the environment. The disproportionate cost in resources of raising cattle is staggering compared to raising grain. Consider these comparisons.[10]

• Approximately half the energy used in agriculture is devoted to raising livestock. It takes the equivalent of 1 gallon of gasoline (tractors, threshers, trucks, refrigeration) to produce 1 pound of grain-fed beef. To produce the yearly average beef consumption of an American family of four requires over 260 gallons of fossil fuel. There are about 75 million households in the United States.

• It takes 8.5 times more fossil fuel to produce 1 calorie of meat protein than to produce 1 calorie of protein from grain for human consumption.

• On average, it takes about 100 times more water to raise a pound of beef than a pound of wheat or corn (table 7.1).

Table 7.1
Gallons of water needed to produce 1 pound of food

Potatoes	60
Wheat	108
Alfalfa	108
Sorghum	133
Corn	168
Rice	229
Soybeans	240
Chicken	419
Beef	11,982

Source: *New Scientist*, February 1, 1997, p. 7.

· Seven pounds of grain are required to produce 1 pound of beef. That amount of grain would provide a daily ration of grain for ten people. However, cattle eat forage crops for much of their lives, and foraging cattle eat plants that humans do not, such as grass. And more than two-thirds of the feed fed to animals in feedlots consists of substances that are either undesirable or unsuited for human food. In essence, the cattle are converting inedible plant materials to human food.
· While 56 million acres of U.S. land are producing hay for livestock, only 4 million acres are producing vegetables for human consumption.

Clearly, there is a sound environmental argument in favor of reducing the consumption of beef. It is a serious drain on fuel and water consumption, as well as a waste of grain and agricultural land. To top it all off, consider that the negative impacts of the by-product, such as water pollution, high blood pressure, heart disease, and some types of cancer, far outweigh the food value of the primary product.

Structure of the Beef Industry
Worldwide there are more than 250 breeds of beef cattle; 61 of them are present in the United States, but fewer than 20 constitute the majority of the genetics used for commercial beef production.[11] Some cattle were developed for meat production, such as the Angus, and others were developed for milk production, like the Holstein. But the unique ability of all cattle to use grazed forages has led to a beef industry characterized by many relatively small producers and wide geographical dispersal.[12]

Most of the scattered and relatively small farms that raise beef cattle send their animals to central locations for succeeding steps in the farm-to-table sequence, and this is where many severe environmental problems surface. From the farms, the cattle are sent to feedlots, which are concentrated in the Great Plains and corn belt. Feedlots with a capacity of 1,000 head or more are a tiny fraction of U.S. feedlots but market the vast majority of fed cattle.[13] Feedlots that hold as many as 32,000 head market 40 percent of the finished cattle. It is much like the situation in grain farming, where most grain is produced by a few megafarms (chapter 3). The cattle industry continues to shift toward a small number of very large feedlots, which are increasingly vertically integrated with the cow-calf and processing sectors to produce the quality of beef American consumers prefer.

In the feedlot the cattle are fed grain and protein concentrates to increase their weight and produce a uniform quality of beef. Two-thirds of the beef cattle are given steroids and hormones to help with weight gain.[14] In the United States, 70 percent of the corn harvest is fed to livestock. And worldwide, nearly 80 percent of all soybeans are used for animal feed.[15] Steer used to live at least four to five years before being slaughtered. Today beef calves can grow from 80 pounds to 1,200 pounds in just fourteen months on a diet of corn, soybeans, antibiotics, and hormones.[16] Depending on their weight when they enter the feedlot, grain feeding can range from 90 to 300 days.[17] Cattle eat about 10 percent of U.S. corn production; other farm animals eat 60 percent.[18] The average weight gain during this "finishing" period is 2.5 to 4 pounds per day; about 7 pounds of feed are needed for each 1 pound of weight gain in the cow.[19] Feedlot cattle are usually ready for market at fourteen to sixteen months of age and weigh between 1,000 and 2,500 pounds, with the females 200 to 300 pounds lighter than the males. The natural life span of cattle is eighteen to twenty-two years, so slaughter at fourteen to sixteen months shortens their life by about 95 percent.[20]

Feedlot rations are generally 70 to 90 percent grain and protein concentrates.[21] The remainder is hay, added for roughage to enable normal rumen activity. Cattle have complex stomachs with four compartments and must have a lot of roughage to digest their food. Because they are ruminant animals, cattle can digest grass, roughage, food by-products, and other materials. Such feed lot rations, which humans cannot eat, are turned into meat. Sheep, goats, deer, buffalo, and camels, among other animals, are also ruminants.

It is possible to keep the animals on their home farms and let them eat only forage crops throughout their lives, but less than 1 percent of the 33 million cattle slaughtered in the United States each year are raised this way. The number of pasture-raised cattle is increasing, however. Between 2000 and 2004, the number of farms raising grass-fed beef grew from 50 to more than 1,000.[22] The number of cattle raised on forage alone is small because American consumers are conditioned to and demand consistency in taste and tenderness. This is easier to obtain when the cattle are "finished" uniformly by being fed grain. Forage-based fattening systems exist in most cattle-raising areas of the United States, but they may yield beef of uneven quality because of seasonal variations in the amount of forage food available. Yearly variations in rainfall make an adequate supply of food even more uncertain. Hence, raising animals on pasture requires more knowledge and skill than sending them to feedlots. In order for grass-fed beef to be succulent and tender, for example, the cattle need high-quality forage, especially in the months prior to slaughter. This requires healthy soil and careful pasture management, which keeps the grass at its optimal stage of growth. Because high-quality pasture is the key to high-quality beef, many farmers who raise animals on pasture refer to themselves as "grass farmers" rather than "ranchers."

Forage-fed cattle produce beef that is healthier for humans than the beef from animals finished with grain in feedlots.[23] The beef from feedlot cattle have, per pound, nearly four times the amount of total fat (marbled beef), four times the amount of saturated fat, three times the amount of transfat, 100 more calories per 6-ounce serving, only one-fourth the amount of heart-healthy omega-3, less than half as much beta-carotene, and much lower amounts of vitamins A and E, and they are usually fed hormones and antibiotics to enhance their growth. There is a high price in decreased health to pay for better taste.

Today's cattle are nutritionally deficient in other ways as well.[24] Analysis of data from 1940 to 2002 revealed that the iron content (at least) in fifteen different meat items has fallen an average of 47 percent; some products showed declines as high as 80 percent. Milk and cheese have been similarly affected. It is not clear whether the loss has resulted from feedlot practices, loss of plant diversity, or depletion of the soil in mineral content or microorganisms. But all of these likely causes are results of industrial farming.

There is an additional problem with feedlot cattle. As noted earlier, the vast majority of crops grown in the United States are used to feed farm animals, not humans. The crops grown for farm animals contain far higher levels of pesticides than crops grown for human consumption.[25] Over 90 percent of the pesticides Americans consume are found in the fat and tissue of meat and dairy products. In 2002, 47 percent of the samples of fat tissues from beef analyzed by the USDA contained pesticide residues.[26]

Another problem in feedlots is the spread of disease. As was true of poultry, disease is a danger when cattle are forced to live in crowded surroundings, and the corporate solution in feedlots is the same: feed the animals antibiotics regularly. The FDA has refused to ban this practice, although there are concerns that the heavy use of antibiotics has resulted in their declining effectiveness against common bacterial infections in humans. *E. coli* caused eight recalls involving almost 182,000 pounds of ground beef in the first ten months of 2006. A few food service groups have shown corporate responsibility by announcing that they will buy only beef and pork that is free of antibiotics, but much more needs to be done.

Large amounts of the antibiotics fed to livestock are excreted in their manure, which is applied to agricultural land to provide nutrients to the crops. The crops are known to absorb the antibiotics, but the amounts and effects are largely unknown. This question is particularly important to organic farmers, because manure is often the main source of nutrients for their crops.

A steer or cow that weighs 1,100 pounds yields about 550 pounds of beef. The 50 percent of the meat animal that does not result in beef is used in by-products. Only 3 percent of the entire animal is wasted.[27] In addition to protein-rich meat, cattle yield many things popular in American culture,[28] including footballs (called "pigskins" but really made of cowhide), basketballs (eleven from one cowhide), shoes (eighteen pairs from one cowhide), camel hair paint brushes (made from the fine hair inside the ears of cattle, not from camels), cosmetics, soap, glue, crayons, margarine, and many other products.

Veal is beef that comes from young cattle. It is particularly tender and whitish in color. The tenderness is created by taking the animal when it is only a few days old and restricting its movement by confining it in a narrow pen for three to eighteen weeks so it can hardly move, much less turn around. The whitish color results from feeding it food that is deliberately low in iron, so the young animal is anemic.[29] When photographs of these tortured creatures first appeared more than twenty years ago, sales of veal plummeted and have never recovered. Before the photos appeared, Americans ate 4 pounds of veal a year on average. Today per capita consumption is about half a pound a year. Interestingly, the decline in sales of veal has not caused most producers to change their treatment of the animals, only to raise fewer of them.

Cattle Waste

More than half the feed given to livestock ends up in manure, and 1.8 trillion pounds of animal waste from cattle, hogs, and other farm animals are produced each year in the United States.[30] Cattle produce 82 percent of it (known to some as "animal processed fiber"), swine 12 percent, and poultry 5 percent.[31] Animal farms produce 86,600 pounds of excrement every second, more than 130 times the amount of waste that people do.[32] According to the EPA, cattle, hog, and chicken waste has polluted 35,000 miles of rivers in twenty-two states and contaminated groundwater in seventeen.[33] When animals graze on pasture and rangeland, their manure is dispersed across a large area, is not concentrated, and decomposes on the soil. It is an excellent fertilizer, rich in nitrogen, phosphorous, and potassium, as well as many minor nutrients and organic matter. The nitrogen and potassium are mostly in the urine, while almost all the phosphorous resides in the feces. However, when thousands of animals are concentrated in a small area like a feedlot, the amount of manure and urine per square foot increases dramatically. Handling and making use of the manure produced in large feedlots is a significant environmental problem. In 2005, 2,000 tons of cow manure caught fire in a feedlot near Lincoln, Nebraska, attributed to heat pent up by decomposing feces inside the pile.[34] It took four months to extinguish the fire by pulling apart the pile, which was 100 feet long, 30 feet high, and 50 feet high. That same year, nearly 3 million gallons of liquefied manure spilled into the Black River in New York State when an earthen storage reservoir gave way.[35] America's 98 million cattle produce 8.2 billion pounds of waste per year. For each 1 pound of steak, a cow produces 53 pounds of feces and urine.[36] Clearly no runoff of fresh water that has passed through feedlots should flow into any waterway.

Over the years, there have been environmental improvements in operations such as feedlots and stockyards, largely forced by federal regulations. The EPA and its state agencies have increasingly focused on feedlots as point sources of pollution and have become stringent and vigilant in their regulations of potential pollution from them. Stockyard locations need to be changed every so often.

A rather unusual use for cattle excretion is promoted by a firm in New Delhi, India, in line with the Hindu ideology that the cow is a sacred animal.[37] The firm sells filtered cow urine, said to be effective for cancer. There also are beauty products made from cow dung and urine that are said to make teeth glisten and also reduce obesity. Some of the company's products are touted as remedies for diabetes, skin diseases, constipation, asthma, arthritis, mosquito repellent, and pimples.

A different approach has been pioneered by Japanese scientists.[38] Using heat and pressure, they have turned cow dung into vanilla extract. The extract will be incorporated into soap and scented candles but not used in food.

Recently an additional harmful side effect of cattle production has been noted. Animal agriculture uses 30 percent of the earth's land for pasture and feed crops, and cattle today produce more greenhouse gases than the world's automobiles.[39] Gases from manure, deforestation to create pastures, and the energy consumed by livestock businesses together account for 18 percent of greenhouse gas emissions by human activities. Livestock account for more than 9 percent of the carbon dioxide derived from human activities, 65 percent of the nitrous oxide, 37 percent of the methane, and 64 percent of the ammonia. These percentages are growing; world meat production is forecast to double by 2050.

Cattle Liberation

In India, cattle are considered sacred because they are a symbol of life-giving Mother Earth. Many Hindus call cows *Go Mata*, or Mother Cow, and killing cattle is forbidden in most of India. Hindus often bow down before the cows while passing them on the streets to offer their respect; others touch the typically mangy animals to feel their heartbeat.

In New Delhi, with a human population of 14 million, 36,000 bulls and cows wander the streets.[40] But their freedom is tempered by a grim existence: most of their diet consists of garbage, and many of the weaker animals lie in the busy streets and are struck by passing cars. To improve the lives of these sacred creatures, cow asylums, called *gaushalas* in Hindi, have been established and are now the home of one-third of the city's cattle.[41] At these shelters, the cattle are fed fresh vegetables and wheat, and during the very hot and humid summer in New Delhi are given a cold shower once or twice a day. The administrator of one of the shelters says that he and his helpers take care of the animals until their last breath. The shelters are supported by donations from pious Hindus.

Europeans have taken a first small step toward improving the preslaughter life of cattle. Veal calves are arguably the most miserable of all farm animals, kept anemic, deprived of straw for bedding, and confined in crates so narrow that they cannot even turn around. Britain banned this system of keeping calves in 1990, and the EU banned it starting in 2007.[42] It persists in the United States.

Milk

There is no finer investment for any community than putting milk into babies.
—Winston Churchill, 1943

Milk production has a significant number of small producers; most operations have fewer than 100 cows. Like grain farms, beef farms, and hog farms, the number of dairy farms has been decreasing, while the size of the farms has been increasing. In 1993, large farms comprised about 14 percent of all dairy operations, housed just over half the cows, and provided 55 percent of the milk produced.[43] By 2000, large farms were 20 percent of operations, had almost two-thirds of the cows, and produced 71 percent of production. Farms with more than 500 milk cows are about 13 percent of the large operations, house 48 percent of the cows, and provide half of the milk produced.[44] Clearly milk production is occurring on a smaller number of larger operations than in the past. The average herd size of specialized dairy farms has increased every year since 1978. The largest have 18,500 cows.[45] Contrary to commonly seen photos of cows standing hock-deep in pasture, thousands of these docile animals spend their lives in barns standing on cement where they are milked automatically. The cows can tolerate the confinement and cement for only a few years, when many develop leg and foot problems. When this happens, they are culled for slaughter along with their sisters whose milk production has fallen and made them uneconomical burdens to the farmer.

Milk and Health

There currently is a major promotional campaign by the milk industry touting the health benefits of their product. Prominent personalities such as movie stars and fashion models are shown with white milk residues around their mouths, the implication apparently being that milk somehow has made them the recognizable people that they are. How valid is the claim that milk promotes health?

Only babies of various animal species drink milk, suggesting that adults of the species do not need it, and each species of mammal produces milk with a unique composition designed to meet the specific needs of its infants. The milk of animals that need to develop a thick layer of insulating fat, such as seals, has a high fat content. The milk of animals that grow rapidly, such as cows, which double their birth weight in fifty days, is rich in protein and minerals. This raises suspicions about the suitability of cow's milk for human infants. Mother's milk is more suited for consumption by human babies than milk from a different mammal. It contains antibodies from the mother's immune system that helps the infant fight off infection and disease. During the first few days after giving birth, a mother releases colostrum, a yellowish liquid that contains less fat and lactose (sugar) and more protein and antibodies than regular breast milk. After about three or four days, colostrum is replaced by a bluish-white milk that is higher in fats and carbohydrates, reflecting the energy needs of a growing baby.[46] Recent research has revealed that human breast milk contains a wide range of protective and beneficial substances for humans not found in substitutes.[47] As usual, nature knows best how to keep us healthy.

In the past sixty years, there has been an upsurge in the use of artificial formula as a replacement for breast milk. As with other commercial replacements for natural products, the effect on human health has been disastrous: for example, twice the risk of dying in the first six weeks of life, five times the risk of gastroenteritis, twice the risk of developing eczema and diabetes, and up to eight times the risk of developing lymphatic cancer.[48] Breast-feeding also helps prevent obesity in infants with overweight or diabetic mothers. One American nutritionist stated, "Breastfeeding is a natural negotiation between mother and baby and you interfere with it at your peril."[49]

In the controversy over the use of baby formula, a $2 billion market, the producing companies show themselves once again to put profits before health. Helmut Maucher, a major corporate lobbyist and honorary chairman of Nestlé, the company that claims 82 percent of the global baby food market, said, "Ethical decisions that injure a firm's ability to compete are actually immoral."[50] Apparently he believes that increasing corporate profits is more moral than protecting the health of newborns. His viewpoint illustrates the validity of the definition of a lobbyist as someone who asks government officials to make laws that will benefit their clients' companies, regardless of how they affect anyone else.

Lactose, a sugar, makes up about 5 percent of milk and gives milk its sweet taste. Between 30 and 50 million Americans have a genetic defect called lactose intolerance, an inability to digest lactose. Such people have low levels of the enzyme lactase, which is necessary to break down lactose. Certain racial and ethnic populations are more affected than others. Seventy-five percent of African Americans and Native Americans and 90 percent of Asian Americans are lactose intolerant. The condition is least common among people of northern European descent.[51] If a person is lactose intolerant and ingests dairy products, the result is stomach cramps, bloating, flatulence, and diarrhea.

The healthfulness of milk and cheese has decreased significantly since data were first collected in the 1930s (table 7.2). For some nutrients, the loss has been catastrophic, with losses as high as 70 percent. The reason for the losses is not clear, but industrial grain farming and its effect on cattle in feedlots may be an important factor, because factory farming and the growth of large feedlots have been the major changes that have affected the raising of cattle since World War II.

An alternative and controversial method of increasing milk production from dairy cows is through the use of recombinant bovine growth hormone (rBGH), intro-

Table 7.2
Changes in the mineral composition of milk and selected cheeses, 1940–2002

	Milk (milligrams)		
	1940	2002	Change
Calcium	120	118.0	−2%
Magnesium	14	11.0	−21%
Iron	0.08	0.030	−62%
Sodium	50	43.0	−14%
Potassium	160	155.0	−3%
Phosphorous	95	93.0	−2%
Copper	0.02	0.01 or less	None remaining
	Cheese		
	Calcium	Magnesium	Iron
Cheddar	−9%	−38%	−47%
Stilton	−10%	−45%	−57%
Parmesan	−70%	−70%	None remaining

Source: Food Magazine, January–March, 2006, p. 10.

duced by Monsanto in 1994, and now administered to about one-third of U.S. dairy cows.[52] Users report production increases of 10 percent (8 to 12 pounds) per day per cow. The FDA approved the use of rBGH in 1993, although it is banned in the European Union and Canada because of cattle safety concerns rather than concerns about the effect on human health. Canadian studies determined that rBGH increases the risk of mastitis by 25 percent, interferes with the cow's reproductive functions, and increases the risk of lameness by 50 percent. Also the abnormally high rate of milk production strains the animal's immune system, and they are more likely to become sick, leading producers to feed the cows increased levels of antibiotics. Traces of these drugs appear in the cows' milk.

Organic Cattle

Their [animals'] interests are allowed to count only when they do not clash with human interests.
—Peter Singer, *Animal Liberation*, 1975

The number of organic cattle is increasing rapidly in the United States, from 17,326 in 1997 to 182,077 in 2005.[53] Forty-eight percent are milk cows, 20 percent are beef cows, and 32 percent are other types of cows. The USDA livestock standards for organic meat and milk specify that antibiotics, wormers, and other modifications cannot be used routinely as preventative measures, and hormones to promote growth are prohibited. The standards include requirements for pesticide-free pasture and access to the outdoors suitable to the natural and behavioral needs of the particular species. Beef and dairy cows, for example, cannot be confined in pens but must periodically be allowed to roam. Organic farmers must maintain organic pasture for the cattle to graze on. Some organic farmers maintain grass-fed-only operations. Others supplement the forage with grain, but organic cattle cannot be fed animal products because of concern animal residues may carry the virus for bovine spongiform encephalopathy (BSE), or mad cow disease, which is transferable to goats and humans at least. Creutzfeldt-Jacob disease, the human form of this deadly brain-wasting disease, had occurred in twenty-three countries and killed 153 people by the end of 2003.[54] At least one organic meat company has capitalized on the growing fear of BSE with the advertising slogan, "Organic Beef—It's What's Safe for Dinner." Organic beef cows totaled 15,197 in 2001 and more than doubled to 36,113 in 2005.[55]

About 1 percent of dairy farms are certified organic. The number of certified organic milk cows jumped from 12,897 in 1997 to 87,082 in 2005, making them the most abundant organic livestock. Organic milk now accounts for more than 3 percent of all milk sold in the United States, and sales are growing at an annual rate of 23 percent, despite the fact that the average cost of organic milk is double that of conventional milk.[56] Overall milk consumption is dropping by 8 percent a year. Organic dairy sales in large supermarkets are increasing 36 percent annually, and dairy sales accounted for more than 11 percent of all organic retail sales in 2000.[57]

Cloning Cattle

Two farmers each claimed to own a certain cow. While one pulled on its head and the other pulled on its tail, the cow was milked by a lawyer.
—Jewish parable

Most of the types of farm animals Americans eat have been cloned: cattle, pigs, sheep, goats, and rabbits. The FDA has determined that meat and milk from cloned animals is safe to eat, a finding that has cleared the way for such products to reach supermarket shelves and for cloning to be widely used to breed livestock.[58] Someday in restaurants you may be asked not only whether you want your steak rare, medium, or well done, but also whether you want it cloned or uncloned. There are now only 1,000 to 2,000 cloned cattle among the nation's 100 million, so there will not be an influx of cloned animal meat anytime soon,[59] and cloning an animal today costs about $15,000, which is much too expensive to make an animal just for its milk or meat. At that price, a hamburger would cost about $100.[60] Cloning is more likely to be used to make copies of prized animals for breeding. For example, scientists announced in April 2005 the creation of dairy cows genetically engineered to resist mastitis, an infection of the udder that costs the U.S. dairy industry more than $2 billion every year through lost productivity.

Surveys show that more than 60 percent of the U.S. population is uncomfortable with the idea of animal cloning for food and milk.[61] The single biggest reason people give is "religious and ethical." Certainly cloning is not necessary for either nutritional or food supply purposes.

Swine

From an ethical point of view, we all stand on an equal footing whether we stand on two feet, or four, or none at all.
—Peter Singer, *In Defense of Animals*, 1985

Pork ranks third in annual U.S. meat consumption, behind chicken and beef, averaging 51 pounds per person annually.[62] Nineteen pounds of this is fresh pork chops, steaks, ribs, fresh ham, and pork parts such as fat backs (fried pork skins), cracklings (crisp rind of roast pork), ears, tails, heads, feet, neck bones, salt pork, chitterlings (intestines), liver, rinds, pork skin, and tripe (stomach). Pork chops and fresh ham account for 41 percent of the fresh pork total. Thirty-two pounds of pork consumption is processed pork (lunch meats, hot dogs, bacon, sausage, smoked ham, and other items); smoked ham and sausage dominate this category. High-income households eat less pork than lower-income homes.[63]

Hogs and pigs, 61 million of them, are raised in about half the states, with Iowa and North Carolina the leading producers.[64] Only 10,000 are raised organically. Hog production in the United States is, like grain and cattle production, increasingly concentrated on fewer and larger farms. The number of hog farms fell by 70 percent between 1990 and 2002,[65] although hog production has remained about the same. The average swine farm has about 600 animals,[66] but "pig cities" have over 1 million animals. Iowa produces the most hogs, about 15 million, with North Carolina second at 10 million.[67] The reason for consolidation is the same: increased productivity and economic efficiency. The detriments of the industrial animal farms are the same as well: poor treatment of the animals and horrendous concentrations of urine and manure. Spokespeople for the swine industry estimate that as many as 20 percent of breeding sows die prematurely from exhaustion and stress due to impacts of restrictive confinement and accelerated breeding schedules on factory farms.[68]

Breeding sows are kept in cages called gestation crates, which are 24 inches across and 7 feet long. These crates virtually immobilize pigs during their pregnancies in metal stalls so narrow they are unable to turn around. Numerous studies have documented crated sows exhibiting behavior characteristic of humans with severe depression and mental illness. Gestation crates are now being eliminated in the European Union, and in the United States, Smithfield Foods announced in 2007 that it will follow the EU's lead. Burger King has announced that by the end of 2007, it would buy 20 percent of its pork from suppliers that do not use sow gestation crates. Apparently continued pressure by animal rights groups is having a beneficial effect for swine as it has for cattle.

Of the 61 million swine in the United States, over 95 percent are continuously confined in metal buildings. Feed is automatically delivered to the animals, which are forced to urinate and defecate where they eat and sleep. Their waste drops into large pits a few feet below their hooves, so that intense ammonia and hydrogen sulfide fumes fill the pigs' lungs and sensitive nostrils. This follows the cutting off of their tails using a wire cutter and without anesthetic. Pigs, like cattle, who also may have their tails "docked" to make milking easier, use their tails to shoo away insects and to communicate, as do dogs. Farmers believe tail removal from hogs reduces tail biting and subsequent infection, although research indicates a three-fold increase in biting of the remaining tail stub.[69] In addition to the toxic fumes, bacteria, yeast, and molds have been recorded in swine buildings at a level more than a thousand times higher than in normal air.

To prevent disease outbreaks (and to stimulate faster growth), the hog industry adds more than 10 million pounds of antibiotics to its feed, more than three times the amount of antibiotics used to treat human illnesses. Veterinarians consider pigs as smart as dogs. Imagine keeping a dog in these conditions: a crowded cage with nothing to chew on, play with, or otherwise occupy its mind.

The USDA's Census of Agriculture statistics reveal that the number of hogs and pigs produced (a pig that weighs more than 250 pounds is called a hog and has a natural life span of eighteen years; a hog has eight to nine piglets per litter and two litters per year) has changed little since 1985, varying irregularly between 48 million and 52 million as pork prices waxed and waned.[70] Although the number of hogs has not changed, the number of hog farms has, decreasing by 85 percent between 1981 and 2003 from 590,000 to 90,000, while the average number of hogs per farm ballooned from 100 to 766. The 2.3 percent of hog farms with more than 5,000 hogs contain 53 percent of the nation's stock, an increase from 32 percent in 1996.[71]

From Recycler to Polluter

Because of the growth of industrial pig farming, the function of hogs has changed over the years. Hogs used to be not only a food source but also a recycling system. Other livestock are primarily herbivores, but hogs are omnivorous, like humans. Households and farming operations collected food scraps and waste products, such as leftovers from dairy processing, and mixed them together with grain to feed, or "slop," the hogs. In this way, household and farm waste was converted into pork, leather, fertilizer, and other useful products. Now raising hogs is not associated with the recycling of waste but with the emission of huge amounts of waste with serious environmental impacts. North Carolina, with the second largest number of hogs, has been a focal point in the battle over hog waste. Ten million animals located in the poorly drained coastal lowlands in the eastern third of the state have generated major water and air quality problems.[72]

The average adult hog produces up to 17.5 pounds of manure and urine per day, ten times the waste of the average adult human.[73] Human waste is produced at 25,000 pounds per second, compared to the 250,000 pounds per second produced by farm animals.[74] On a farm with 6,000 hogs, this amounts to 50 tons of raw sewage produced each day. Hog waste, unlike human waste, does not go to water treatment plants because it would be too expensive to process like human waste. In pig factories the excretions fall through slatted floors into a basement, where they are periodically flushed into giant outdoor pits called lagoons, which may cover up to 12 acres and are prone to leakage.[75] A study done at North Carolina State University estimated that as many as half of the existing lagoons are leaking badly enough to contaminate groundwater.[76] Typically water is used for the flushing process, so the resulting slurry is up to 97 percent liquid.[77]

Smithfield Foods, the world's biggest pork producer and packer with 26 percent of the market, produces over 11 million hogs each year, slaughtering 80,000 every day.[78] Ninety percent of McDonald's bacon comes from them. Smithfield operates in thirty-six states and has an environmental record that is tough to beat for its disregard of the environment.[79] Its hog facilities spray untreated liquid manure with its four hundred dangerous substances onto croplands or pastures, which quickly become saturated. The waste seeps downward into groundwater or is carried by rain into nearby streams or lakes. The volume of animal waste from the industrial hog farms is much greater than is needed as fertilizer for crops grown on the farms, so manure-containing lagoons are rarely empty.

In 1997 the EPA found 6,900 violations of the Clean Water Act at a Smithfield plant in Virginia.[80] In a twenty-three-month period between 1999 and 2001, Smithfield companies and partners were fined at least fifty-five times. A lagoon rupture at one of its facilities spilled 1.5 million gallons of liquefied hog feces and urine into nearby wetlands. Smithfield has poisoned thousands of miles of waterways and killed billions of fish. A 25 million gallon spill from one 8-acre pig waste storage area killed 10 million fish in the Neuse River in North Carolina in 1995.[81] Smithfield is not alone in its violations of federal clean water rules. A Cargill pork factory in Missouri killed 53,000 fish in the Loutre River when it dumped hog waste.[82] In Iowa, Minnesota, and Missouri, which account for more than a third of hog production, recorded animal waste spills rose from 20 in 1992 (killing at least 55,000 fish) to more than 40 in 1996 (killing at least 670,000 fish).[83]

Environmental problems with swine waste have not abated. Spills have continued, as have other environmental calamities. In 2005 in Nebraska fourteen environmental violations of state laws were reported and included, in addition to spills from hog waste lagoons, dumping dead hogs in the lagoons. In 2007 in Indiana a tanker carrying 6,000 gallons of liquefied hog manure tipped over, spilling half the waste. In 2004 in Missouri breaks in irrigation pipes spilled 26,000 gallons of hog waste onto the owner's property and into a neighbor's pond. In addition to these calamities by hog producers and transporters, coastal North Carolina is a common site of hurricane landfalls, an occurrence whose deluge of water swamps the manure-filled lagoons, sending it into waterways and coastal ecosystems.

Liquid hog manure and urine mixtures have an odor that defies description. Neighboring farmers choke, vomit, and faint from the gases that emanate from hog operations.[84] The smell cannot be removed from skin or clothing, even with the strongest soap. The odors can be so strong that they nauseate people flying in airplanes as high as 3,000 feet above industrial pig plants.[85] Numerous studies show that factory farm workers and downwind neighbors are unusually susceptible to lung disease, eye infections, nosebleeds, gastrointestinal illness, depression, and even brain damage. Every year pig factory workers become seriously ill and die from deadly gases such as hydrogen sulfide, methane, and ammonia that emanate from liquid manure pits.[86] Inside the pig buildings, the ammonia and other gases from manure irritate animals' lungs to the point where over 80 percent of the pigs have pneumonia when they are slaughtered.[87] The fumes are so strong that the animals quickly die from asphyxiation if the ventilation systems fail.

Odors from hog farms have had legal recognition since 1610 in England, and forty-four states in the United States now regulate odors on hog farms.[88]

Genetic Engineering and Enviropigs

Genetic engineering has recently created a conundrum for organic swine farmers and environmentalists.[89] Pig manure, like most other solid animal excretions, contains phosphorous that during rain runs off into surrounding water bodies, stimulating algal growth, depleting the oxygen supply in the water, and killing the fish. In the 1980s, phosphorous pollution killed all aquatic life in a 25-mile-long fjord in Denmark.

Recently scientists have used genetic engineering to create a pig whose manure contains very little phosphorous. The animal has been dubbed Enviropig by its creators, but is considered a disguised Frankenpig by Greenpeace. The scientists constructed a novel DNA molecule that, when planted in a pig embryo, gives the Enviropig the ability to secrete a phosphorous-extracting enzyme in its saliva. The new molecule enables Enviropig to extract all the phosphorous it needs from grain alone, without the phosphorous supplements that farmers now use. This reduces the phosphorous content of their manure by up to 75 percent, clearly an environmental plus. Pigs can also be engineered to digest natural grasses and hay, as cattle and sheep do, reducing the need to grow energy-intensive corn to feed pigs.

The Enviropig is one of many new technologies that poses difficult questions for environmentalists and organic food enthusiasts. Should they remain categorically opposed to genetically modified foods even at the expense of the environment? What happens when philosophy collides with pragmatism?

Another recent genetic modification is cloned pigs that make their own omega-3 fatty acids, potentially leading to fatty bacon that is good for the heart.[90] The saturated fats and cholesterol will still be there, but the genetic manipulation might help balance the bad with a bit of good. Also in the pipeline are cows that make omega-3s in their milk and chickens that contain them in their eggs. It is getting harder to predict whether our great-grandchildren will be eating any natural food at all.

Sheep

Bah, bah, black sheep, have you any wool?
Yes sir, yes sir, three bags full.
—Children's nursery rhyme

There are more than 1 billion farmed sheep in the world, but only 6.2 million of them are in the United States, marking a continual decline from 56.2 million in 1942. Most of the nation's 68,000 sheep farms are located in western states. Texas and California are the largest sheep-producing states.[91] Fewer than 5,000 sheep are raised organically.

Sheep are raised for both meat and wool. Historically in the United States, lamb and mutton have been viewed as by-products of wool production, even though meat production from sheep accounts for about three-quarters of the revenue on sheep farms.[92] Lamb and mutton production varies with the price of wool. If wool prices are high, fewer sheep are slaughtered and lamb and mutton production falls.

In the list of jobs immigrants perform that no American wants, sheep herding probably ranks near the top.[93] There are about 825 shepherds who work on the nation's sheep farms; most of them are immigrants on three-year work visas from Peru, Chile, Bolivia, and Mexico. They live with the sheep from late March until fall in pastures far from town or even a paved road. For weeks on end, visits by the boss are the shepherd's only contact with humans. During lambing season when the ewes give birth, they work twelve- to sixteen-hour days, seven days a week. They live in small trailers, most of them without toilets, heat, or potable water. Salaries are about $1,000 per month.

Mature female sheep (ewes) can breed until about seven years old, are pregnant for five months, and give birth to one or two lambs. If they are not slaughtered, they live about fifteen years. The average sheep farmer raises 86 sheep, but as with cattle, large operations are increasing at the expense of smaller ones. In 2002 there were 74,000 sheep farms, but the 150 farms with over 5,000 sheep produced nearly one-quarter of the U.S. sheep population.[94] Sheep producers manage their flocks on pasture and range forage until they weigh 60 to 80 pounds, when they are sent to feedlots to be fattened and finished for slaughter. The average lamb carcass weighs 68 pounds, a sharp increase from 55 pounds twenty-five years ago. Per capita annual consumption of sheep has decreased during the past forty years from 5 pounds to 1 pound as beef consumption and then poultry consumption increased.[95]

Most sheep meat is sold as lamb and comes from animals less than fourteen months old. Mutton comes from older animals and is less desirable to American consumers. Legs and loins are the most popular cuts, and ethnic Americans are the major consumers. Much of the less desirable cuts go to the pet food industry or are exported. Imports, nearly all from Australia and New Zealand, supply one-third of our sheep meat.[96]

8

Seafood: The Killing Fields

How inappropriate to call this planet Earth, when clearly it is ocean.
—Arthur C. Clarke, 1990

Nearly 71 percent of the earth's surface is covered by oceans teeming with life, with an average depth of about 12,000 feet. The near shore and shallow part of the ocean, where most commercial fishing takes place, is called the continental shelf. It slopes seaward from the beach at a few feet per mile off the Atlantic coast and extends to a depth of about 425 feet, perhaps 50 miles off the coast. At this depth the seafloor slope increases to 200 to 300 feet per mile down to a depth of 5,000 to 10,000 feet, from where the slope changes gradually and merges with the deep sea-floor. Along the U.S. Pacific margin, the continental shelf is very narrow, perhaps only 15 to 20 miles wide, because of active mountain building near the coast.

Availability of Fish

I wiped away the weeds and foam,
I fetched my sea-born treasures home.
—Ralph Waldo Emerson, "Each and All," 1834

Life in the ocean extends to its greatest depths, and as a potential food source for humans, the ocean is unrivaled. With a surface area of 139.5 million square miles and an average depth of 2.3 miles, it has a volume of 320.8 cubic miles. In comparison, life on land is restricted to a thin zone perhaps a thousand feet thick, the height to which birds fly. Go much higher, and the cold temperatures and decreased amount of oxygen make life impossible. "Life volume" near the land surface is only 11.5 million square miles (57.5 million square miles times 0.2 miles), less than 4 percent of the ocean's volume of life. From the standpoint of potential human food supply, the ocean is twenty-eight times more valuable than the land. Actually the figure of twenty-eight times is an underestimate, because about 65 percent of the raw weight of finfish is eaten, compared to 50 percent of the raw weight of cattle, chicken, and pigs and 40 percent of sheep.[1] This difference exists because fish are supported by water and therefore do not have to put as much of their growth energy into bones, so more of their weight is edible. As saltwater livestock, fish are unrivaled.

The world fish catch increased steadily from 19 million tons in 1950 to 88 million tons in 1988, but net change has been near zero since then and stands at 90 million tons in 2003, fifteen years later.[2] Three-quarters of the catch was from ocean waters and 25 percent from fresh waters. The freshwater commercial catch is all from outside the United States, as freshwater fishing is essentially a recreational activity for Americans. Americans caught 5.4 million tons of the ocean catch. Some analyses conclude that the world catch has decreased by 660,000 tons each year, a fact that has been concealed because of politically motivated overreporting of fish catches by China.[3] But whether the ocean catch has stagnated or declined, there is no doubt that seafood consumption by Americans has exploded during the past twenty-five years, sustained by sharp increases in fish farming (aquaculture). Seafood consumption in 2004 was 16.6 pounds per person per year, an all-time high.[4] Children do not eat much fish, so the number per person understates consumption by adults. Nevertheless, Americans still eat nearly six times more meat than fish. Seventy-eight percent of America's seafood is imported.[5]

When the eating habits of other countries are considered, Americans are not big eaters of seafood. According to the FAO, about 1 billion people worldwide rely on fish for at least 30 percent of their animal protein, and as the number of people who live near the coast continues to increase, so does the consumption of seafood.[6] Sixty percent of the human population now lives on or near the coast. Also, studies show that as incomes rise, people consume more fish on average.

Marine organisms, many of them tasty to the human palate, occur at all oceanic depths and in all body sizes (table 8.1), from very small (plankton and algae) to very large (sharks and whales), although the largest concentrations of marine animals occur on the continental shelves at shallow depths, where their food is more abundant.

The world ocean has no natural boundaries, no readily observable natural demarcation lines such as the Rio Grande or the Himalaya or the boundary between land and sea. Who owns the water and the life within it? Clearly it is a valuable resource that all nations might seek to claim as their own, or at least all nations with a coastline might claim. Such an unmarked, unowned area is called a commons, a term brought into common parlance in 1968 by the late ecologist Garret Hardin.[7] Every user wants to maximize use, typically without regard for the long-term health of the commons, and the commons is eventually destroyed. The guiding principle of users of a commons is: if I don't increase my use of it, someone else will. The ocean is a commons, as are the air we breathe and outer space. Use of the air does not deplete

Table 8.1
Top fish species harvested worldwide

Species	Total annual harvest
Herring, sardine, anchovy	22.5
Carp, barbel, other cyprinids	17.3
Cod, hake, haddock	8.4
Tuna, bonito, billfish	6.1
Oyster	4.5
Shrimp, prawn	4.3
Clam, cockle, arkshell	4.3
Squid, cuttlefish, octopus	3.2
Salmon, trout, smelts	2.6
Tilapia, other cichlids	2.2
Scallop, pecten	2.0
Mussel	1.7
Crab, sea-spiders	1.3
Flounder, halibut, sole	1.0
Shark, ray, chimaeras	0.8

Source: UN Food and Agriculture Organization.

it but may poison it, as industrialized nations do. Outer space can be cluttered with human space junk, which is increasing almost yearly. And the ocean can be emptied of all its palatable sea creatures. Although the seafood catch is commonly described as harvesting, it is not a harvest in the sense that collection of corn, wheat, or soybeans is a harvest. The grain is planted each year; we actively renew it. However, we add nothing to the ocean; we simply scoop up what it produces, and if we scoop up too much, reproduction will be unable to restore the "crop." What we do in the sea is better described as hunting rather than harvesting.

In the eighteenth century, it became accepted that three nautical miles (3.45 land miles) offshore was the limit of sovereignty of a coastal nation, a distance probably based on the length of the English unit of measure called a league when Great Britain was the dominant sea power.[8] The commons was the area beyond the 3-mile limit. But as technology increased and knowledge of the ocean and its potential food, oil, and mineral riches increased, some nations arbitrarily extended their sovereignty to 4 or 5 nautical miles; many claimed 12 miles, and others claimed 200 miles. To overcome this problem of ocean ownership, the United Nations enacted the Law of the Sea Treaty in 1982, which came into force in 1996.[9] It placed about

one-third of the ocean, the 200-mile zone nearest the continents, under various national jurisdictions. This is the limit for exclusive control over the management of
fisheries. The rest of the ocean remains a commons, used by all but protected by
none.

Even this generous 200-nautical-mile limit to fisheries exclusivity did not solve all
problems of seafood ownership. Fish are migratory, making property rights uncertain, particularly in the outer parts of the exclusivity zone. For example, about 10
percent of the productive Grand Banks off Newfoundland is beyond Canada's 200-
mile exclusivity zone. Large fishing vessels from many other nations fished there
in the 1980s and 1990s and largely depleted the fish catch over the entire Grand
Banks. The formerly dominant fauna has been reduced to just a few percent of its
long-term averages and been made commercially extinct. The number of spawning-
size cod in 1992 was 1 percent of that in 1962.[10]

Another problem that makes the 200-mile limit imperfect is ocean temperature.
Fish can sense temperature variations, and as the temperature of near-surface waters
changes seasonally, a nation may "own" the fish in the summer but not in the winter. Some migrations are occurring in response to global warming and may be permanent. For example, fish in the North Sea are migrating to cooler waters (farther
north or deeper) to escape the warming seas.[11] In a twenty-five-year study, researchers found that of thirty-six fish species studied, twenty-one have moved: fifteen to
cooler latitudes and six to cooler depths. Whether these species will be replaced by
commercially valuable species that prefer warmer water is unknown. If cod is
replaced with, for example, red mullet, there will be no commercial loss. But if the
cod are replaced with jellyfish, the loss to commercial fishers would be immense.

The United States, which has the longest coastline in the world (6,053 miles for
the contiguous forty-eight states, increasing to 13,443 miles when Alaska and
Hawaii are included), has not ratified the Law of the Sea Treaty.[12] Many members
of Congress believe that putting the UN in charge of the unlegislated two-thirds of
the ocean will be harmful to American interests because of strong anti-American or
anti-Western attitudes by many UN member states.

Ocean Fisheries Management

You did not kill the fish only to keep alive and to sell for food, he thought. You killed him for
pride and because you are a fisherman.
—Ernest Hemingway, *The Old Man and the Sea*, 1952

There are between 15,000 and 20,000 species of fish in the ocean,[13] and until the
last few decades of the twentieth century, few people, even those in professions focused entirely on fisheries, worried about managing the fish in the ocean. Virtually
everyone subscribed to the view of the nineteenth-century naturalist Jean-Baptiste de

Lamarck: "Animals living in … the sea waters … are protected from the destruction of their species by man. Their multiplication is so rapid and their means of evading pursuit or traps are so great, that there is no likelihood of his being able to destroy the entire species of any of these animals."[14] Modern fishing technology has made Lamarck's view wrong. It is now possible to completely empty the sea of its seafood riches.

In the 1850s forty-three schooners from a single port in Massachusetts plied part of the Nova Scotian shelf for cod.[15] Crews dropped lines, each with one hook, over the sides of the ships and jiggled their bait along the seafloor to entice the big predatory fish. Although the combined fleet used fewer than 1,200 hooks, they hauled in more than 7,000 tons of cod from part of the shelf each year.[16] That was about 560 tons more cod than the ninety modern ships fishing the entire Nova Scotian shelf today.[17] Twentieth-century fishing reduced the tonnage of adult cod in the North Atlantic to only 4 percent of what it had been in 1852.[18]

An explosion of fishing technologies occurred during the 1950s and 1960s as military hardware was adapted to serve the commercial fishing industry. Radar allowed boats to navigate in total fog and darkness, and sonar (acoustic fish finders) made it possible to detect schools of fish deep under the ocean's opaque blanket. Electronic navigation aids such as satellite positioning systems turned the trackless sea into a grid so factory ships could return to within 50 feet of a chosen location, such as sites where fish gathered and bred. The hunting vessels now receive satellite weather maps of water-temperature fronts, indicating where fish will be traveling.

There are about 30 million professional fishermen worldwide, but 50 percent of the fish caught at sea are captured by only 1 percent of the boats,[19] the massive ships that are more like floating factories than the boat used by Ernest Hemingway's protagonist in *The Old Man and the Sea*. Partly because of subsidies of one kind or another given by nations to their fishing fleets, there are now more than 23,000 fishing vessels of more than 100 tons, as well as 3 to 4 million small ones scouring the ocean in increasingly deeper waters.[20] The estimates of worldwide subsidies to the fishing industry range from $15 billion to $50 billion per year, with the European Union the largest subsidizer. At least $2.5 billion is provided by the U.S. government to subsidize fishing in the North Atlantic each year, supporting incomes and paying portions of boat fuel and equipment bills. Without subsidies, the world's fishing industry would be bankrupt.[21] Government subsidies distort the fishing industry in the same way that farming subsidies distort agricultural output: they reward unsustainable fishing and encourage overcapitalization, such as increasing the size of fleets.

Not only have the number and size of fishing vessels increased but so have their capabilities to catch fish. Commercial fishing vessels now deploy gear of enormous proportions, beyond the imagining of naturalist Lamarck 200 years ago. There are transparent or blue-colored submerged longlines 80 miles in length with thousands of hooks baited with squid. Longlines used to catch ten fish per hundred hooks; now they are lucky to catch one.[22] The longlines are complemented by bag-shaped bottom-scraping trawl nets, large enough to engulf twelve jumbo jetliners, and 40-mile long drift nets. While the longlines and gill nets snag the large predatory fish such as shark, swordfish, and tuna, the trawlers literally scrape the bottom clean, harvesting an entire ecosystem along with the catch of the day. This effect was first noticed in 1376 when fishermen complained to King Edward III that trawls were "destroying the flowers of the land beneath the water there, and also spat of oysters, mussels, and other fish upon which the great fish are accustomed to be fed and nourished."[23] Nets dragged along the seafloor scoop up everything in the way, the sub-sea equivalent of collecting the entire farm when the goal is to bring in a bushel of apples.

Although counting the numbers and types of fish in the oceans is as much art as science, experts believe that the variety of species in the world's oceans has dropped by as much as 50 percent in the past fifty years, mostly because of commercial fishing and habitat destruction.[24] According to fishery experts, only 10 percent of large fish such as tuna, swordfish, marlin, cod, halibut, and flounder are left in the sea.[25] In the early twentieth century, harpooned swordfish were routinely 300 pounds apiece. Swordfish caught on longline hooks by the mid-1990s averaged less than 90 pounds, barely big enough to reproduce.[26] More than half the seafood caught in 2002 was either herring or carp.[27] In the North Atlantic fishing grounds, the salmon population decreased from 590,000 individuals in 1971 to 40,000 in 1998; Arctic cod from 2.7 million metric tons to 0.3 million; bluefin tuna from 55 million metric tons to less than 20 million.[28] An average bluefin tuna weighs about 80 pounds, so there would be about 550 million fish. A more recent estimate places the number in the western Atlantic at only 25,000 fish.[29] Regardless of which estimate is correct, it is clear that industrial fishing has scoured the global ocean. Eighty to 90 percent of the fish in some populations are removed every year.[30]

And then there is the by-catch of shark, dolphin, seal, sea turtle, albatross, and anything else unfortunate enough to be lured toward the fishing nets. More than 100,000 seabirds are killed each year just in the Chilean sea bass fishery around Antarctica.[31] This collateral damage, which is generally discarded, makes up 25 to 30 percent of the total "fish" catch.

One small positive aspect of bycatch is that after rising through the 1900s, the number of shark attacks on surfers has dropped 30 percent since 2000 because of a global decline in shark populations. It now averages four per year.[32] Fishing fleets worldwide kill an estimated 26 to 73 million sharks per year, a decline that is having damaging ecological effects on other sea life. The abundance of twelve of the fourteen species the sharks prey on has increased dramatically, and these species

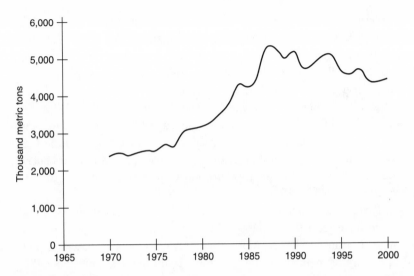

Figure 8.1
Ocean fish catch by the United States, 1970–2000. *Source*: Earthtrends, Earthtrends.wri.org/
pdf_library/country_profiles/Coa_cou_840.pdf.

are wiping out other marine organisms. For example, the number of cownose rays
has swelled tenfold. Because these organisms shred the sea grass that houses crabs
and clams, North Carolina was forced to shut down its century-old bay scallop fishery in 2004.

The total world fish catch rose steadily for half a century but leveled off in 1989
at about 88 million tons, rose to 92 million tons in 1994, and has fluctuated ran-
domly between 88 and 96 million tons since then.[33] The catch by U.S. fishers has
declined by nearly 20 percent since peaking in 1987 (figure 8.1). Open ocean commercial fishing has greatly reduced fish stocks in all major fisheries. Research indicates that almost every time a new species becomes a target or boats move into a
new fishing ground, populations nosedive within ten to fifteen years.[34] According
to a 2005 report from the FAO, 16 percent of world fish stocks are overexploited,
52 percent are fully exploited (they are fished at their maximum biological productivity), 21 percent are moderately exploited and could support modest increases in
fishing and harvests, 7 percent are depleted, and 1 percent are recovering from depletion. Only 3 percent are underexploited. This means 68 percent of the world's
marine food resources are fished to the hilt or beyond. Expansion would come at
the price of extinction.[35] If the earth's oceans were a human being, they'd be rushed
to the hospital, admitted to the intensive care unit, and listed in grave condition.

Staple stocks like North Atlantic cod and haddock have been fished to near extinction. Cod can weigh more than 200 pounds but now are rarely found above

20 pounds; the average is 4 to 7 pounds.[36] The number of spawning-sized cod in eastern Canada in 1992 was found to be around 1 percent of that in 1962,[37] and in 1992 the Canadian government was forced to impose a moratorium on cod fishing on the Grand Banks. But the cod had not recovered by 2004, possibly because many of the young survivors from the fishing holocaust are swallowed by predators before they can reproduce. There also appears to have been an increase in illegal cod fishing, concealed as bycatch.[38] In Georges Bank, the Cape Cod fishing grounds to the south of the Grand Banks, cod stocks collapsed in the mid-1990s, and the U.S. government closed thousands of square miles of fishing grounds and limited fishers' days at sea. Nevertheless, the cod population around Cape Cod dipped another 25 percent between 2001 and 2004.[39]

In the Fraser River near Vancouver, Canada, in 2004, only about 524,000 salmon appeared to have returned to the spawning grounds, barely one-quarter of the number that made it in 2000.[40] In Alaska, the average size of a spawning pink salmon has dropped by 30 percent in the past forty years because only those thin enough to squirm through the mesh of a gill net survive to reproduce.[41] This change is genetic, so it is not clear that larger salmon will ever return.

Until recently it was thought that all eggs and larvae have the same odds of survival, regardless of their parents' size. But it now appears that the biggest fish are the most valuable for maintaining the population.[42] A red snapper 2 feet long produces more than 200 times as many eggs as females that are two-thirds her size. Larvae from older and larger black rockfish are bigger, grow more than three times as fast, and can survive without food for twice as long as larvae from younger females. But commercial fishers target big fish rather than small ones because bigger fish yield higher profits. There is more marketable meat per landed fish. Also, big fish tend to have a milder flavor and fewer bones per pound than small fish do. Large fish species, which tend to have long maturation periods, have suffered disproportionately. Swordfish maturity does not occur until age five or six, bluefin tuna must be three to eight years before they reproduce, halibut take twelve years, and most rockfish must be well into their teens.[43]

Highly migratory species such as tuna, swordfish, and wild salmon are under serious threat. Swordfish lines do not discriminate between babies and behemoths, and with most of the big fish gone, fishers are catching and eating many juveniles, keeping them from reproducing and replenishing the stocks. Most salmon now comes from land-based fish farms.[44] The seafood riches of the world ocean are being mined out most rapidly in the Atlantic, but the Pacific Ocean fishing fleet has been affected as well. In coastal Oregon, whole fishing fleets are sitting idle. Pacific cod and halibut fishing were closed in 2003 in an effort to save four species of rockfish. They are not a target of fishers yet, but they are caught and discarded as bycatch in such numbers that they are down to 4 percent of their former numbers in many areas.[45] As has been amply demonstrated on land, humans are very good at killing things. We may soon find our seafood restricted to jellyfish and plankton stew. Human predation knows no limits. From the viewpoint of the web of life, humans may be the most dangerous and lethal threat on earth.

Fishers have countered the loss of preferred fish by switching to species of lesser value, usually those lower on the food web. This is one reason that global fish catches have not declined dramatically in the past twenty-five years, despite severe overfishing. Of course, predation of sea creatures lower on the food chain robs larger fish, marine mammals, and seabirds of food. During the 1980s, five of the less desirable species made up nearly 30 percent of the world fish catch but accounted for only 6 percent of its monetary value.[46] Now there are virtually no untapped marine fish that can be exploited economically. Human predation has exceeded the sustainable yield, the amount of fish that can be removed from the sea without harming the species.

The American Fisheries Society, an organization of 10,000 fisheries scientists and managers, in 2000 published a list of marine fish stocks at risk of extinction in North America.[47] The list, which contains eighty-two species and subspecies considered vulnerable, threatened, or endangered with global extinction, includes Atlantic salmon and halibut and several species of shark, sawfish, sturgeon, Pacific coast smelt, cod, rockfish, and grouper. Yet despite the grim report, funding for ocean research, which was 7 percent of the total federal research budget twenty-five years ago, is just 3.5 percent today, and the decline is continuing.[48] Space research is more glamorous and saleable to Americans than ocean research. Everyone has heard of the National Aeronautics and Space Administration (NASA), but few nonscientists know of JOIDES, the Joint Oceanographic Institutions for Deep Earth Sampling ship that has logged over 250,000 miles since 1978 drilling holes in the ocean floor and recovering more than 85 miles of sediment and rock cores. And fewer still are aware of the research vessels that traverse and sample the oceans to uncover its secrets. Sampling rocks on Mars has more public appeal than determining the food resources of the ocean. Our ignorance of the food chains and the ecology of ocean life has resulted in nearly emptying the sea of its visible life. There are simply too many boats catching too many fish.

However, all is not bleak, at least in U.S. waters. There are signs that some fish stocks are recovering in some areas.[49] A report in 2004 by the National Oceanic and Atmospheric Administration (NOAA), the federal agency responsible for fisheries maintenance in U.S. waters, was encouraging. There are 894 federally managed fish stocks, and of these, 76 stocks are currently classified as overfished, a drop of 10 from 2002. But 60 fish stocks are experiencing overfishing and need better management. At present forty-six management plans are in place. In 2003 four fish stocks were fully rebuilt, and overfishing practices were stopped for five species.

The turnarounds were accomplished by NOAA in cooperation with regional fishery councils, commercial fishers, environmental groups, and coastal states.

Economic theory predicts that when the available amount of a wanted product decreases, the price of the product will increase. This is certainly true for seafood. Between 1970 and 2000 the price of seafood increased one-third faster than the consumer price index (CPI), with invertebrates such as shrimp, lobster, clam, and scallop increasing eight times faster than the CPI.[50] The combination of decreasing seafood catch and increased demand for seafood has caused the increase. However, Americans are unfazed by the price rise. Per capita consumption of seafood increased by 12 percent between 2001 and 2004 to a record 16.3 pounds.[51] Shrimp consumption has shown the biggest increase in recent years, surpassing tuna in 2001.[52] Because of increased consumption of seafood, the United States is increasingly dependent on imports, with more than half its supplies coming from overseas.

Fish and Pollution

Here in the United States we turn our rivers and streams into sewers and dumping grounds ... and exterminate fishes, but at last it looks as if our people were awakening.
—Theodore Roosevelt, 1913

As noted in chapter 2, a considerable amount of the fertilizer and pesticides farmers spray on their fields runs off laterally into streams and downward into groundwaters, and eventually to the ocean. The well-known dead zone off the mouth of the Mississippi River results from excess fertilizer. However, there are other sources of water pollution, and evidence has appeared that freshwater fish are being affected by endocrine-disrupting chemicals being released into the environment from waste storage facilities on livestock farms. The amount of estrogen coming out of pigs and cows is estimated to be more than ten times higher than the amount the human population puts out.[53] Scientists at the National Institutes of Health have found that male fish in contaminated rivers have been feminized by endocrine disrupters to the extent that only 20 to 30 percent of them are able to release fish sperm (milt). Those that do produce milt produce up to 50 percent less than do normal male fish. Also, the ability of these sperm to fertilize eggs and produce viable offspring is reduced.[54]

In 2004, federal scientists reported finding egg-growing male fish in Maryland's Potomac River, and in 2007 they were discovered in Boulder Creek, Colorado.[55] The investigators believe the abnormality may be caused by pollutants from sewage plants, feedlots, and factories. The effluent from the sewage plants comes from humans discharging natural estrogens and the synthetic estrogens found in contraceptive pills directly into sewage. Upstream of the wastewater treatment plant along Boulder Creek, the ratio of males to females is 50–50. Below the plant, females outnumber males by five to one, and 10 percent of the fish were neither male nor female but had sexual characteristics of both.

In the United Kingdom, 33 percent of the fish in the rivers and streams are intersex (having characteristics of both males and females) because of endocrine-disrupting chemicals in amounts of only parts per trillion released from sewage efflu-ent.[56] In waterways that receive large inputs of effluents from sewage treatment works, all of the male fish are intersex to varying degrees. Industrial chemicals dis-charged into the rivers may also be involved. It is likely that many chemicals in the environment, possibly interacting with one another, cause this condition in fish.

Similar results have been found in the Pacific Ocean off the coast of southern Cal-ifornia, apparently a result of the discharge of sewage effluent into the ocean near Los Angeles.[57] Nearly 1 billion gallons of treated and filtered sewage are released into the ocean every day through three underwater pipelines. Eleven of eighty-two fish caught near the pipelines had ovary tissue in their testes. Two related studies found that two-thirds of male fish near one of the pipelines had egg-producing qual-ities.[58] The likely feminizing culprit is the chemical oxybenzone, which occurs in sunscreen and mimics estrogen's chemical makeup. Apparently it is washed off tanned bodies in the shower, passes unchanged through sewage works, and settles on the seabed, where it is eaten by bottom-feeding fish. No one knows whether eat-ing the amounts found in the fish will cause reproductive problems in humans.

Sex-altering chemicals are only a small sample of the pollutants found in the tissues of freshwater and saltwater organisms.[59] Researchers believe that antidepres-sant drugs in the water are depressing freshwater wild mussel populations and have found these drugs in the brains of fish captured downstream of sewage treatment plants. Off the Florida coast, flame-retardant chemicals are accumulating in fish. Prey fish such as perch average 43 parts per billion (ppb) of these chemicals, sharks average 750 ppb in their fat, and dolphins contain 1,190 ppb. Some sharks and dol-phins had concentrations as high as 4,200 ppb. In laboratory tests, these chemicals impair hormonal and reproductive function and disrupt fetal development.

PCB concentrations in the top predators in these studies were even more dra-matic. Sharks average 25,800 ppb in their fat, and bottlenose dolphins 162,000 ppb in their blubber. The investigators believe that the concentrations of both the flame-retardant chemicals and the PCBs are doubling every two to four years in the sharks and dolphins.

Pollution far below the level recognized as dangerous for aquatic life has altered fish behavior in American lakes.[60] Just a few parts per billion of heavy metals such as copper and zinc cause salmon to lose their sense of smell, so they cannot smell chemicals that warn of a nearby predator. In the lakes, the heavy metals also made minnows unable to recognize their eggs; the fish ate them instead of protecting them.

Leeches lost their ability to smell food, and zooplankton lost their ability to evade predators. The contamination in the lakes was much too weak to kill the organisms outright, but their populations were clearly suffering. These new data are another previously unrecognized example of the harm that human activities are having on other living organisms.

No one knows what effects the pollutants in seafood may have on humans, but it is clear they are harmful to the freshwater and open-ocean sea creatures in which they are accumulating. The long-term effects have yet to be determined but certainly they cannot be good.

Crustaceans and Shellfish

Shrimp boats are acoming, their sails are in sight.
Shrimp boats are acoming, there's dancing tonight.
—John Phillips, "Shrimp Boats Are Coming," 1950s

Shrimp, crab, clam, scallop, oyster, and lobster are popular seafood among Americans, and consumption has increased dramatically over the years, and the sight of aficionados slurping clams and oysters "down the hatch" and the sound of cracking lobster or crab shells is common in Red Lobster and similar seafood restaurants.

Shrimp

In the contest for the tastiness of seafood without fins, it is no contest among Americans. Shrimp wins hands down, with per capita consumption skyrocketing from 2.2 pounds in 1990 to 4.2 pounds in 2004, a 90 percent increase over the fourteen-year period, most of it since 1996.[61] Shrimp was 25 percent of total seafood consumption in 2004.[62] Approximately 80 percent of the shrimp is eaten in restaurants.[63]

Per capita consumption of shrimp in 2004 was six times greater than consumption of crabs, nine times greater than clams, and more than twelve times greater than scallops.[64] Oysters and lobsters lag far behind. The United States is the world's biggest consumer, and the vast majority of the shrimp we eat is imported, about half from China and Thailand.[65] Imports totaled 227 tons in 2004, between 40 million and 80 million shrimp, depending on their size. About half of the imports are not ocean shrimp but are cultured on land in saltwater ponds.

Saltwater shrimp live on soft mud bottoms at depths of 30 to 1,500 feet and are harvested in the United States by a fleet of more than 20,000 vessels of various sizes. In 2004, 83 percent of America's domestic ocean shrimp were harvested from the Gulf of Mexico, 8 percent from the South Atlantic coast, 8 percent from the Pacific coast, and 1 percent from New England. The Gulf shrimp are mostly from Texas and Louisiana; South Atlantic shrimp are mostly from North Carolina; Pacific shrimp come from Washington and Oregon; and New England shrimp are found mostly in the Gulf of Maine, a cold-water variety not viable south of New England, as contrasted to the warm-water varieties harvested from the southern and southeastern coasts.

The method used to harvest shrimp is an ecologic disaster because of the amount of bycatch, which is five to ten times the amount of shrimp.[66] These unwanted creatures are tossed back into the ocean dead or dying. Typical organisms are fish, crustaceans, turtles, corals, and assorted plants that make up seafloor communities. The harvesting of shrimp accounts for one-third of the world's discarded catch while producing less than 2 percent of global seafood.[67]

Crab

Crabs are mostly marine creatures and live in all oceans, from their shallowest parts to the greatest depths of the ocean. The annual catch of crabs is about the same weight as shrimp, but crabs are mostly shell, and shrimp are not. There are thousands of species, many with familiar names such as blue crab, Dungeness crab, and horseshoe crab. Chesapeake Bay is the nation's largest producer, and it is estimated that one-third of blue crabs come from these estuarine waters.[68] Crab pots are the preferred technique for catching crabs, with the bait being menhaden, eel, or herring. Both hard-shell crabs and "soft-shelled" ones are harvested, the soft-shelled ones being crabs that have just completed one of several molting phases.

The blue crab, so named because of brilliant blue claws, is a pugnacious predator. The 2006 catch of blue crabs in Chesapeake Bay was only 50 million pounds, far below the high of 111 million pounds in 1994 and near the historic low of 48 million pounds in 2003.[69] Causes for the low catch include a decrease in oxygen supply because of pollution of the estuary by nutrients from agricultural areas around it and the loss of submerged aquatic vegetation that young crabs need for shelter and food during their development. Management plans are in place to stem the trend of decrease in the crab population, but reducing the runoff of excess nutrients from the land will be difficult.

Clam

Chesapeake Bay is also a major producer of clams, cherrystone clams being a favorite of shellfish eaters. Clams are distributed in the western Atlantic from the Gulf of St. Lawrence south to Cape Hatteras, and commercial concentrations are found primarily off New Jersey, the Delmarva Peninsula, and on Georges Bank. In the mid-Atlantic region, they are found from the beach zone to a depth of about 200 feet.[70] Beyond 130 feet, however, their abundance is too low to be commercially useful. Clams reach harvestable size in about six years and have a maximum size of about 9 inches. The edible part of a clam may consist of the muscles that operate the shell; the siphon, or neck; and the foot, which the clam uses to propel itself through sand.

Harvesting is done with a hydraulic clam dredge, and the catch varies between 24,000 and 35,000 metric tons. The allowable catch is regulated by the mid-Atlantic Fishery Management Council in cooperation with state authorities.

Clams are particularly vulnerable to toxins in their environment, which can be abundant in bays where runoff from pollutants on the land surface is high. The Georges Bank region has been closed to the harvesting of clams since 1990 due to the risk of paralytic shellfish poison.[71]

In 2005, it was reported that clams contain high levels of a chemical known to boost libido by raising testosterone production.[72] It is not yet known whether the oyster, a close relative to the clam and long reputed to enhance sexual drive, also contains high levels of this chemical.

Scallop

Most bay scallops come from China, where they are farm-raised on suspension nets rather than being harvested by dredging the seafloor.[73] This is an ecologically sounder method, as it avoids the dredging that destroys entire seafloor communities.

Scallops in the western Atlantic Ocean have about the same distribution as clams but at shallower depths. Commercial concentrations are found at depths between 130 feet and 325 feet. The U.S. catch has been rising rapidly since catch management was instituted in the 1980s to protect the species. The catch in 2003 was 23,900 million tons, up from only 10,000 in 1999, with Massachusetts (48 percent) and Virginia (31 percent) the major harvesters.[74] The U.S. fishery is managed by the New England Fishery Management Council.

Lobster

Lobsters can live to a hundred years in the wild, but few do because of the tastiness of their flesh. We have come a long way since the Pilgrims pronounced them "fit only for pigs" and middle-class Americans in the nineteenth century snubbed them as food for the poor because of their abundance. Lobsters are cold-water animals, live along the eastern coast of the United States from Labrador south to Cape Hatteras, prefer rocky areas, and are locally abundant in coastal regions within the Gulf of Maine and off southern New England. The common way of catching them is in traps, although they sometimes occur as bycatch in dredging for shrimp. The annual catch is about 30,000 tons and is managed by the New England Fishery Management Council.

Maine's state government does not limit the number of licenses it issues, so without a controlling mechanism, the famous Maine lobster might vanish. To preserve the lobster catch, lobstermen practice exclusion through a system of traditional fishing rights. Acceptance into the lobster fishing community is essential before someone can fish, and thereafter the person admitted can extract lobsters only in the territory held by that community. This has resulted in a sustainable harvest and higher catches of larger, commercially valuable lobsters by fishermen in the exclusion communities.

Lobsters live in deep water in the fall and return to shallower depths in the spring. They can go without eating for as long as six months. They are caught in wooden cratelike traps or pots baited with fish and shellfish, which are sunk to the ocean floor and marked with a buoy. The lobsters that are caught and kept usually weigh from 1 to 5 pounds; those larger than 5 pounds are considered excellent egg producers and are generally thrown back to sea to replenish the population. The largest lobster on record was taken in 1977 off Nova Scotia and weighed 44 pounds.[75]

Lobsters are able to regrow lost limbs and antennae, indicative of a primitive nervous system compared to humans or other types of animals. They can "drop" a claw and go off like nothing happened. Physicians and biochemists would like to understand how lobsters are able to regenerate lost body parts.

Oyster

Oysters, best known for their reputed aphrodisiac powers, have been a favorite of food lovers for thousands of years.[76] The virility legend stems from mythology. Aphrodite, the Greek goddess of love, sprang forth from an oyster shell and promptly gave birth to both Eros and the English word *aphrodisiac*. As described by Diane Brown, author of *The Seduction Cookbook*, "Oysters are so sensual just in their nature. They have that slippery, slurpy sensation when you eat them that makes them very seductive."[77]

Americans eat more oysters than any other people in the world.[78] Their rough, irregularly shaped shells are unattractive to human eyes, and perhaps to the oyster as well, as an oyster is able to produce both sperm and eggs during its life span.[79] No need for romance among members of this supposedly aphrodisiac creature.

Unlike most other seafood, the taste of oysters varies greatly depending on the type of algae fed on and the salinity in the area where they live.[80] Like wine aficionados, oyster connoisseurs take pride in distinguishing the different tastes that result from the various harvested regions. To distinguish the oysters, both the Atlantic and Pacific oyster frequently take on the name of the region from which they came. The East Coast yields the majority of oysters in the United States, and the various varieties have names such as Chesapeake, Blue Point, Long Island, and Chincoteague, names that are designated on menus in expensive restaurants. Oysters on the Atlantic coast occur from the Gulf of St. Lawrence southward to the Gulf of Mexico at depths of 8 to 25 feet.

Unfortunately for both ocean ecology and oyster lovers, oyster harvests in Chesapeake Bay, one of America's chief harvest areas, declined from 53,000 tons in 1880 to 10,000 tons in 1980 to 100 tons in 2003, from several million bushels in 1948 to 1 million bushels in 1984 to only 26,500 bushels in 2004,[81] a decline traced to over-harvesting, parasites, and environmental pollution by agricultural nutrients, toxic pesticide runoff, and siltation. A traveler commented in 1701 that "the abundance of oysters is incredible. There are whole banks of them so that the ships must avoid them.... They surpass those in England by far in size ... they are four times as large. I often cut them in two, before I could put them in my mouth."[82] This is certainly not the case today.

Healthy oysters in estuaries such as Chesapeake Bay consume algae and other waterborne nutrients by filtering up to 5 quarts of water per hour, a function barely noticeable in the bay today because of their near extinction. In 1880 there were enough oysters to filter all the water in the bay in three days; by 1988 it took more than a year for the remaining oysters to accomplish the same task.[83] The inflow of nitrogen and phosphorous from the agricultural areas in the Chesapeake Bay watershed has so depleted the water of oxygen that a dead zone has developed in the central part of the bay that covers up to one-third of the bay, decimating the mussel population (*mussel* is the general name for saltwater bivalves). The Chesapeake Bay Program partners have had the Oyster Management Program in place since 1994 in an attempt to resuscitate the oyster population. The program has had little success so far: the oyster population is only 1 percent of what it was in 1880, the dawn of the region's oyster industry.

Aquaculture

Underfishing is a sin of omission, overfishing a crime against humanity.
—Pedro Ojeda Paullada, 1983

Aquaculture, sometimes called mariculture when applied to ocean species, is the production of food through the controlled growth and harvesting of plants and organisms that live in water. Using low-cost equipment and simple methods, aquaculture has the potential to supply more protein-rich foods than traditional land farming and to reduce our dependence on depleted stocks of ocean and freshwater fish, crustaceans, and shellfish (figure 8.2).

Aquaculture originated in China thousands of years ago and is now the world's fastest-growing source of food, increasing at a rate of 10 percent a year, from 6.2 million tons in 1983 to 17.8 million tons in 1993 to 59.4 million tons in 2004 (figure 8.3).[84] Forty percent of the entire global seafood market is now supplied by aquaculture.

Aquaculture is entirely responsible for the increase in fish harvests that has occurred since the late 1980s.[85] Ninety percent of aquaculture still takes place in

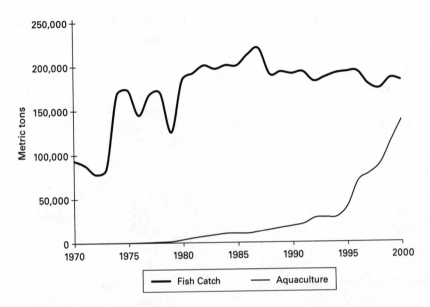

Figure 8.2
Freshwater catch and aquaculture production in the United States, 1970–2000. *Source*:
Earthtrends, Earthtrends.wri.org.

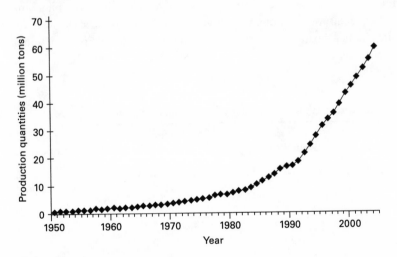

Figure 8.3
Aquaculture production, 2004. *Source*: *Environment*, April 2007, p. 38.

Asia. Between 1990 and 2003,[86] the aquaculture industry in the United States grew at an average compound rate of more than 9 percent a year, and the production of farmed fish is expected to nearly double by 2020.[87] Freshwater aquaculture totals more than half of farmed seafood. Already, 50 percent of world fish production is farmed[88]; half of all fresh and frozen seafood Americans eat is farmed.[89] Commonly raised fish are salmon, cod, halibut, perch, bass, and tuna; shellfish include clams, oysters, and scallops; crustaceans include crab, lobster, and shrimp. Almost 90 percent of the shrimp consumed each year in the United States is imported, and about half of the imported shrimp is farm-grown.[90] More than 260 species of seafood are cultivated.

The percentage of seafood resulting from marine harvests has decreased every year since 1980 (when it was 93 percent) because of the devastation wrought on world fish populations.[91] For example, salmon are extinct across almost half their historic range in the continental United States and are threatened throughout much of their remaining range in California, the Pacific Northwest, and New England (table 8.2).[92] Overfishing and dam construction, which blocks the salmon from reaching their spawning grounds, are largely responsible for the decline. Most of the salmon we eat today have lived in captivity for most of their lives, compared to just 6 percent in 1988. Similarly, 40 percent of the shellfish and 65 percent of freshwater fish have lived mostly in farm environments.[93] The fact that world seafood supplies have not dropped more precipitously is due almost entirely to the phenomenal growth in aquaculture. Although high-value carnivorous species such as salmon dominate the net worth of the aquaculture industry, production in volume is dominated by freshwater, predominantly vegetarian, fish.[94]

Table 8.2
Estimated historic and current salmon populations, by region (1,000 fish)

Region	Historic number of fish	Current number of fish	Current/ historic (percent)
Alaska	175,160	187,470	100 or more
British Columbia	65,556	24,800	36
California	3,060	278	9
Puget Sound	20,036	1,600	8
Oregon Coast	3,074	213	7
Columbia River	13,072	221	2
Washington Coast	3,935	72	2 or less

Source: D. R. Montgomery, Geology, Geomorphology, and the Restoration Ecology of Salmon., *GSA Today*, November 2004, p. 4.

Aquaculturalists hatch eggs from carnivorous fish such as salmon in freshwater and grow the fish for a year in tanks before transferring them to giant plastic cages or pens of netting suspended near shore in bays or estuaries. They feed the salmon pellets composed primarily of fish meal, vegetable matter, and vitamins; after three years, they harvest and sell the fattened fish. The harvesting process consists of a ship with a vacuum cleaner type of hose that sucks the 10-pound salmon from their cages or pens and deposits them onto a metal slide. They are then stunned by a punch machine, killed, and dumped into the ship's hold.[95]

Aquaculture is often regarded as an environmentally benign add-on to the world's seafood stocks. For herbivorous species such as carp and shellfish, this is correct, and these seafood varieties account for about 90 percent of world aquaculture production.[96] They feed by eating algae and by cleaning the water of plant plankton, ecologically sound activities. But for the high-value carnivorous species such as salmon, halibut, trout, tuna, and cod, the benefit of farmed fish is not evident.

Up to 80 percent of the ingredients in the feeds for these fish is fish oil and fish meal, which are made from small ocean species such as anchovy, herring, eels, sardine, menhaden, or the remnants of fish processing.[97] In 1948, only 7.7 percent of global marine landings were used for fish meal and fish oil. Today about 37 percent is reduced to feed, eliminating an important source of human sustenance. Fish meal and fish oil supply essential amino acids that are deficient in plant proteins and fatty acids not found in vegetable oils. They also provide energy, which is important because fish tend to convert carbohydrates to energy inefficiently. It requires up to 7 tons of smaller fish for each 1 ton of farmed salmon produced, clearly not a sustainable process.[98] One of every three fish that is caught now is used to feed farmed carnivorous fish.[99]

In Britain, retailer Marks & Spencer conducted a survey among its shoppers in 2003 and found that sustainability of fishing was named as their main environmental concern, ranking above genetically modified foods.[100] The head chef at one of London's seafood eateries insists on sustainable fish for quality reasons. He says that fish line-caught by small fishermen are fresher and firmer, although they are more expensive. He will not buy bass or cod unless they were line-caught.

If the number of farmed fish continues to rise at its current rate and the supply of oil and meal stays the same, as it has for the past dozen years, demand will outstrip the supply of oil by 2010.[101] If those projections are extended, fish meal will face the same problem by 2050.[102] Currently fish farms use about 40 percent of the world's supply of fish oil and 31 percent of its fish meal.[103] To circumvent this looming problem, feed companies are turning to larger fish such as mackerel and whiting for fish feed, causing further pressure on natural stocks. Companies have started going after krill, tiny crustaceans found predominantly in Antarctica.[104] This attack on the web of life is at the base of the food chain, which has serious ramifications for all life in the ocean. Research is under way to determine how to keep carnivorous farmed fish healthy on a more vegetarian diet. At present, the cost of the nutritional supplements that would be needed makes a vegetarian diet for these fish is not cost-effective. And putting farmed fish on a vegetarian diet reduces the amount of omega-3 fatty acids in fish flesh, the compounds that reduce the risk of heart disease in people who eat them.

Salmon

Farm fish are known to escape from pens in all salmon aquaculture areas of the world due to storms, marine mammal predation, and human error. There is chronic low-level leakage of fish from pens, as well as periodic, catastrophic escapes of thousands or tens of thousands of fish at one time.[105] More than 1 million Atlantic salmon were reported to have escaped from farms in Washington and British Columbia between 1992 and 2002.[106] In Scotland, more than 1 million salmon escaped from fish farms between 2000 and 2005.[107] As much as 40 percent of Atlantic salmon caught by fishers in parts of the North Atlantic Ocean are of farmed origin.[108]

Studies have been made of the effect escaped farmed salmon have on the wild salmon population in the ocean. It was found that farmed fish were exceedingly short-lived in the wild, surviving on average only 2 percent as long as their wild counterparts. Perhaps this is analogous to what happens to house pets abandoned by their owners. The salmon created by interbreeding between farmed and wild populations survived only 27 to 89 percent as long as their fully wild cousins; the farmed salmon had shortened considerably the life span of the salmon population.[109] Such a shortened life span could devastate the beleaguered Atlantic salmon, which has already been hit by pollution and overfishing. Wild stocks are extinct, endangered, or vulnerable in more than half the 2,000 salmon-spawning rivers around the North Atlantic.[110]

One of the reasons people eat fish is their widely touted health benefits, particularly the protection they offer against heart disease. The benefit comes from the content of omega-3 fatty acids in their flesh. But the fat on farmed salmon is very different from their wild cousins.[111] Farmed salmon have four times more body fat than wild salmon (15 percent compared to 4 percent), and this fat contains 200 percent more artery-damaging saturated fat and only one-half to one-third the amount of omega-3 found in wild salmon. Clearly shoppers should buy wild salmon rather than the farmed variety, but it is not always possible to do so. Not only is there a lack of labeling in stores, but commonly the salmon is mislabeled. *Consumer Reports* found that 43 percent of salmon labeled "wild" in supermarkets were actually farm-raised.[112] The mislabeling is not only a health concern, but a budgetary concern as well: wild salmon costs twice as much per pound as farmed salmon.

This is only the beginning of the alterations humans are making in the salmon population. Genetic modification is already being used in fish farming. Scientists are investigating a salmon that carries a gene inducing the production of a growth hormone that makes the fish grow bigger and faster.[113] GM salmon can grow more than seven times larger than their wild relatives. Research indicates that in a head-to-head battle for food, normal Coho salmon lose out to their genetically engineered cousins.[114] Not only do the aggressive GM salmon gobble up most of the feed when raised in tanks with ordinary salmon, they also eat their weaker competitors, often leading to dramatic population crashes with only one or two GM fish left in tanks that originally held fifty salmon.[115] Currently no GM fish are being commercially farmed, but based on the experience with grain farming, GM fish may soon be for sale at supermarkets. Farm-raised salmon form the bulk of the salmon eaten by Americans today.

As with all other industrial farming on land or sea, there is the problem of waste. Fish grown in offshore cages, as salmon frequently are, concentrate large quantities of untreated nutrients, harmful chemicals, paracitides, pesticides, spawning hormones, and waste such as uneaten food and dead fish. A moderately sized salmon farm of 200,000 fish releases an amount of fecal matter equal to the untreated sewage from 65,000 people.[116] The nitrogenous waste produced in 2000 by farmed salmon in Norway was roughly equal to the sewage produced by Norway's 4 million people.[117]

Organic matter tends to accumulate underneath salmon cage operations, creating a dead zone that might extend 100 to 500 feet in diameter beyond the farms,[118] a zone much like the better-known dead zone near the mouth of the Mississippi River. A study published in 2004 found numerous persistent organic pollutants at a level several times higher in farmed salmon than in wild salmon.[119] Because these pollutants accumulate in animal fat, animal protein in farmed fish feed was suggested as the likely source.

Salmon researchers recommend that farm-raised Atlantic salmon should not be eaten more than once a month, farmed Pacific salmon no more than four times monthly, and wild salmon up to eight times a month.[120] The U.S. government has not yet endorsed these recommendations. Moreover, the fish may transmit diseases such as sea lice, which eat salmon flesh, to wild stock. Scientists believe that sea lice present in escaped farmed fish led to a 98 percent decline in pink salmon populations near Vancouver Island in 2002.[121]

Shrimp

Shrimp farming is only a minor part of American aquaculture today because of the high cost of coastal land. Farmed shrimp in the United States contribute no more than 1 percent of the world's farmed shrimp. But shrimp aquaculture is a major industry in Southeast Asia, where it causes significant destruction of coastal mangrove forests and wetlands that protect coastlines and serve as nurseries for local fish. Mangrove destruction can cause a decline in local fish supplies that exceed the gains from shrimp production. Experts disagree on the percentage of worldwide destruction of mangroves and wetlands that is due to shrimp farming, but the major causes are urban development, rice production, grazing, and tourism.[122]

Shrimp are grown in a controlled environment. The eggs or larvae are either gathered from the natural environment or grown in hatcheries after being taken from female brood stock and raised to maturity in shallow ponds. The food fed to the developing shrimp is high in fish meal, like salmon food, and so puts more pressure on oceanic fisheries.

The overuse of antibiotics that is so common in industrial animal farming on land is also a problem in fish aquaculture. A report by the American Society of Microbiology in 1995 singled out the use of antibiotics in aquaculture as potentially a leading cause of the evolution of antibiotic-resistant bacteria in humans.[123] And the shrimp are already becoming resistant. A 2004 study in Sweden found that 77 percent of bacteria in farmed shrimp were resistant to one or more antibiotics.[124]

Mercury in Fish

When someone is chronically ill, the cost of pollution to him is almost infinite.
—Anonymous congressional staff member, 1974

Although emissions of mercury into the environment have decreased significantly in recent years and seem destined to be reduced further during the next decade, there is still much cause for concern.[125] The main sources of mercury in the environment are emissions from the smokestacks of America's 1,100 coal-burning power plants (41 percent of the total), municipal waste incinerators (29 percent of the total), medical waste incineration, and chlorine production.[126] Deposition is highest in the East and Midwest, where coal-burning plants are located. No federal laws require existing power plants to control their mercury emissions, although technologies are available that remove almost all the mercury at the stack level.[127] Much of the mercury stays airborne for up to two years and spreads around the globe, returning to earth with rain or snow; the nations of the world receive our mercury, and industrialized nations in Europe and elsewhere send us theirs.

In the atmosphere, the mercury can react with chlorine that was emitted from volcanoes. Also, some mercury is emitted from power plants as a water-soluble mercury-chlorine compound formed when mercury reacts with chlorine in coal. Precipitation quickly washes the mercury-chlorine compound into lakes, rivers, and oceans, where bacteria take it up, attach it to carbon atoms, and convert it into the powerful neurotoxin methylmercury, which then enters the gills of fish.

The methylmercury passes up the food chain, increasing in amount in each higher organism, and reaching the highest levels in large predatory fish species such as swordfish and tuna. A top predatory fish such as tuna can have sequestered in its flesh methylmercury levels that are 1 million times higher than the water it swam in.[128] From the predatory tuna, the mercury compound gets into other predatory species such as humans. Mercury levels in tuna from top California sushi restaurants are, on average, double the FDA limit.[129] Sushi from New York City was also found to contain excessive levels of mercury. Eating sushi has become the new Russian roulette.

Because of the high concentrations of methylmercury in large predators such as tuna, swordfish, marlin, king mackerel, tilefish, and shark, the FDA in 2001 advised pregnant women and nursing mothers to severely limit their intake of these fish. The FDA estimates that about 630,000 infants are born every year in the United States with unsafe levels of mercury in their blood.[130] Shark, swordfish, and marlin have levels of mercury five to seven times higher than that of canned tuna and two to four times higher than that of fresh tuna.[131] Albacore tuna contains more mercury than other kinds.[132]

Freshwater fish, like their marine cousins, also sequester mercury. Levels are so high that forty-eight states have declared their freshwater fish unsafe to eat. (The other two states have not tested their fish.) Measurements in American lakes show that the number of lake acres under advisory for mercury pollution has increased each year, from 3 million in 1993 to 13 million in 2003.[133] The polluted lakes cover 12 million acres. Freshwater fish caught almost anywhere in the country are contaminated with mercury. More than half the fish in 260 freshwater lakes across the United States have levels of mercury higher than the limit EPA recommends for consumption by women of childbearing age and small children.[134] In 2003, every state but Wyoming and Alaska issued warnings about mercury.

Although women of childbearing age and small children need to be careful about fish consumption, for others the health benefits of eating fish far exceed the risks, according to reports from the National Academy of Sciences and the Harvard School of Public Health. The healthy fats in fish, particularly omega-3 fatty acids, protect against heart disease and possibly against diabetes, cancer, and some mental disorders. In a study of more than 8,000 pregnant women and their children, researchers found that children whose mothers ate less than 12 ounces of seafood a week were about 45 percent more likely to fall into the lowest 25 percent in IQ.[135] Interestingly, fish do not actually produce omega-3 fatty acids; they capture it from the food chain and accumulate it in their tissues, another example of the interconnectedness of the web of life.

Antibiotics in Fish

To live by medicine is to live horribly.
—Carl Linnaeus, *Diaeta Naturalis*, circa 1750

It is common practice in the industrial aquaculture industry, as in the cattle industry, to use large amounts of antibiotics to prevent infection. These antibiotics are often not biodegradable and remain in the farm waters for long periods of time, encouraging the growth of bacteria resistant to the antibiotic. The properties that make the bacteria resistant can then be transferred to humans who eat the fish. A report from New York Medical College published late in 2006 cautions that "if we don't curb the heavy use of antibiotics in aquaculture, we will ultimately see more and more antibiotic pathogens emerging, causing increased disease to fish, animals, and humans alike."[136]

9

Fruits and Vegetables: Plants to Cherish

Cheerfully adorn the proudest table,
Since yours is to bear the glorious label—
"Richest in vitamines!"
—Rose Fyleman, "To an Orange," twentieth century

There is no clear botanical distinction between a fruit and a vegetable. We think of a fruit as a minor but edible part of a vine, bush, or tree, such as a grape, a blueberry, an apple, or a cherry. We cannot eat the whole plant with its leaves, branches, stem, and roots. And fruit normally grows above ground level. The plant that yields the fruit does not need to be replanted each year; a peach tree produces peaches this year, next year, and in years after that.

A vegetable is a plant that can be eaten almost in its entirety and includes the leaves (lettuce), stems (asparagus), roots (carrot), and tubers (potato) of various plants. Vegetables are not sweet and sugary, as many fruits are, and are not acidic like citruses. Furthermore, they must be replanted each year, and they grow at or below ground level. But there are exceptions. We all think of watermelon as a fruit, and it does grow on a vine. But it must be replanted annually and grows at ground level, so technically it is a vegetable. String beans and green peas in their pods grow on a vine, as do grapes, so perhaps they should be classed with the fruits. Lest you think the distinction to be only academic, it is worth noting that in a court case dealing with fruits and vegetables, the U.S. Supreme Court in 1893 declared the tomato to be a vegetable and therefore subject to import taxes.[1] Fruits at that time were not subject to import taxes, so other countries would have been able to flood the market with lower-priced tomatoes. Who would have thought that classifying tomatoes had economic ramifications? Perhaps we should settle for the distinction between a fruit and a vegetable similar to the oft-quoted description of pornography: I may not be able to define it, but I know it when I see it.

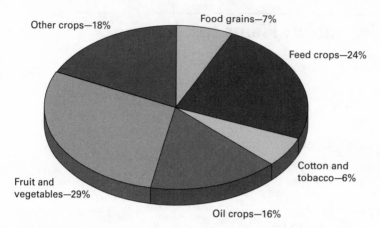

Food grains—7%

Other crops—18%

Feed crops—24%

Cotton and tobacco—6%

Fruit and vegetables—29%

Oil crops—16%

Figure 9.1
U.S. farm cash receipts for crops, average 2002–2004. *Source*: U.S. Department of Agriculture.

Production and Consumption of Fruits and Vegetables

Let it please thee to keep in order a moderate-sized farm, so that thy garners [storehouses] may be full of fruits in their season.

—Hesiod, *Works and Days*, circa 720 B.C.

I am better off with vegetables at the bottom of my garden than with all the fairies of the Midsummer Night's Dream.

—Dorothy Leigh Sayers, *Lord, I Thank Thee*, 1943

Fruits and vegetables form only a minor part of American crop area. In 2002 they were produced on 13 million acres, only 3 percent of harvested cropland, an increase of only 10 percent since 1994. Nevertheless, because of their relatively high retail value, fruits and vegetables provide nearly one-third of farm cash receipts (figure 9.1). Retail prices for them have increased by 73 percent since 1980, in large part because they are not subsidized.[2] Four percent of vegetables and 2 percent of fruit is produced organically. Excepting dried peas and lentils, fruits and vegetables are not eligible for federal subsidies.

Per capita consumption of fruit and tree nuts has been essentially unchanged in recent years, increasing only 1 percent between 1992–1994 and 2002–2004. In contrast, the 2005 per capita consumption of vegetables in the United States increased 21 percent since 1982.[3] Nevertheless, in the United States in 2005, 400 times more land grew hay for livestock than grew vegetables and melons for people.[4]

Americans eat about 60 percent more vegetables than fruit,[5] mostly as fresh vegetables and french fries (eaten primarily in fast food restaurants). Oranges, apples, and grapes are the most popular fruits; potatoes, tomatoes, and corn are the most consumed vegetables (table 9.1). The number of varieties of fruit is decreasing, as was the case for grain. Of the 7,100 fruit varieties that were known to exist in 1800, only 1,000 exist today.[6] The reason for the decline for fruit varieties is the same as it is for grain: the growth of large industrial farms that grow only a few, high-value crops.

Similarly, the number of fruit farms has been decreasing. Between 1997 and 2002 grapes were the only major fruit that increased in farm number, probably in response to the huge increase in wine consumption in the United States, from 1.94 gallons per person in 1997 to 2.83 gallons per person in 2002, an increase of 46 percent in only five years. In 2003 it increased another 5 percent, to 2.98 gallons.[7] Nearly all domestic grapes come from California, and more than half of U.S. grape production is used to make wine. Chardonnay is the American favorite (table 9.2).[8]

Increased production of fruit is concentrated in the hands of a few producers. Fruit farm sizes increased between 1997 and 2003. Only 2 percent of fruit and nut farms are larger than 500 producing acres, but they produce 51 percent of total fruit and nut revenue.[9] Fruit farm operators who are sixty-five years of age or older account for about a quarter of total U.S. fruit revenue.[10] These trends point to a future with even fewer farms that are large enough to compete in world markets, many with lower labor costs than American farms.

The consolidation trend of vegetable farms mirrors that of fruit farms. With some exceptions, vegetable farms are largely individually owned and relatively small, with two-thirds harvesting fewer than 25 acres.[11] However, relatively few farms account for most of the vegetables produced. In 1997, about 11 percent of vegetable farms had sales in excess of $500,000, yet they accounted for about 70 percent of vegetable acreage.[12]

Fruits and vegetables may be classified according to two major end uses: fresh market or processing. Processing can be further subdivided into canning, freezing, juicing, and dried/dehydrating. Sixty to 70 percent of fruit production is processed (orange juice from oranges, wine from grapes), and about 50 percent of the vegetable harvest is processed (tomatoes and potatoes.)[13] Although a few commodities are suitable for either end use (apples, grapes, broccoli, cauliflower, asparagus), growing for processing is distinct from growing for the fresh market. Most vegetable varieties grown for processing are better adapted for mechanical harvesting and often lack characteristics desirable for fresh market sale. In contrast, most fruit varieties grown for processing are still hand-harvested, and there is a year-round need for fruit pickers.

Table 9.1
Top fruit and vegetables by average per capita consumption, 2002–2004

Vegetables and melons	Per capita consumption (pounds)	Fruit and tree nuts	Per capita consumption (pounds)
Potatoes, all	135.0	Oranges, all	81.3
Tomatoes, all	89.5	Apples, all	46.6
Sweet corn, all	26.6	Wine grapes	30.0
Lettuce, head	22.4	Bananas	26.2
Onions, all	21.5	Grapes, excl. wine	19.2
Watermelon	13.6	Pineapples, all	13.3
Carrots, all	11.9	Grapefruit, all	10.6
Cucumbers, all	11.1	Peaches, all	9.5
Lettuce, romaine/leaf	10.9	Lemons, all	7.4
Cantaloupe	10.5	Strawberries, all	6.6
Cabbage, all	9.2	Pears, all	5.6
Broccoli, all	8.0	Tangerines and tangelos	3.8
Snap beans, all	7.4	Avocados	2.7
Bell peppers, all	6.9	Limes	2.6
Dry beans, all	6.3	Mangos	2.0
Celery, all	6.2	Cherries, all	1.6
Chili peppers, all	5.7	Olives	1.3
Squash, all	4.5	Plums and prunes, all	1.3
Sweet potatoes, all	4.4	Almonds	1.1
Mushrooms, all	4.1	Papayas	0.9
Others	29.4	Others	8.4
Total	445.1	Total	282.0

Note: Data are in fresh-weight equivalent. "All" refers to all uses, fresh and processed.
Source: *Fruit and Tree Nuts Yearbook* and *Vegetables and Melons Yearbook*, USDA, Economic Research Service.

Table 9.2
Percentage of major grape varieties crushed for wine production in California, 2005

Chardonnay	17.1%
Cabernet sauvignon	12.5
Zinfandel	10.4
Thompson seedless	10.2
Merlot	9.8
French colombard	7.0
Rubired	3.9
Syrah	3.4
Sauvignon blanc	2.7
Chenin blanc	2.2
Pinot noir	2.2
Others	18.5

Source: California Department of Food and Agriculture, *Final Grape Crush Report, 2005 Crop*, 2006.

Harvesting and Farm Workers

The laborer is worthy of his reward.
—1 Corinthians 9:9

Harvest workers are mostly migrant workers and are hired both locally and from other areas and countries to help get the crop in on time, before the quality begins to decrease. The need for many manual laborers to do the picking of apples and other fruits and vegetables has created social and legal problems concerned with the treatment of the workers and their movement to the United States from Mexico, an important source of farm labor at harvest time. Seventy percent of U.S. farm workers are illegal immigrants.[14] Their willingness to work long hours for low wages has enabled California to sustain its agricultural production, despite the loss since 1964 of more than 9 million acres of farmland.

Fruit and vegetable growers rely on a thriving market in labor. This has been, until recently, one of the main reasons that there was no serious effort to stop the immense flow of illegal immigrants crossing the border from Mexico to Arizona. Mexico is a reliable supplier of 16 percent of our imported oil, and Mexico regards the northward flow of their citizens as a safety valve for the country's unemployment problem. The U.S. government is concerned about irritating this southern neighbor, but in 2006 was forced by internal political pressures to stem the flow of illegal immigrants by building a fence along the Rio Grande.

About 85 percent of fruit and vegetables produced in the United States are hand-harvested or cultivated.[15] Hand-picking accounts for about 50 percent of total production costs, compared with only 16 percent when mechanical harvesting is used.[16] Commercial hand-harvesting of fruit and vegetable crops is hard, tedious, and time-consuming work. Because of this, and the typical low pay for such work, the supply of workers available for hand-harvesting is decreasing steadily, and shortages are occurring. Without enough workers when needed for a few weeks each year, part of most hand-harvested crops will be lost. Maintaining access to an affordable labor pool is a top concern for the fruit and vegetable industry.

Although some crops, such as hard-shelled nuts, corn, and potatoes, can be picked by machine without fear of damage, squishy produce, such as tomatoes, blueberries, cherries, strawberries, grapes, eggplant, and asparagus, require careful handwork. About twenty-five fruit crops and twenty vegetable crops lack feasible mechanical harvesting options.[17] Crops harvested by mechanical means are mostly used for processed products, in which bruising and other fruit damage is not apparent. Consumers of fresh produce want perfect-looking apples and peaches.

The need for hand labor in harvesting will not decrease in the foreseeable future. The government's policy for research and development of new machines to make harvesting more efficient was stated by Secretary of Agriculture Bob Bergland in 1979: "I will not put federal money into any project that reduces the need for farm labor." This policy ended public funding for research and development projects focused on reducing the cost and increasing the labor productivity for harvesting horticultural crops.[18]

Health Benefits of Fruits and Vegetables

Look to your health; and if you have it, praise God, for health is the second blessing that we mortals are capable of; a blessing that money cannot buy.
—Izaak Walton, *The Compleat Angler*, 1653

Much of the benefit derived from fruits revolves around their antioxidant properties. Antioxidants are substances that inhibit oxidation. Oxygen is essential to animal life, but its use in the body creates chemicals known as free radicals. Free radicals steal electrons from molecules in bodily cells, damaging them in a process called oxidation. Oxidation damages the cell membranes and the genetic material (DNA) in cells in much the same way as the oxygen in the air turns a cut apple brown or exposure to air makes butter go rancid. In this way free radicals affect the rate at which we age, produce cataracts and other eye diseases, contribute to arthritic joint inflammation, and damage brain cells to promote Parkinson's or Alzheimer's disease. They are also implicated in cancers and in increasing the ability of low-density lipoprotein cholesterol (the bad type of cholesterol) to stick to artery walls.[19] Antioxidants in fruits neutralize free radicals and supplement the antioxidants produced naturally by the body. The more people engage in harmful activities such as smoking, excessive alcohol consumption, or excessive exercise, the more antioxidants they need. Other things that increase the need for antioxidant supplements are pollution, solar radiation, and X-rays.[20]

The best sources of antioxidants are fruits, vegetables, tea, and red wine. The more brightly colored the fruit or vegetable is, the richer it is in antioxidants.[21] Research has revealed the best antioxidant foods to be blueberries, cranberries, blackberries, prunes, raisins, raspberries, strawberries, and spinach, all of them strongly colored. Other good sources are orange-yellow foods, such as carrots, squash, pumpkin, apricots, tomatoes, and pink grapefruit.

Apples

Remember Johnny Appleseed,
All ye who love the apple;
He served his kind by Word and Deed,
In God's grand greenwood chapel.
—William Henry Venable, *Johnny Appleseed*

We have all heard the adage, "An apple a day keeps the doctor away," which comes from an old English adage, "To eat an apple before going to bed, will make the doctor beg his bread."[22] Certainly nothing will keep physicians away indefinitely, but eating apples might delay the time when you need them. Research indicates that apples, especially apple skins, are rich in antioxidant compounds called flavinoids, particularly quercetin, which has the best health-promoting capabilities.[23] Quercetin is found most abundantly in apples, onions, tea, and wine. In addition, one medium apple contains 20 percent of the recommended daily fiber intake, and the roughage or dietary fiber found in apples is beneficial for stimulating regular bowel movement. Roughage has also been shown to reduce the occurrence of intestinal disorders, including diverticulosis, hemorrhoids, and possibly some cancers of the digestive and intestinal tract.[24] Apples have cholesterol-lowering effects, which helps prevent arterial blockage and consequent heart attacks and strokes. They may also reduce blood pressure levels. In 2004, it was found that apple juice may improve thinking ability and protect against damage that contributes to age-related brain disorders such as Alzheimer's disease.[25]

Pesticides in Apples

Apple trees are susceptible to a host of fungal and bacterial diseases and insect pests, and nearly all commercial orchards spray poisons up to eighteen times per crop to maintain high fruit quality, tree health, and high yields. Apple trees provide an example of the frequency and importance of these artificial poisons to the fruit tree farmer. Many of these poisons are found on the apples consumers buy in their local supermarkets and are not easily removed by washing (chapter 2). Apples are one of the most contaminated fruits in the supermarket.

Organic Apples

Because of the high level of pesticide contamination found on apples grown by industrial agriculture, they are best bought from organic growers. The United States is apparently the world leader in organic apple production, but there are no data for China, the dominant apple producer. In 2003, 14.5 percent of fruit and vegetable acreage was certified organic; two-thirds were vegetables and one-third fruit. In 2005 apples were grown organically in the United States on 12,515 acres, 3 percent of total apple acreage. Washington led the way in 2005 with 53 percent of organic apple production, followed by California with 27 percent and Arizona with 8 percent.[26] The 2005 acreage of organic apples was an increase of 50 percent from 2000. As with grain and meat products, consumption of organic fruit is increasing rapidly.

Grapes

Take your sharp sickle and gather the clusters of grapes from the earth's vine, because its grapes are ripe.
—Revelation 14:18

Grapes are a versatile fruit and have a variety of uses: juices (crushed grapes), raisins (dried grapes), jellies (sugared grape juice), and wines (fermented grapes). Grape production in the United States in 2003 was about 6.5 million tons, and the yield was 6.8 tons per acre on 953,000 acres.[27] Eighty-six percent of production in 2004 was from California. Grapes destined to be made into wine were about 51 percent of the total, with red wine about 20 percent more popular than white. Future raisins were about 37 percent, and table grapes were only 12 percent.[28] In terms of per capita consumption, the USDA says that in 2004, Americans ate an average of 8.5 pounds of fresh grapes, drank 8 quarts of wine, ate 1.5 pounds of raisins, drank 1.8 quarts of grape juice, and ate 0.2 pounds of canned grapes.[29] Clearly, Americans appreciate grapes.

About 10 percent of America's grapes are produced organically. The percentage is higher than for most other fruits because of consumer preference. Grape spraying is particularly dangerous because of the use of methyl bromide. When a new grape vineyard is started, the barren soil is sterilized with this compound, which is acutely toxic to humans and wildlife. In addition, it evaporates and contributes to ozone depletion and resultant increase in skin cancers.

As the grape plants grow, the methyl bromide is followed by a coterie of insecticides, fungicides, and herbicides to kill practically everything except the plants. Exposure to the sulfurous fungicides is the major cause of reported farmworker illnesses in California. However, grapes sold in stores test low in pesticide residues. About 10 percent of California's grapes are grown organically.[30] Pesticide residues are a problem with imported grapes and occur on 86 percent of those tested.[31] Thirty-five different pesticides were found. Thirty-six percent of the samples contained two pesticides, 27 percent had one, 24 percent had three, and 10 percent had four. Only 14 percent of the grape samples tested were free of pesticides. The United States is a net importer of grapes, with Chile supplying two-thirds of our imports. Mexico supplies nearly all the rest.[32]

Wine is the most popular alcoholic drink in the United States according to a 2005 Gallup poll: wine 39 percent, beer 36 percent, and liquor 26 percent.[33] Wine can be made from the fermented juice of many fruits besides the ever-popular grape. Apples, apricots, berries, pears, and even flowers such as dandelions can serve as the raw material. Of the approximately 10,000 different grape varieties, 180 are made into wine.[34]

Berries

Wife, into the garden and set me a plot
with strawberry roots, the best to be got.
Such growing abroad among thorns in the wood,
well chosen and picked, prove excellent good.
—Thomas Tusser, *Five Hundred Points of Good Husbandry*, 1557

Like grapes, berries are subject to frequent pesticide applications for control of insects and disease pathogens. These applications are particularly hazardous for field workers because of the plant coverage required for effective pest control. Frequently all parts of the plant must be covered, including the undersides of the leaves and all surfaces of the berries. This often requires the use of very small droplets and the use of air to promote penetration of the canopy and deposit of the pesticide. This produces many aerosols, which attack the eyes and skin and can be inhaled and penetrate deep into the lungs.

Many fields are fumigated with highly toxic materials before the berries are planted in order to reduce the population of pests such as nematodes, bacteria, fungi, and viruses before they can attack the young plants. Fumigation requires special training and involves injection of a gas or liquid into the soil and covering with a plastic sheet to prevent the pesticide from escaping too soon.

Strawberries

Strawberries are the most popular berry in the United States. Production has doubled since 1989.[35] We produce more than a quarter of the world's supply.[36] A survey in 1998 revealed that 94 percent of American households consume strawberries.[37] Strawberry consumers agree with the seventeenth-century writer William Butler that "doubtless God could have made a better berry, but doubtless God never did." Wild strawberries grow everywhere in the United States, largely from seeds sown by birds. When birds eat the wild berries, the seeds pass through them intact and in reasonably good condition.[38] The germinating seeds respond to light rather than moisture and therefore need no covering of earth to start growing.

The volume of commercial strawberries produced in the United States is only 20 percent of the volume of grapes, but the berries have 50 percent of the value of grapes, that is, a strawberry is more than twice as expensive as a grape. There are more than 600 varieties of strawberries.[39] More than 90 percent of the strawberry crop is harvested for the fresh market, and 25 percent is frozen for the processed fruit market.[40] Fresh strawberries are hand-picked and rushed to coolers, where huge fans pull out the field heat, and they are then shipped within 24 hours in refrigerated trucks or air-freighted to their final destination. Strawberries selected for processing are gently washed, sorted, and frozen quickly to ensure that the best flavor and appearance are preserved.

Pesticide sprays on strawberries are particularly dangerous to consumers, as their soft, fleshy character allows the poison to easily penetrate the fruit, unlike fruits with a hard or stiff skin, such watermelons, bananas, oranges, or even grapes. When the ease of penetration is added to the fact that strawberries are dowsed with poisons six times during and after the growing season, the decision to eat industrially grown strawberries should not be taken lightly.

Blueberries

The production of blueberries is highly concentrated in North America, with the United States producing 55 percent of the world's crop.[41] Canada supplies 28 percent, Poland 10 percent, and the rest of the world only 7 percent. In 2004, 228 million pounds of blueberries were cultivated, only 10 percent the amount of strawberries.

Recent research has revealed that blueberries, particularly the wild variety, are exceptionally rich in nutrients and number 1 in antioxidant activity. We have heard in recent years of the cardiac benefits of red wine, but blueberries deliver nearly 20 percent more of the free radical fighters than wine does. It is noteworthy that wild blueberries are nearly 50 percent richer in antioxidants than the cultivated variety.[42]

Vegetables

Life expectancy would grow by leaps and bounds if green vegetables smelled as good as bacon.
—Doug Larson, botanist

Vegetables rival fruit as a source of nutrients but are not well liked by American consumers. Even young children seem to have an inborn aversion to eating vegetables, an antipathy that remains in many people as they grow older. This is most unfortunate, as these plants are extraordinarily nourishing. They are excellent sources of fiber, vitamins A and C, potassium, calcium, and iron, and they are low in calories. Legumes (beans, peas, lentils) are a good source of complex carbohydrates, are high in protein content, and can be used to some extent as meat substitutes.

Potatoes

What I say is that, if a fellow really likes potatoes, he must be a pretty decent sort of fellow.
—A. A. Milne, children's author

Although we no longer emulate the Incas in worshiping the potato, this tuber is the leading vegetable crop in the United States, forming 29 percent of our vegetable consumption—about 130 pounds annually per person.[43] More than half of the annual consumption is processed rather than fresh (fast food french fries and potato chips). Second among our favorite vegetables is the tomato, at 90 pounds per year. Planted potato acreage has declined slowly for the past 80 years, but the yield per acre has increased steadily at an annual rate of 2 percent since 1950, so production has continued to trend upward.[44] Potatoes are the most important vegetable crop in the nation, accounting for 17 percent of all vegetable and melon cash receipts in 2002 (lettuce 13 percent, tomatoes 10 percent).[45] Pound for pound among U.S. crops, potatoes are topped only by wheat flour in importance in the American diet.[46] The potato is the best package of nutrition in the world, rich in minerals, vitamins, calories, and protein, and it is virtually fat free (at least until it is smeared with butter and buried in bacon bits and sour cream).

Pesticides in Potatoes

Potatoes are the most pesticide-intensive U.S. crop because of the heavy use of soil fumigants to suppress parasitic nematodes and disease-causing organisms. Pesticides are sprayed on them thirteen times a season. Prominent among the fumigants is methyl bromide, soon to be phased out because it damages the ozone layer.[47] According to the federal government and the Environmental Working Group, pesticides were found on 79 percent of potatoes tested.[48] Half of the samples contained one pesticide, nearly a quarter had two, and 6 percent had three or more. Potatoes were among the dozen most consistently contaminated types of produce sold in supermarkets. The potatoes were all washed and prepared for consumption before being tested, so it is apparent that washing does not remove the pesticides.

Genetic Engineering and the Potato

In 2005, Japanese scientists inserted a human enzyme into potatoes and rice that breaks down a range of common herbicides, pesticides, and industrial chemicals.[49] Critics say the idea of eating a potato containing human genes smacks of cannibalism. Perhaps of more consequence is the concern of Joe Cummins, professor emeritus at the University of Western Ontario in Canada: "The human enzymes put into rice are responsible for causing most kinds of human cancer by activating certain pollutants into forms that attack the genes causing mutation."[50] No one knows whether the human enzyme inserted into the potato has the same effect as the one put into rice.

Produce and Food-Borne Illness

Each season has its own disease, its peril every hour!
—Reginald Heber, "At a Funeral," eighteenth-century poet

In the latter part of 2006, fresh produce was the cause of several widespread episodes of food-borne illness, first from spinach, then from tomatoes, and then from green onions and lettuce.[51] The number of produce-related outbreaks of food-borne illness has been increasing in recent years, from about forty in 1999 to eighty-six in 2004, according to the Center for Science in the Public Interest. According to the Centers for Disease Control and Prevention, 76 million Americans, one-quarter of the population, are sickened, 325,000 are hospitalized, and 5,000 die each year from something they ate. Americans are more likely to get sick from eating contaminated produce than from any other food item. In 2007, 66 percent of shoppers were confident that the food they buy at the grocery store is safe, down from 82 percent in 2006, according to the Food Marketing Institute.

Several factors have contributed to the sharp rise in outbreaks, an important factor being the industrialization and centralization of the food system. The fast food industry's demand for uniform products has encouraged centralization in every agricultural sector. Fruits and vegetables are now grown, packaged and shipped like industrial commodities. The Taco Bell distribution center in New Jersey, the likely source of the green onion and lettuce contamination, supplies more than 1,100 restaurants in the Northeast. Other factors favoring an increase in food-borne illness are the increased consumption of fruits and vegetables, improved reporting of outbreaks, and an aging population whose weakened immune systems are more susceptible to food-borne illness. Unlike meat, much produce is eaten uncooked, and live and harmful organisms can enter the body unimpeded.

The outbreak from spinach was caused by a strain of the bacterium *E. coli* O157:H7, whose benign cousins live harmlessly in human intestines. The harmful strain produces toxins that destroy the intestinal lining, leading to bloody diarrhea, kidney failure, and occasionally death. The *E. coli* attack from spinach killed three and sickened 204 in twenty-six states and Canada and was traced to cattle feces near a California spinach field and to wild pigs that roamed through it. The tomatoes, tainted by the Salmonella bacterium, infected at least 183 people in twenty-one states. The contaminated green onions originated in California and sickened about 200.

The food safety regulatory system is the shared responsibility of local, state, and federal agencies. In some cases, the federal government has delegated the responsibility for ensuring food safety to states and municipalities, which are often able to respond more quickly to localized public health problems. Approximately 80 percent of food safety inspections are made at state and local levels. For example, the states have the primary responsibility for ensuring the safety of milk and the sanitary operation of restaurants.

Although it is not possible to make every meal risk free, germ free, and sterile, there is no question that improvements in the government's food inspection system can be made. Food safety oversight in the United States is poorly and inadequately managed. Responsibility for it is split among fifteen agencies in the federal government, operating under at least thirty statutes. The USDA has 7,600 inspectors and is responsible for the safety of meat, poultry, and processed egg products, whereas most of the rest—80 percent of the food supply—are the responsibility of the FDA. But the FDA's Center for Food Safety and Applied Nutrition has seen its budget and staff cut over the past decade, and the number of inspections for food safety has fallen sharply for decades.

Operating funds dropped from $48 million in 2003 to an estimated $25 million in 2007, and the number of full-time inspectors fell from about 2,200 to 1,962 in that same time frame. The number of food safety inspections conducted by the FDA has decreased 90 percent from 35,000 in the 1970s to 3,400 today. The FDA in fact inspects less than 1 percent of the food it is responsible for, much of which is imported (figure 9.2).

Their problem is illustrated by seafood imports. Six and a half million tons of seafood (seventy-nine percent of consumption) are imported into the United States every year from 160 countries, but only eighty-five inspectors work primarily with seafood. As noted by the director of the FDA's food safety arm, "We have 60,000 to 80,000 facilities that we're responsible for in any given year. Explosive growth in

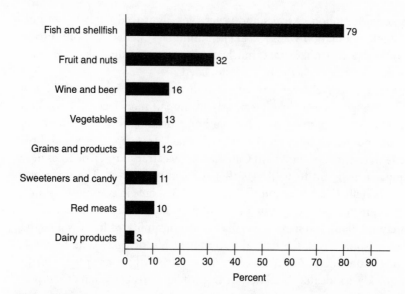

Figure 9.2
Import shares by volume are highest for fish and shell fish, 2000–2005. *Source*: USDA Economic Research Service, *Amber Waves*, February 2008.

the number of processors and the amount of imported foods means that manufacturers have to build safety into their products rather than us chasing after them."[52] The FDA has so few inspectors at ports and domestic food production plants that most domestic plants get a visit from an inspector only once every five to ten years. The amount and value of imported food increased by 50 percent between 2002 and 2006 alone, and it is financially and logistically impossible to test every food product that enters the United States. Only a program that stations American inspectors at producing and shipping points in the countries that supply imported food has a chance of controlling the cleanliness of the food we eat.

The rapid growth of imported food products is a major cause of America's inadequate inspection system (figure 9.3). More than 130 countries, many of them in the Third World and with substandard inspection systems, ship food items to the United States. In 2006, inspectors sampled just 20,662 shipments out of more than 8.9 million that arrived at American ports: only 2.3 percent.[53] As noted by former FDA commissioner William Hubbard, "The public thinks the food supply is much more protected than it is. If people really knew how weak the FDA program is, they would be shocked." In reference to imported food, he said, "The word is out. If you send a problem shipment to the United States it is going to get in and you won't get caught, and you won't have your food returned to you, let alone get arrested or imprisoned."[54]

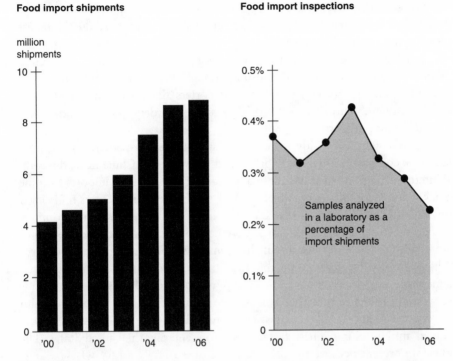

Food import shipments

Food import inspections

Figure 9.3
Growth and inspection rates of food imports. *Source*: U.S. Food and Drug Administration.

An example of the dangers of uninspected food products from Third World countries occurred early in 2007 when wheat flour from China was found to be contaminated with melamine, an industrial toxin.[55] It made its way into pet food, chicken feed, pig feed, and fish feed used in aquaculture. Investigators suspect that Chinese exporters boosted their profits by using cheap, unprocessed, low-protein flour and adding melamine, which gives false high-protein readings.

Other Chinese imports caught at American ports include dried apples preserved with a cancer-causing chemical, mushrooms laced with illegal pesticides, and frozen catfish laden with banned antibiotics and a disinfectant powder that has been banned in China for the past five years because it is a suspected carcinogen. China supplies a rapidly increasing percentage (currently 5 percent) of all catfish sold in the United States and 22 percent of America's total fish imports. The Chinese government's own reports express alarm that many rivers in which their fish are cultured are so contaminated with heavy metals from industrial by-products and pesticides that they are too dangerous to touch, much less raise fish in.[56] Industrial and urban sewage in the water forces farmers to use chemicals to keep the fish alive.

These contaminated products were among 107 food imports from China that the FDA detained at U.S. ports in April 2007, along with more than 1,000 shipments of tainted Chinese dietary supplements, toxic Chinese cosmetics, and counterfeit Chinese medicines. Inspection records reveal that China has flooded the United States with foods unfit for human consumption for years. And for years, FDA inspectors have simply returned to Chinese importers the small portion of those products they caught, many of which turned up at U.S. borders again, making a second or third attempt at entry.

According to Robert B. Cassidy, a former assistant trade representative for China, "So many U.S. companies are directly or indirectly involved in China now, the commercial interest of the United States these days has become to allow imports to come in as quickly and smoothly as possible."[57] As a result, the United States finds itself "kowtowing to China," even as that country keeps sending American consumers adulterated and mislabeled foods. Carol Foreman, a former assistant secretary of agriculture, adds that it is not just about cheap imports: "Our farmers and food processors have drooled for years to be able to sell their food to that massive market. The Chinese counterfeit. They have a serious piracy problem. But we put up with it because we want to sell to them."[58] This urge to compromise with China on health issues is reflected in an FDA directive issued in June 2007 that was designed to prevent banned antibiotics in Chinese farmed fish from entering the United States. The antibiotics are present because the fish are being raised in a country whose waterways are tainted by sewage, pesticides, heavy metals, and other pollutants. The new FDA rule requires that laboratories test five types of farm-raised seafood from China to document that the seafood is safe. However, the certifying laboratory can be located in any country, including China, which sounds a bit like asking the fox to guard the henhouse.

By halting the import of certain seafood products from China because of "filth" and illegal contaminants, the FDA has given new significance to the "Made in America" label. A Gallup Poll in July 2007 revealed that 51 percent of American shoppers now say they "make a special effort to buy items produced in the United States," and 74 percent are very or somewhat concerned about food from China. Sixty-one percent are concerned about food from Mexico. But it is not only Third World countries that send contaminated food to the United States (table 9.3).

The seriousness of inadequate inspections was emphasized in 2004 by Tommy Thompson, former secretary of health and human services, who said, "For the life of me I cannot understand why the terrorists have not attacked our food supply because it is so easy to do."[59]

Because of the sharp rise in imports of food, which doubled between 1996 and 2006, and the substandard inspection systems in many countries, there should be

Table 9.3
Food imports denied by the FDA, selected countries, July 2006–June 2007

	Number of refused food shipments	Most frequent food violation and counts		Total value of food imports, 2006
Mexico	1,235	Filth (mostly on candy, chilis, juice, seafood)	385	$9.8 billion
India	1,187	Salmonella (spices, seeds, shrimp)	256	1.2
China[a]	899	Filth (produce, seafood, bean curd, noodles)	287	3.8
Dominican Republic	821	Pesticide (produce)	789	0.3
Vietnam	385	Salmonella (seafood, black pepper)	118	1.1
Indonesia	344	Filth (seafood, crackers, candy)	122	1.5
Japan	305	Missing documentation (drinks, soups, beans)	143	0.5
Italy	273	Missing documentation (beans, jarred foods)	138	2.9
Denmark	87	Problems with nutrition label (candy)	82	0.4

[a] Does not include Hong Kong.
Sources: *Food and Drug Administration; U.S. International Trade Commission.*

mandatory country-of-origin labeling on food products. It is clear that Americans want this, and shoppers need to be informed about where their food comes from. In fact, such a law was passed as part of the 2002 farm bill and was scheduled to take effect in 2004, but enforcement has been blocked, largely by lobbyists for the meat industry, which opposes the law. The law required country-of-origin labeling on beef, pork, lamb, fresh fruits and vegetables, seafood, and peanuts. Only seafood labeling has taken effect.

Opposition to enforcement of the law was led by Henry Bonilla, a Republican member of the House of Representatives (who was defeated for reelection in 2006). His congressional district is a large cattle ranching region, and he was the top congressional recipient of campaign funds from the livestock industry in 2006 and 2004 and ranked second in 2002. He received the money from the livestock industry in 2004 as a reward for pushing through in 2003 a two-year delay in implementation

of the mandatory labeling law, and in 2005 he again succeeded in delaying implementation until September 2008. The five-year delay in implementing mandatory
labeling is another example of a public good being thwarted by well organized
and funded industries and their congressional lobbyists.

Vegetarianism

The average age (longevity) of a meat eater is 63. I am on the verge of 85 and still work as
hard as ever. I have lived quite long enough and am trying to die; but I simply cannot do it. A
single beef-steak would finish me; but I cannot bring myself to swallow it. I am oppressed
with a dread of living forever. That is the only disadvantage of vegetarianism.
—George Bernard Shaw, 1940

He is a heavy eater of beef. Me thinks it doth harm to his wit.
—William Shakespeare, *Twelfth Night*

Vegetarians are people who do not eat meat (including fish); perhaps one-third of
them are vegans.[60] Vegans not only eschew meat but also do not use any products
derived from animals, including edible products such as milk, eggs, and honey.
Maintaining health on a vegan diet, although possible, requires immense education
about food and nutrition and thus is not possible for most people. Vegans also prohibit the use of nonedible products such as leather shoes, wool clothing, silk, and
pillows stuffed with chicken feathers or goose down. They use only soap made of
vegetable oil instead of animal fat and avoid products containing ingredients that
have been tested on animals. Veganism is a philosophy and way of living that seeks
to exclude, as far as possible and practical, all forms of exploitation of, and cruelty
to, animals for food, clothing, or any other purpose. For example, horseback riding
may be prohibited because it is not "natural" for a horse to carry a person on its
back. The American Vegan Society Web site says, "It is not mere passiveness, but a
positive method of meeting the dilemmas and decisions of daily life. In the western
world, we call it Dynamic Harmlessness."

There are many subdivisions of vegetarians and vegans. Lacto-ovo vegetarians do
not eat meat, fish, or fowl but do eat dairy and egg products. Ovo-vegetarians omit
dairy products but eat eggs. Lacto-vegetarians omit eggs but not dairy products.
Vegans divide into two groups: those who do not use any animal products and
those who refuse foods derived from animals but use products that are not edible,
such as leather or chicken feathers.

Polls conducted by the Harris organization in 2000 and 2003 suggest that 2.5
percent of Americans are vegetarians, people who engage in a "non-violent diet."[61]
A study at Arizona State University reported that, sight unseen, salad eaters were
rated more moral, virtuous, and considerate than steak eaters.[62] Twice as many

women are vegetarians as men. The list of famous people who are or were vegetarians includes not only hippies from the 1960s but also Buddha, Socrates, Plato, Leonardo da Vinci, Percy Shelley, Leo Tolstoy, Mohandas Gandhi, and George Bernard Shaw, as well as the inventors of the graham cracker (Sylvester Graham) and cornflakes (John Kellogg).[63] Interestingly, people with high IQs are more likely to be vegetarians.[64]

There are three main reasons people decide to become vegetarians: health considerations, concern for the environment, and compassion for animals. Less common reasons are economics (meat is expensive), world hunger issues, and religious beliefs.

Health Considerations

Appropriately planned vegetarian or vegan diets can be nutritionally adequate and offer health benefits. Vegetarian diets contain lower levels of saturated fat and cholesterol and higher levels of carbohydrates, fiber, magnesium, potassium, boron, folate, and antioxidants such as vitamins C and E. Vegetarians have a lower body mass index than nonvegetarians, as well as lower blood cholesterol and lower rates of death from heart disease.[65] During World War II when meat and dairy products were strictly rationed, the rate of heart disease plummeted. In the United States today, someone dies of heart disease every 45 seconds.

Vegetarians also have lower blood pressure and lower rates of hypertension, adult-onset diabetes, diverticulitis, gallstones, dementia, and prostate and colon cancer.[66] As Albert Einstein said, "Nothing will benefit human health and increase the chances for survival on Earth as much as the evolution to a vegetarian diet." In support of this belief is the result of a twenty-year study of vegetarian Seventh Day Adventists that indicated a 3.6 year survival advantage for them.[67] It has also been reported that the very oldest people tend to have a large proportion of vegetarians among their ranks.[68] And meat producers acknowledge that vegetarian diets can be healthy and have responded to the call for leaner food. The National Pork Board says that compared with twenty years ago, pork is on the average 31 percent lower in fat and 29 percent lower in saturated fat, and it has 14 percent fewer calories and 10 percent less cholesterol.[69]

In addition to the naturally harmful effects of meat consumption, there are those that humans have added. Meat is the leading source of pesticides consumed in the American diet. The relative contributions are meat, 55 percent; dairy products, 23 percent; vegetables, 6 percent, fruits, 4 percent; grains, 1 percent.[70] With regard to the federal government's highly touted meat inspection program to detect, among other things, pesticide contamination, it is worth noting that fewer than 1 out of every 250,000 slaughtered animals is tested for toxic chemical residues.

Concern for the Environment

Most people do not think of diet as an environmental issue, but it is.[71] Meat production is highly inefficient. By growing grain that is fed to livestock, which is subsequently fed to humans, producers end up with much less food than they would have by feeding grain and other plant products directly to humans. More than 70 percent of the grain grown in the United States is eaten not by humans but by livestock.[72] Sixteen pounds of grain and soybeans are needed to produce 1 pound of feedlot beef. An acre of farmland can produce 40,000 pounds of potatoes, 50,000 pounds of tomatoes, or 250 pounds of beef.[73] A person on a meat-based diet needs twenty times the amount of land needed to feed a vegetarian. The huge percentage of grain produced for livestock contributes to the depletion of topsoil. And growing food for livestock uses so much fuel for farm machinery and for the manufacture of fertilizers and pesticides that grain might be considered a by-product of petroleum. According to People for the Ethical Treatment of Animals, producing a single hamburger patty uses enough fossil fuel to drive a small car 20 miles and enough water for seventeen showers.[74] Americans eat about 38 billion hamburgers each year.[75]

Of all the water used in the United States, more than half goes to livestock production.[76] The amount of water used in the production of an average steer is enough to float a U.S. naval destroyer.[77] It takes 40 to 100 times more water to produce 1 pound of meat than to produce 1 pound of wheat.[78] Beef production uses more water than is used to grow America's entire fruit and vegetable crop.[79] And then there is the problem of animal excrement, considered in detail in chapters 2 and 7. Pollution of streams and groundwater by runoff from feedlots occupied by thousands of animals is common. Many hundreds of thousands of fish, and probably millions of them, have died from pollutants. As noted in chapter 8, the EPA has determined that livestock waste has polluted more than 35,000 miles of rivers and contaminated groundwater in dozens of states.

One example of pollution by animal excrement is on the Eastern Shore of Maryland, where 1 billion chickens are raised each year.[80] The chicken feed contains antibiotics and arsenic, which is added to the feed to kill harmful organisms that breed in the crowded conditions in which the chickens are raised. The arsenic ends up in chicken excrement, which is sprayed on local agricultural fields as fertilizer for soybeans and corn to feed the chickens, in such quantities that the arsenic is leaching into nearby streams. There is arsenic in the drinking water on the Eastern Shore, possibly originating from this source.

The Union of Concerned Scientists in 2004 named meat eating as the consumer choice with the second-worst environmental impact (after cars) on global warming, air and water pollution, and habitat alteration. Britain's Environment Agency reached a similar conclusion.[81]

Compassion for Animals

The treatment of farm animals destined for the dinner tables of Americans can only be described as barbaric. Hardly any of our 300 million citizens ever come face to face with such cruelty, and it is safe to say that few of us would want to. We want to eat meat, lots of it, and regard the efficient raising and slaughter of animals as a necessary prerequisite. It is much like the way those who believe in capital punishment view executions. They may believe that a convicted child killer deserves to die but probably have no desire to witness the hanging, electrocution, or lethal injection. Few humans enjoy watching the maltreatment or killing of anything with a central nervous system. And the "cuter" or closer on the evolutionary ladder the victim is to us, the more distasteful it seems. Who wants to eat a hamburger while thinking we are consuming a ground-up cow who had its throat slit by a machete or had its head beaten in with a sledgehammer? It is ghastly and, if the American Dietary Association is to be believed, unnecessary as well to human health.[82]

Some groups of Americans go to great lengths to try to get meat eating out of the public's mind. People for the Ethical Treatment of Animals has offered money to various towns to change their name. Slaughterville, Oklahoma, named after former resident Jim Slaughter, turned down $20,000 by refusing to rename itself Veggieville. Refusals had previously been made by towns named Fishkill, Rodeo, and Hamburg.[83]

Are Humans Designed to Eat Meat?

Let us permit nature to have her way; she understands her business better than we do.
—Michel de Montaigne, sixteenth century

For most people, deciding to adopt a vegetarian diet is based on health considerations or is an ethical choice. But there is an alternate way of approaching the question of whether it is the right thing to do: Are human anatomy and metabolism characteristic of herbivores, carnivores, or omnivores? What has evolution designed humans to eat? In what ways related to food consumption do carnivores differ from herbivores?[84]

Carnivorous animals such as lions, dogs, and cats have unique characteristics that set them apart from all other animals. Carnivores have large mouths in relation to their head size. Big mouths are better for grabbing, killing, and dismembering prey. Herbivores and humans have small mouths. In this respect we resemble herbivores. The preferred method of eating for carnivores is to swallow the food whole, while herbivores and omnivores extensively chew their food before swallowing.

Carnivorous animals have sharp claws and powerful jaws, and the chewing apparatus of a carnivore is rich in pointed canine teeth to pierce tough hide and to spear and tear flesh. The head and pointed teeth of perhaps the most deadly carnivore ever to appear on earth, the dinosaur Tyrannosaurus Rex, are monstrous in the extreme. Alligators and crocodiles are modern examples. Carnivores do not have molars (flat back teeth), which vegetarian animals need for grinding their plant food.

The digestion of plant food starts in the mouth with an enzyme in the saliva. Herbivore and human saliva contain this enzyme; carnivore saliva does not. Plant food must be chewed well and thoroughly mixed with the saliva in order to be broken down. For this reason, herbivores (and humans) can use a slight side-to-side motion of their jaws to grind their food, as opposed to the exclusively up-and-down motion of carnivores. Unlike grains, flesh does not need to be chewed in the mouth to pre-digest it; it is digested mostly in the stomach and intestines. A cat, for example, can hardly chew at all. Humans have four canine and twelve molar teeth, suggesting an omnivorous eating habit with an emphasis on plant food.

Carnivores have short intestinal tracts, about three times their body length. This is because flesh decays rapidly, and the products of this decay quickly poison the bloodstream if they remain too long in the body. So a short digestive tract evolved for rapid expulsion of putrefactive bacteria from decomposing flesh. Complementing this adaptation are stomachs with ten times more hydrochloric acid than non-carnivorous animals (to digest fibrous tissue and bones). Like carnivores, humans secrete hydrochloric acid in their stomachs; herbivores do not. An herbivore's intestines are twelve times its body length. A human's tract is four to four and a half times its body length, clearly closer to carnivores.

Interestingly, studies have shown that a meat diet has an extremely harmful effect on plant eaters.[85] Meat-eating animals such as dogs have an almost unlimited capacity to handle saturated fats and cholesterol. However, if a half-pound of animal fat is added daily over a long period of time to a rabbit's diet, the animal's blood vessels become caked with fat, and atherosclerosis develops after two months. Human digestive systems, like the rabbit's, are not designed to digest meat, and the more of it they eat, the more caked their arteries become.

Both herbivores and humans suck water into their mouths; carnivores lap it up with their tongues. Both herbivores and humans sweat through pores in their skin; carnivores pant and sweat only through their tongue. There are many other lines of physiological evidence that can be brought to bear in the argument over whether humans are more like herbivores than carnivores. Some favor carnivores, others herbivores.[86] On balance, I would give an edge to the herbivores, so I recommend going heavy on vegetables and fruits and light on meat. These are also the federal government's recommendations, which are based on nutritional considerations rather than anatomical ones.

It also is possible to raise a few psychological points, briefly referred to earlier. It is obvious that our natural instincts are not carnivorous. Most people clearly prefer other people to kill and butcher their meat for them and would be sickened if they had to do the killing and slaughtering themselves. A human will do it to survive, but it certainly is not a preferred way of spending time for most of us (not like picking blueberries or peaches). Instead of eating raw meat as all flesh-eating animals do, humans cook it (which, of course, animals cannot do) and disguise it with sauces and spices (another thing animals cannot do) so that it bears no resemblance to its raw state. A carnivore such as a cat will salivate with desire at the smell of a piece of raw flesh but not at the smell of fruit or vegetables. As a vegetarian publication points out, "If a man could delight in pouncing on a bird, tear its still-living limbs apart with his teeth, and suck the warm blood, one might conclude that nature provided him with meat-eating instinct. On the other hand, a bunch of luscious grapes makes his mouth water, and even in the absence of hunger he will eat fruit because it tastes so good."[87]

10

Food Processing: What Is This Stuff We're Eating?

It's quite clear that consumers are now almost more aware of the packaging than the product.
—John Elkington, *EcoSource*, 1990

Go to your cupboard, pull out a package of something, and read the list of ingredients. Now ask yourself, Which items in this list have I never heard of or have no idea what it is? You may be surprised at the number of unknown words you have just read, words that, for all you know, might be the ingredients for glue or synthetic road-deicer rather than for something you would want to eat. If you check out some of these words in a book on the chemistry of food, you will find that they are assorted chemicals added to primary ingredients, such as wheat, milk, or meat, intended to add to or cover smells, color, stabilize, emulsify, bleach, texturize, soften, preserve, sweeten, or add more flavor to what nature provided. The FDA lists approximately 2,800 international food additives and about 3,000 chemicals that are deliberately added to our food supply.[1] When considering the number of chemicals used in the process of growing and processing food, we consume between 10,000 and 15,000 added artificial chemicals in a day of eating.[2] Most of the food we eat has been processed in some way and is very different from the way we ate 350 years ago or even 100 years ago. Capitalism and advancing technology have changed America's eating habits.

Many of the chemicals added to the raw food from farms are not listed on food packages. The FDA does not require food additives Generally Regarded as Safe (GRAS) to be put on an ingredients label. All that is required are the words *artificial flavor* or *artificial coloring* or *natural*. However, not everything that is natural is good for you, or even desirable. Alcohol, tobacco, marijuana, and cocaine, for example, are all natural, as are insect parts, arsenic, mercury, and cattle feces.

The average American eats his or her body weight in food additives each year, about 150 pounds of them.[3] About 10 percent of this is flavoring agents, preservatives, and dyes, many of them considered GRAS by the FDA. However, it is the food industry, not the FDA, that provides the evidence to support the GRAS claim. Having chemical manufacturers provide their own research to support a claim that a chemical is safe is the same as letting a cigarette manufacturer do the research to show cigarettes are safe. This seems less than advisable, but in the case of food, it is mandated by the high cost and time-consuming nature of testing. The FDA has neither the time nor the money to do the testing.

The Processing Industry

The contrast between the machine and the pastoral ideal dramatizes the great issue of our culture.
—Leo Marx, *The Machine in the Garden*, 1964

There are 22,000 different food items in today's supermarkets,[4] and nearly 100 percent of them have been processed to a greater or lesser degree. And many thousands of new products are introduced each year—21,000 in 2003. During processing, the prevention of contamination is a central objective. Most processing to prevent contamination emphasizes microbial control, eliminating bacteria that may already be present and preventing the growth of new organisms. Techniques used include refrigeration, freezing, drying, adding chemicals or harmless but predatory organisms, pasteurizing, sterilizing, irradiation, and fermenting.[5] Of lesser importance from the viewpoint of safety, but of great importance from the viewpoint of product salability are color, flavor, texture, and general appearance, including packaging. All of these may be produced or changed by processing, with the nature of the processing determined by the character of the input foodstuff, which may be grain, animal parts, or animal products such as milk.

When you eat fresh fruit or vegetables, you have little or no contact with the food processing industry. Other than the waxing of store apples, the sizing of eggs, the shelling of nuts, the boning of fish, or the bags of chopped greens, you can get by without processing services. Actually the apples need not be waxed, the eggs sized, the nuts shelled, the fish boned, or the greens chopped for you. But they are, and we buy them. In general, the food manufacturing industry is the standard intermediary between grain farmers, livestock producers, dairy farmers, and fishers, and the ultimate consumer. Various specialized industries process raw fruits, vegetables, grains, meats, fish, and milk into finished products. The production of most of these products requires industrial services. Few of us want to buy milk and the necessary equipment to make our own cottage cheese, yogurt, butter, ice cream, cream cheese, or sour cream. We want to eat these products and do not have the knowledge, time, or financial resources to replace the industry that produces them for us. Similarly, turning the grain farmers' produce into flour, corn flakes, bread, cake, crackers, or potato chips is not how Americans want to spend their time, although a tiny percentage do bake their own bread or make their own peanut butter. And turning cattle and hogs into T-bone steak, bacon, baby-back ribs, turkey bologna, or sugar-cured ham is a job we prefer others to do for us. Clearly the food processing industry is a necessary part of modern America's food culture. It is regulated by the USDA.

Thirty-four percent of food manufacturing workers are employed in plants that slaughter and process animals, and another 19 percent work in establishments that make bakery goods. In 2004 the food manufacturing industry employed 1.5 million people in about 29,000 establishments, 89 percent of which employed fewer than 100 workers.[6] More than half of the establishments had fewer than ten workers. As noted in earlier chapters, in grain farming and livestock operations, the few largest farms produce most of the products. In the food processing industry, only 2 percent of establishments employ 500 or more workers, but they account for 36 percent of all jobs.[7] In the United States, large size has economic advantages.

Meat

Speciesism is the belief that we are entitled to treat members of another species in a way in which it would be wrong to treat members of our own.
—Peter Singer, *Philosophic Exchange*, 1974

Just as the trend is for fewer but larger farms to produce the vast majority of livestock, the trend in the meat processing industry is toward fewer but much larger meat processing plants, owned by fewer companies. This has tended to concentrate employment in a few locations. For example, in 2004 only five states—California, Illinois, Iowa, Pennsylvania, and Texas—employed 24 percent of all workers in animal slaughtering and processing.[8] Meat processing is the most labor-intensive food processing operation, and killing and cutting up the animals we eat has always been bloody, hard, and dangerous work. Animals are not uniform in size, and slaughterers and meatpackers must slaughter, skin, eviscerate, and cut each carcass into large pieces, usually by hand using large, suspended power saws. They also clean and salt hides and make sausage. Meat, poultry, and fish cutters and trimmers use sharp hand tools to break down the large primary cuts into smaller sizes for shipment to wholesalers and retailers. These workers use knives and other hand tools to eviscerate, split, and bone chickens and turkeys. Automating these processes is difficult.

Because of volume requirements and the minimal level of training of slaughterhouse workers, the accident rate in the meat processing industry is high. In 2003 there were 12.9 injuries per 100 workers in animal slaughtering plants, the highest rate in food manufacturing.[9] Many of these accidents were severe, such as the loss of fingers or other human body parts. The problem of worker safety when working with sharp and fast-moving machinery and hand-held equipment is made more difficult by the nature of the workers. Slaughterhouse owners are no different from other employers, and they try to hire the cheapest labor they can find. Increasingly these are immigrants, many of them in the country illegally, whose undocumented status, language difficulties, and lack of education compound the risks of the job. Existing federal laws are too weak or too riddled with loopholes to be effective.[10] Meat slaughterhouses are dangerous places to work.

According to the U.S. Department of Labor, nearly one in three slaughterhouse workers suffers from illness or injury every year, compared to one in ten workers in other manufacturing jobs.[11] The rate of repetitive stress injury for slaughterhouse employees is thirty-five times higher than that of other manufacturing jobs. Because of the difficult and dangerous nature of slaughterhouse work, half of the workers quit within ninety days of being hired.

The process of turning an obese steer or cow into steaks for human consumption is a ghastly process that can be done only by either easily inured individuals or individuals too poor or uneducated to have other options. It is dehumanizing work. Turning an 800-pound cow or even a 5-pound chicken into tenders for the supermarket checkout or fast food restaurant counter is by its nature demanding, physical labor in bloody, greasy surroundings. There is an old saying that if slaughterhouses had glass walls, everyone would be a vegetarian.

Cattle are deliberately made obese in feedlots using fortified grain to maximize their commercial value. When they are deemed fat enough, they are shipped to a slaughterhouse perhaps a hundred miles away, where they will walk up a ramp through a door to the kill floor. The animals pass over a bar, their legs on both sides, and the floor slowly drops away so that they are carried on the moving bar, which acts as a conveyor belt. A short distance later the animal passes under a man on a catwalk holding a pneumatic device called a stunner. However, it does more than stun. It injects a metal bolt the size of a thick pencil into the animal's brain, which should kill it.

At that point, chains are attached to the (one hopes) dead animal's rear legs, and it is lifted up by the chains that are attached to an overhead trolley. The animal is then drained of blood by a person who sticks a long knife into it to cut the aorta. From there the corpse goes through a series of stations to clean it and remove the hide. Many steps are taken to make sure the feedlot manure that often cakes the hide of the animal when it enters the kill building does not contaminate the animal's flesh, our future sirloins and hamburger. Another possible source of contamination is the stomach contents of the animal. Contamination by bacteria is an ever-present danger. If manure gets on some meat and then that meat is ground up with lots of other meat, the whole lot of it can be contaminated. Evisceration and tying off the intestines must be competently done.

Unfortunately, incompetence occurs all too often.[12] A nationwide study by the USDA in 1996 found that 7.5 percent of the ground beef samples taken at processing plants were contaminated with Salmonella, 11.7 percent were contaminated with *Listeria monocytogenes*, 30 percent were contaminated with *Staphylococcus aureus*, and 53.3 percent were contaminated with *Clostridium perfringens*. All of these pathogens can make people sick; food poisoning by *Listeria* generally requires hospitalization and proves fatal in 20 percent of the cases. In the USDA study 78.6 percent of the ground beef contained microbes that are spread primarily by fecal material. As Eric Schlosser says, "There is shit in the meat."[13]

Meat processing has changed in recent decades. A hamburger bought in 1970 probably contained meat from one steer or cow, which would have been processed at a local butcher shop or small meat-packing plant. Today a typical fast food hamburger patty contains meat from more than 1,000 different cattle, raised in as many as five different countries. It may look like an old-fashioned hamburger, but it is a fundamentally different thing.[14] Meat from a greater number of animals increases the possibility that the hamburger will be contaminated with bacteria, fecal matter, or some other undesirable material.

It is difficult to guarantee cleanliness (or safety) when slaughterhouses process 300 or 400 cattle or 1,250 swine an hour, and when the industry averages a turnover rate of employees between 75 percent and 100 percent a year.[15] It would seem that a well-trained and stable workforce would be necessities in a slaughterhouse factory.

Slaughterhouses for hogs or chickens are little different from those for cattle, processing one thousand hogs per hour or thousands of broilers per hour,[16] with the workers pulling and cutting with sharp hooks, knives, and other implements. Contamination of chicken carcasses in the slaughterhouse is little different from the contamination in the cattle slaughterhouse. A study by *Consumer Reports* in 1998 revealed that 74 percent of store-bought chickens were contaminated with Campylobacter or Salmonella (or both), bacterial contaminants responsible for thousands of deaths and millions of sicknesses in the United States each year.[17]

According to the USDA, contaminated meat causes approximately 70 percent of all food-borne illnesses.[18] In the United States, food-borne illness sickens 76 million people, causes 325,000 hospitalizations, and kills 5,000 people every year.[19]

Food Additives

It is not clear that society will lose all that much by some reduction in the number of chemicals introduced.

—J. Clarence Davies, Senior Fellow, Resources for the Future

There is a saying that people eat first with their nose, then with their eyes, and in the end with their mouth. The aroma and the flavor, the looks and the taste, are all essential for enhancing the appeal of food. As someone has remarked, if you are what you eat, it seems the average consumer consists mostly of thickener, water, salt, and sugar. Food colors and flavor enhancers have been used since ancient times, although with nowhere near the types, amounts, and intensities of modern processing techniques. Food manufacturers have become experts at manipulating the human desire to eat. A large part of America's obesity problem can be attributed to this expertise.

Food additives play an important part in the food supply, ensuring that the food is safe, has a long shelf life, and meets the needs or wants of consumers. Processing nearly always means the addition of manufactured chemicals to the basic foodstuff that came from the farm. There are no corn chip bushes or Cheerios vines on a farm, and spaghetti trees have yet to be discovered. No cow is made of bologna, liverwurst, or frankfurter, no part of a chicken is a McNugget, no chicken has a boneless breast, and no part of a hog is made of sausage, spam, or has sugar-cured bacon as part of its anatomy. These are all products whose immediate origins involve considerable processing, an important part of which is a careful selection of additives. The additives are necessary not only to produce the unnatural products that we enjoy, but also to retard spoilage as the products travel across the United States in trucks. For example, sulfur dioxide is added to meat products such as sausage meat to prevent the growth of microorganisms. Additives under the rubric of expanded shelf life technologies can double or even triple the shelf life of a fresh loaf of bread, from several days to ten or more. Apples treated with SmartFresh stay firm and tasty three times longer than untreated apples.[20] However, only 1 percent of the weight of the additives used in food is to inhibit the growth of harmful microorganisms.[21] The other 99 percent are added to entice consumers to eat more.

Some substances used as food additives occur naturally, such as ascorbic acid (vitamin C) in fruit and lecithin in egg yolks, soybeans, peanuts, and corn. The human body cannot distinguish between a chemical naturally present in food and that same chemical present as an additive.[22] However, the body may be aware that it is receiving much larger amounts of a naturally occurring substance than it can satisfactorily deal with. Typically the body's long-term tolerances for added chemicals are not known.

One universal addition to meat and fish is water. A solution of chemicals called polyphosphates is injected into the meat on the production line, which acidifies the meat so that it absorbs water. The water gain can be substantial, and the process is commended in advertisements to producers and butchers on the grounds that it allows them to sell water in the guise of meat.

Some food additives have more than one function.[23] For example, acidity regulators help to maintain a constant level of acid in food. This is important for taste, as well as to influence how other substances in the food interact. Chemical reactions can vary as the level of acidity changes, and the growth of microorganisms is affected by acid content. Antioxidants may be added to retard or prevent rancid flavors from developing in fats and oils when they are exposed to oxygen. Flavor enhancers are added to make foods tastier so consumers buy more and eat more of them. Preservatives retard bacterial growth. Thickeners increase the viscosity of spaghetti sauce or ketchup to give them the desired consistency. The list of possible additives in processed foods is endless, as chemists are continually devising new substances to improve the palatability of food. Intolerance to an additive does not depend on whether the food additive is derived from a natural or synthetic source.[24]

A recent addition to food preservation is carbon monoxide in meat packages so that the meat will stay red for weeks.[25] The process, used in factory-wrapped (or case-ready) meat, replaces most of the oxygen in the package with other gases. These include tiny amounts of carbon monoxide, which react with the pigment in the meat, producing a red color. Retailers favor this addition; they lose large amounts of money ($1 billion in 2003) because consumers shy away from meat that has turned brown from exposure to oxygen, although it might still be fairly fresh and perfectly safe. People and organizations opposed to the carbon monoxide addition say that the process, which is also used to keep tuna rosy, allows stores to sell meat that is no longer fresh and that consumers would not know until they opened the package at home and smelled it. The label on the package does not indicate whether the meat has been laced with carbon monoxide. Food processors say hamburger can be sold for twenty-eight days after leaving the processing plant and solid cuts for thirty-five days, whether or not the meat has remained red.

Some of the natural or artificial foods we consume have lists of additives that no human being can comprehend. For example, compare an old-fashioned strawberry milk shake with one bought from a fast food outlet.[26] The old way contained milk, cream, sugar, ice cream, and strawberries. The drink bought at a local fast food outlet today contains milk fat, nonfat milk, sugar, sweet whey, high-fructose corn syrup, corn syrup, guar gum, mono- and diglycerides, cellulose gum, sodium phosphate, carrageenan, citric acid, sodium benzoate, red coloring #40 and artificial strawberry flavor (amyl acetate, amyl butyrate, amyl valerate, anethol, anisyl formate, benzyl acetate, benzyl isobutyrate, butyric acid, cinnamyl isobutyrate, cinnamyl valerate, cognac essential oil, diacetyl, dipropyl kentone, ethyl acetate, ethyl amylketone, ethyl butyrate, ethyl cinnamate, ethyl heptanoate, ethyl heptylate, ethyl lactate, ethyl methylphenylglycidate, ethyl nitrate, ethyl propionate, ethyl valerate, heliotropin, hydroxyphenyl-2-butanone, x-ionone, isobutyl anthranilate, isobutyl butyrate, lemon essential oil, maltol, 4-methylacetophenone, methyl anthranilate, methyl benzoate, methyl cinnamate, methyl heptine carbonate, methyl naphthyl ketone, methyl salicylate, mint essential oil, neroli essential oil, nerolin, neryl isobutyrate, orris butter, phenethyl alcohol, rose, rum ether, y-undecalactone, vannilin, and solvent).

Color plays an important part in people's response to food. If you doubt this, ask yourself whether you would buy eggs with green or purple yolks or perhaps black butter. For this reason, almost anything found in a supermarket today has a dash of something for eye appeal. It may be natural, such as the caramel color in most soda, which comes primarily from heated sugar. Alternatively, the color additive may be synthetic, such as the petroleum-based dye known as FD&C Red No. 40, shorthand for 6-hydroxy-5-[(2-methoxy-5-methyl-4-sulphophynl)azo]-2-naphthalenesulfonic acid].[27]

Adverse reactions to food additives occur in a small proportion of the population. However, more people are intolerant to common foods such as peanuts, milk, or eggs than to food additives. One additive that has caused considerable concern is MSG (often called hydrolyzed vegetable protein on labels) classified as GRAS since 1959 by the FDA and deemed safe by the American Medical Association in 1992, but claimed by some environmental groups to be harmful. MSG is used as a flavor enhancer in a variety of foods prepared at home, in restaurants, and by food processors. Many scientists believe that MSG stimulates receptors in the tongue to augment meat-like flavors, tricking the brain into believing that the food tastes better than it does, so you will eat more of it.[28] In Japan, MSG was first sold under the brand name Ajinomoto, which literally means "essence of taste."[29] MSG is one of the most extensively studied food ingredients in the food supply; hundreds of scientific studies and numerous scientific evaluations have tried to ascertain its effect on the body. The conclusion of the research is that MSG is a safe and useful flavor enhancer in the amounts that people consume in processed foods, although some people do have short-term reactions, such as numbness, burning, tingling, facial pressure, tightness, chest pain, headache, or nausea.

Food Processing and Nutrition

Technology is a queer thing. It brings you great gifts with one hand, and it stabs you in the back with the other.
—C. P. Snow, 1971

The nutrient content of food can be affected by any of the variety of things that happen to it during its growing, harvesting, storing, and preparing.[30] Crops are fertilized, and fertilizer tends to reduce the content of vitamin C in the crop while increasing its protein content. Cereals are ground to remove the fibrous husks or bran, but the bran contains most of the plant's dietary fiber, B-group vitamins, phytochemicals, and some minerals. Loss of the bran coating is the reason white bread is less nutritious than whole wheat bread. Molasses or cocoa may be added to white bread by the bread processor to give the bread the appearance of whole wheat bread.

Grains from the farm are typically refined, a process that means the opposite of "improved." George Orwell would be pleased. Humans have been refining grains since at least the Industrial Revolution, favoring white flour (and white rice) even at the price of lost nutrients. Refining grains has two objectives: extending shelf life because the nutrients lost renders the grains less nutritious to pests as well as people and making the grains easier to digest because the fiber that slows the release of their sugars is removed. Refining processes are one of the many ways that food processors contribute to the growing epidemic of obesity in the United States.

Consider the transformation of whole wheat flour into the more popular white bread. At the grain mill, long rollers with hundreds of metal teeth repeatedly crush raw wheat berries, sifting and separating them between large screens and eventually stripping the strongly flavored nutrient- and fiber-packed germ and bran from the starchy, bland endosperm. The resulting white flour contains barely any fiber, vitamins, or minerals, the building blocks of healthy food. One slice of white bread has 65 percent less fiber, magnesium, and potassium than whole wheat bread. The bran alone in whole wheat bread gives it twenty times more antioxidant power.[31]

Refined grains have been linked to a lowering of insulin sensitivity and unhealthy spikes in blood sugar, a precursor of type 2 diabetes. Twenty percent of the typical American diet now comes from refined grains: bread, pasta, doughnuts, potato chips, muffins, crackers, and other processed grain products.

When the bread label says "fortified," it means that some of the nutrients that were lost during milling have been artificially replaced. However, it is not possible to replace everything that was removed. The fiber added to some fortified products is often in the form of resistant starch, which may not be as beneficial as the original fiber that was removed. No one knows which nutrients are most important.

Before a food is canned or frozen, it is usually blanched by heating it quickly with steam or water. The water-soluble vitamins are easily destroyed by blanching, and cooking in water has the same effect. Canned food is heated inside the can to kill any dangerous microorganisms and extend the food's shelf life. This may affect the taste and texture of the food, making it less appealing. Water-soluble vitamins are particularly sensitive to high temperatures.

Drying foods, such as apricots or cranberries, can reduce the amount of vitamin C they retain but can concentrate other nutrients, particularly fiber in plant foods. Dehydrating foods also makes them more energy dense, which may contribute to weight gain. If a dehydrated food is reconstituted and cooked with water, further nutrients are leached from the food and lost in the cooking water. Most vegetables are peeled or trimmed before cooking to remove the tough skin or outer leaves. The bulk of nutrients, such as vitamins, tend to lie close to the skin surface of most vegetables. It is healthier to eat the skin of a baked potato or a cucumber than to throw it away.

Clearly the nutrient value of any food, including those grown organically, is likely to be altered by processing, and vegetables are particularly susceptible. Some processing is done at home by consumers; simply cooking or combining a food with other foodstuffs to create a recipe is also a form of food processing. The loss of nutrients when cooking depends on the temperature, duration of cooking, and the nutrient involved. Some vitamins are more stable and therefore less affected by processing than others. Water-soluble vitamins (B-group and C) are less stable than fat-soluble vitamins (K, A, D, and E) during food processing and storage. The least stable vitamins are folate, thiamin, and vitamin C. Most stable are niacin (vitamin B_3), vitamin K, vitamin D, biotin (vitamin B_7), and pantothenic acid (vitamin B_5). Cooking in water also causes the loss of the more water-soluble minerals, such as sodium, potassium, and calcium. Eating broccoli or spinach raw is more nutritious than cooking it first.

Use fresh ingredients whenever possible, as nutrients may deteriorate with age. Improper storage methods can also destroy nutrients in fresh foods. Keep cold foods cold, and store leftovers in airtight containers. Microwave, steam, roast, or grill vegetables rather than boil them. If you boil the vegetables, save the nutrient-rich water for soup stock or gravies, or give it to your dogs to drink. They find it tastier than tap water.

Dairy Processing

Sometimes it seems like environmentalism's main accomplishment has been to transform a simple trip to the supermarket into an occasion for anguished self-criticism, and a shopping list into the measure of one's relationship with Mother Earth.
—Bill Gifford, *City Paper*, 1991

Milk most often means the nutrient fluid produced by the mammary glands of female mammals, but it can also refer to the white juice in a coconut or to nonanimal substitutes such as soya milk, rice milk, or almond milk. Although milk from mammals such as goats or sheep is popular in non-Western countries, virtually all milk in the United States is from cattle and is processed in a dairy facility to generate many milk-based products. As noted in chapter 9, the most popular of these are cheese (38 percent), milk and cream (33 percent), butter (13 percent), and ice cream (9 percent). The remaining 7 percent is processed into a wide variety of products, including dry whole milk and nonfat dry milk, condensed milk, skim milk, buttermilk, yogurt, and sherbet. The cheese category can be subdivided into cottage cheese, cream cheese, and the hard sliceable cheeses such as American, muenster, swiss, mozzarella, and several others. In North America most dairies are local companies.

Milk

As all Americans over the age of fifty who may remember milk in glass bottles are aware, standing fresh milk has a tendency to separate into a high-fat cream layer on top of a larger skim milk layer. The cream is often sold as a separate product with its own uses. In the United States a blended mixture of half cream and half milk, called half and half, is sold in smaller quantities and commonly used to fatten coffee. Modern commercial dairy processing techniques remove all the butterfat from the milk, then add back an appropriate amount depending on which product is being produced on that particular line. Whole milk contains about 3.25 percent butterfat, low-fat milk about half this amount, and skim milk about 0.1 percent fat. (In New Zealand in 2001, a cow was discovered that had a rare gene mutation that causes it to produce skim milk. Naturally produced skim milk should soon be available). A decrease in the amount of fat in milk increases the percentage of protein.

Cartons of milk are sold today in supermarkets in homogenized form. Milk is homogenized by mechanically reducing the fat globules to a size that stabilizes them in solution so the cream will not separate. It is then heated to at least 145°F for at least 30 minutes or to 161°F for at least 15 seconds (pasteurized) to kill even the most resistant harmful bacteria and then cooled for storage and transportation. Pasteurized milk is perishable and must be kept cold by suppliers and consumers; it is transported in refrigerated trucks. Dairies print expiration dates on each container, after which stores remove any unsold milk from their shelves. However, the milk remains drinkable for several days after the expiration date on the carton.

Pasteurization is required in the United States to kill bacteria that occur in the milk, but the process is not without nutritional drawbacks.[32] Raw milk contains enzymes that permit raw milk to digest itself. These enzymes are destroyed by the pasteurization process, which makes digestion in the stomach difficult and the calcium in the milk largely indigestible. It is noteworthy that American women, who consume lots of milk products, suffer the world's highest incidence of osteoporosis. Studies have failed to associate drinking large amounts of milk with a decrease in bone fractures in either men or women.[33]

As described in chapter 9, dairy cattle in the United States spend their lives indoors and so receive no solar ultraviolet-B radiation. The resulting milk is deficient in vitamin D, which must be added in the processing plant. Milk today also commonly has flavorings added to it for better taste or to provide a new product for the grocery shelf. Until relatively recently, the only flavoring in milk was chocolate.

Cheese

Turophilia, a love of cheese, is a common affliction in the United States. As a result, more than 400 types of cheese are produced, and the United States produces more than 30 percent of the world's supply.[34] In 2003 each American ate 30.8 pounds of cheese. About one-third of America's milk production is used to make cheese, and consumption has been increasing for more than fifty years.[35] There now exist farmstead cheeses, defined as an artisan cheese produced on a farm using only milk from the farm's herd, similar to an estate wine produced from a winery's own grapes.[36] California, the number 1 cheese-producing state, has ten farmstead cheesemakers, up from only three in 1997. In 2006 the state eclipsed Wisconsin in total cheese production.

Wisconsin produces 26.5 percent of domestic cheese, and employs 33 percent of all cheese manufacturing workers. Popular types of natural cheeses include unripened ones (cottage cheese, cream cheese), soft (brie, camembert), semihard (brick, muenster, roquefort, stilton), hard (colby, cheddar, gruyere), blue veined (blue, gorgonzola), cooked hard cheeses (swiss, parmesan), and pasta filata (stretched curd, such as mozzarella, provolone). Then there are processed cheeses such as American cheese (a mixture of cheddar and colby), which in 2004 accounted for 42 percent of total cheese production,[37] and various cheese spreads, which are made by blending two or more varieties of cheese or blending portions of the same type of cheese in different stages of ripeness. All cheeses are high in cholesterol and saturated fat, and the harder the cheese is, the worse it is for blood vessels.

The modern manufacture of cheese consists of four basic steps: coagulating, draining, salting, and ripening.[38] Processed cheese manufacture has the additional steps of cleaning, blending, and melting. The method is different for each variety of cheese. Initially the milk is homogenized to ensure a constant level of fat. A centrifuge, which skims off the surplus fat as cream, may be used first to obtain the fat levels appropriate for different varieties of cheese. Pasteurization is next.

The milk is then coagulated, or clotted, by physical and chemical modifications to the milk, leading to the separation of the solid part of the milk (the curd) from the liquid part (the whey). The modifications include the addition of harmless, active bacteria and an enzyme called rennet that may be of animal, fungal, or bacterial origin. The rennet reacts with the protein to coagulate the milk. Most of the fat and protein from the milk stay in the curd, but nearly all of the lactose and some of the minerals, protein, and vitamins escape into the whey. The curd is cut into variously sized cubes and cooked at various temperatures to produce different varieties of cheese. The proto-cheese is drained to separate the curd from the whey, and the curd is dried and pressed. As many of us recall, Little Miss Muffet ate both curds and whey.

For some cheeses, special procedures may occur before, during, or after draining. Blue-veined cheeses are seeded with penicillium powder prior to drainage. Cooked hard cheeses are stirred and warmed to complete separation of the whey, which may be used to make whey cheese and other products.

Knitting, or transforming, the curd allows the accumulating lactic acid to chemically change the curd; knitting also includes salting and pressing. This step leads to the characteristic texture of different cheeses, such as provolone, mozzarella, cottage cheese, or cream cheese. Finally, the cheese is ripened or cured, giving the different varieties of cheese their characteristic and unique textures, aromas, appearances, and tastes through complex physical and chemical processes that are controlled by adjusting temperature, humidity, and duration of ripening. Curing may involve adding yeast and bacteria (brick, brie, camembert), molds (blue cheese), or gas-forming organisms (swiss cheese). For all cheeses, the purpose of ripening is to allow beneficial bacteria and enzymes to transform the fresh curd into a cheese of a specific flavor, texture, and appearance. Cottage and cream cheeses are not ripened and usually have a bland flavor and soft body.

Swiss cheese was in the news in 2002 because the USDA created new guidelines that regulated the hole size of domestically produced Swiss cheese.[39] It reduced the hole size by 50 percent because new cheese-slicing machinery got jammed on larger holes. The Swiss objected to the guidelines and insist that Emmentaler cheese, as the Swiss call it, must have large holes.

Butter
Butter is made by churning fresh or fermented cream or milk and consists of butterfat surrounding minuscule droplets of water and milk proteins. Salt, flavorings, or preservatives are sometimes added. The color of the butter depends on the animal's feed and is sometimes manipulated with food colorings.

Unhomogenized milk and cream contain microscopic globules of butterfat surrounded by a membrane of emulsifiers and proteins. (An emulsion is a liquid in which droplets of fat are floating.) The fat consists mostly of chemicals called triglycerides, which many physicians believe are as damaging to human blood vessels as cholesterol. The membrane that coats the fat globules serves to prevent the fat in milk from coagulating. Making butter involves agitating pasteurized cream to damage the fat globule membranes, allowing the fats to come together and separate from the other parts of the cream.[40] Butter is essentially the fat of the milk. Churning produces small grains of butter floating in the water-based part of the cream, which is buttermilk. The buttermilk is drained away, and the butter grains are pressed and kneaded to form them into a solid mass. Commercial butter is about 80 percent butterfat and 15 percent water. Whipped butter is produced by incorporating nitrogen gas, which does not react with the butter. Using normal air would promote oxidation and rancidity.

Margarine

Margarine is not a milk product but is commonly used in place of butter. Twice as much margarine as butter is consumed in the United States.[41] Margarine is made from a wide variety of animal or vegetable fats and is often mixed with skim milk, salt, and emulsifiers. It is an emulsion of water in oil and is high in saturated fat and transfat, an unhealthy fat produced when the vegetable oils that are the base of margarine are chemically modified so they solidify. The process involves adding hydrogen atoms to the oil molecules. Partial hydrogenation yields transfats; complete hydrogenation yields saturated fats. Americans consume four to five times more saturated fat than transfat in their diet. The federal government now requires the transfat content of packaged foods to be listed on labels.

Ice Cream

The manufacture of ice cream consists of a sequence of seven steps: blending of the mix ingredients, pasteurization, homogenization, aging, freezing, packaging, and hardening.[42] The mix of ingredients consists of milk, cream, sugar (12 to 16 percent), and flavorings. Federal law mandates that anything labeled ice cream contain at least 10 percent milk fat, and regular ice creams have 10 to 16 percent; some premium brands have as much as 20 percent of this high-calorie component.[43] Pasteurization destroys any pathogenic bacteria and reduces the number of spoilage organisms. Homogenization reduces the size of the fat globules, increases the smoothness of the ice cream, makes it easier to whip, and increases resistance to melting. Aging overnight improves the whipping qualities of the mix and increases the viscosity or thickness of the ice cream.

Following aging, liquid flavors, fruit purees, or colors are added, and the mixture is dynamically frozen. This process consists of whipping the mixture as a large amount of air is pumped into it and it passes through a freezing apparatus. Without the air, which occupies 15 to 50 percent the volume of ice cream, the ice cream would be similar to a frozen ice cube. The finest ice creams have between 15 percent and 30 percent air. Cheaper ice creams contain more air, which decreases the density of the product and thus decreases transportation costs and provides a less solid product in a quart of ice cream. After the dynamic freezing, particulate materials such as fruits, nuts, candy, cookies, or other additives may be inserted into the ice cream. When kept in the freezing compartment of a refrigerator at a temperature less than about −13°F, ice cream is stable indefinitely without danger of ice crystal growth.

Ice milk contains less fat than ice cream, and sherbet contains even less. French ice cream is enriched with egg yolks.

Packaging

Waste is not just the stuff you throw away, of course, it's the stuff you use to excess.
—Joy Williams, *Esquire*, 1989

Proper packaging of food is essential to make sure the food remains wholesome during its journey from processor to consumer; packaging also increases shelf life, an important consideration for food producers and marketers.[44] Packaging makes food easier to handle, protects it from extremes of temperature during transport, locks out microorganisms and chemicals that could contaminate the food, and helps prevent physical and chemical changes and maintain the nutritional qualities of food. It also prevents criminal poisoning of food by terrorists or blackmailers. Packaging is not cheap; it accounts for 10 to 50 percent of the price of food.[45]

Milk is stored in opaque containers to prevent vitamins from being destroyed by light. Oil is packaged in containers that are impermeable to oxygen because oxygen makes fats become rancid. Conversely, plastic wraps permeable to oxygen are used with fruits and vegetables to allow them to "breathe" and with meats to ensure that they will maintain the vibrant red color that consumers associate with freshness. Metal (sheet steel sometimes with an internal coating of tin, enamel, or vinyl) and glass containers are used in canning because these materials can withstand the high temperatures and changes in pressure during the processing of their contents. An advantage of glass is its transparency, which enables the consumer to see the product inside. On the negative side, glass is not impact resistant and is relatively heavy, which increases transportation costs.

Plastic is lightweight and unbreakable and has become the dominant material used in food packaging. Most plastics used in food packaging are heat resistant so that they can survive high-temperature sterilization processes. There are many types of hard and flexible plastics used in food containers and wrappings, and they can be made into a wide variety of shapes, as well as the thin films that are used as bags and wraps. It is difficult to imagine a world without these petroleum-based plastics. Roughly 7 to 10 percent of the petroleum consumed in the United States is used to manufacture plastics and fibers.[46] One common type of nonflexible plastic used in food packaging is known as PET, a useful abbreviation for its chemical name, polyethylene terephthalate. We see it as containers for soft drinks, bottled water, mayonnaise, and ketchup. Americans throw away 2.5 million plastic bottles every hour, and less than 3 percent are recycled.[47]

The most popular flexible plastic wrap, constituting about two-thirds of all food packaging wrap,[48] is made from low-density polyethylene, sometimes with a chemical abbreviated as DEHA added to make the wrap more flexible and to stick. The plastic's adherence to itself keeps air out of the wrap's contents, increasing its life span and keeping bacteria and germs out of the food. DEHA has been accused of leaking into the food it surrounds, particularly high-fat foods such as meats and cheeses, and of causing a variety of gene mutations and illnesses. Although trace amounts of the chemical may indeed pass from the wrap into the food, particularly during microwaving, there is no evidence that the amounts are large enough to cause harm to humans.[49] Microwaving should be done using only plastic wrap or plastic containers that say they are microwave safe.

The most recent development in packaging is to embed aromas in the plastic packaging material.[50] Plastic bags and bottles will soon attract consumers by, for example, the smell of chocolate, fresh bread, or oven-baked cookies emanating from packages of the products. An Israeli scientist has discovered how to make food products smell like flowers. Several studies have shown that pleasant scents encourage shoppers to linger over a product, increase the number of times they examine it, and in some cases increase their willingness to pay higher prices too. The aromas last about a month on the supermarket shelf. Other packaging innovations soon to appear include a flat electronic display that can be applied to boxes like a label, allowing for tiny lights, miniature games, or flashing messages such as BUY ME! NEW TASTE SENSATION, or possibly RECOMMENDED BY BRAD PITT AND ANGELINA JOLIE.

Other high-tech packaging innovations aim to educate or instruct consumers. Recognizing that consumers often damage fresh fruit by squeezing it to determine its ripeness, the pear industry has started using a label that changes color to show consumers when pears are ready to be eaten. Companies are also developing labels for meat and seafood that will change color if the package was not stored at the right temperature at any point along the delivery chain.

The amount of plastic packaging of food has increased over the decades and now accounts for one-third of an average household's waste. This has contributed to a glut in landfills. However, the glut has resulted as much from demographic factors as from the food packagers themselves.[51] The decrease in the size of households and their increasing number has resulted in more people buying smaller portions of food and thus more packaging. Higher living standards in the United States have led to more consumer goods and to the transportation of exotic foods over long distances, requiring a large amount of packaging to maintain freshness. The growth and expansion of cities has created longer distances between food producers in rural areas and consumers in urban areas and has led to a demand for packaging. There has been an increase in households in which both partners work, which has led to a higher demand for convenience foods, which invariably have a lot of packaging.

Glassworks and plastic producers fight for increasing sales of their packaging materials, but it seems increasingly that glass packaging does not have a bright future. The only significant advantage of glass over plastic is that it is inert and there is no migration of possibly harmful chemicals into the food. But because the amounts leaked by plastics are currently considered negligible, this apparent advantage loses most of its significance.

Paper is not frequently used in packaging. However, when paper is coated with plastic or other materials to make it stronger and impermeable to water, it can be used more widely. Reinforced paper-based flour and sugar containers are common, as are paperboard for packaging frozen foods and egg cartons. Shipping cartons are usually made of corrugated cardboard.

In recent years, environmental concerns have influenced food packaging. Companies that make packaging materials are trying to develop packaging that is recyclable, biodegradable, or more compact so it will use less landfill space, as well as to eliminate unnecessary packaging.[52] And recycling of plastic bottles and aluminum cans is promoted widely in American cities.

Labeling

Where is the wisdom we have lost in knowledge?
Where is the knowledge we have lost in information?
—Thomas Stearns Eliot, *The Rock*, 1934

Labels on hard-shell packages have become more informative in recent years, in part because of new governmental regulations. If consumers are interested they can get a good idea of the healthfulness, or lack of it, of the processed food they are buying.[53] However, whether the consumer is really being informed by much of the labeling is questionable. Humorist Dave Barry wrote,

When I purchase an item at the supermarket, I can be confident that the label will state how much riboflavin is in it. The United States government requires this, and for a good reason, which is: I have no idea. I don't even know what riboflavin is. I do know I eat a lot of it. For example, I often start the day with a hearty Kellogg's strawberry Pop-Tart, which has, according to the label, a riboflavin rating of 10 percent. I assume this means that 10 percent of the Pop-Tart is riboflavin. Maybe it's the red stuff in the middle. Anyway, I'm hoping riboflavin is a good thing; if it turns out that it's a bad thing, like "riboflavin" is the Latin word for "cockroach pus," then I am definitely in trouble.[54]

Nutritional information is not required on restaurant menus, but starting in 2006, McDonald's fast food restaurants, in an attempt to counter its growing image as a major cause of the obesity epidemic sweeping the United States, introduced such information on its food packages. The information consists of icons and bar charts displaying how McDonald's menu items relate to daily recommendations for calories, protein, fat, carbohydrates, and sodium. The fat percentage does not have a breakdown into saturated, transfat, and unsaturated, however. (Perhaps this is because there is little unsaturated fat in a McDonald's hamburger.)

Can a consumer rely on the content weights given on processed food packages? In 2004, the USDA tested ninety-nine single-serving processed foods to determine whether the amounts printed on the food packages in stores were accurate.[55] The results indicated that only thirty-seven of the ninety-nine labels were accurate: fifteen were underweight and forty-seven overweight. Nearly half the time the food processor is giving consumers more than their money's worth. Federal regulations require most of the foods tested to be within 10 percent of the weight printed on the package, but many of the food labels exceeded this amount. The three heaviest groups of products were breakfast cereals, sliced bread, and prepackaged breakfast items, and they varied by as much as 72 percent from what their labels reported. Lightweight foods such as precooked bacon and bologna missed their mark by as much as 25 percent.

In addition to nutritional information and package weight, the package may have other information printed on it, sometimes in easily visible bold lettering. Some of this labeling is meaningful and significant, such as "USDA Organic." But most of these labels are not legally defined and are meaningless—for example, "fresh," "nonpolluting," "no chemicals," "ozone friendly," "allergy tested," "environmentally friendly," "farm fresh," "wholesome," "original," and "natural." Advertisers are clever, literate people. The instant coffee I drink in the morning is described on the label as "bold and intense," differing from their companion blend characterized as "rich and smooth." Both descriptions of my morning blend are certainly more inviting than a "weak and wimpy" coffee but are equally meaningless. Advertisers have an inexhaustible supply of loosely defined but pleasant-sounding adjectives. As columnist Melvin Madocks says, they are "the carnival barkers of life who misuse language to pitch and con and make the quick kill."

Food Distribution

Our national flower is the concrete cloverleaf.
—Lewis Mumford, *The Culture of Cities*, 1938

After food is grown, processed, and packaged, it enters an extensive distribution network that transports food products from the farmer or the food processor to retail outlets across the country. Transport can be by truck, train, or plane, but given the inadequacies of America's rail system and the high cost of air freight, most food is moved by truck. Commercial agricultural transporters move 95 percent of livestock, over 90 percent of all fruits and vegetables, and nearly 70 percent of domestic grain movements.[56] Even perishable food can be transported great distances in refrigerated trucks, a necessity in a large country such as the United States.

Distribution networks satisfy consumer demand for variety, making available, even in remote areas, foods that are not locally grown or processed. The same network also distributes imported food from the seaport to locations hundreds or even thousands of miles away in America's interior. Although food distribution is essentially unseen by the average consumer, it plays a vital role in ensuring the availability of even the most basic foodstuffs.

Some large grocery store chains have the resources to buy food products directly from the processors, transport them, and store them in warehouses until they are needed at the store. But for the declining number of independent groceries in the United States, food wholesalers fulfill these roles. Small groups of retailers may join together with a cooperative wholesaler that sells only to the member-owners. Some food is sold directly to a retail store, particularly foods such as bread and dairy products that must be delivered fresh every day or every few days.

In these ways, food makes its way to retailers such as restaurants, fast food outlets, supermarkets, and convenience stores. In the United States, supermarkets are the predominant type of food retailer, and when the food reaches this destination, the seller must consider factors such as positioning, branding, and pricing. Positioning refers to location in the store—near the entrance or toward the rear; top, middle, or lower shelf; sold in the standard manner or as a drawing card or "loss leader." Branding refers to whether the label is the store's brand, a local labeled product, or a nationwide brand.

Pricing strategy is not unrelated to store location and branding. Marketers position the items they most want to sell on the shelves between knee and shoulder height, the "strike zone" in baseball terminology. The highest markup items and the most expensive name brands are likely to be placed at chest height to make it easy for customers to grab and toss them in their shopping cart. The main areas where people are likely to walk—the paths to milk and bread—are usually strewn with high-priced land mines. Should a higher-priced sugary breakfast cereal with a figurine of Bugs Bunny in it be located where a five-year-old can see and reach it? Where should the store place the overly priced private-label avocado dip or the low-priced cans of oily tuna? These are important decisions in the high-volume, low-profit margin grocery business, and the store manager's job may depend on making the most customer-savvy decisions.

Buying the Food and Eating Well

Americans can eat garbage, provided you sprinkle it liberally with ketchup, mustard, chili sauce, Tabasco sauce, cayenne pepper, or any other condiment which destroys the original flavor of the dish.
—Henry Miller, American writer 1891–1980

American shoppers take for granted supermarket shelves groaning under the weight of a vast array of foods. Store aisles glisten, and soft music guides shoppers past pyramids of shiny, attractive-looking produce. There are plums in November and strawberries in February and luscious-looking vegetables year round. And there are never-ending rows of eye-catching packages touting lower fat, lower sodium, reduced sugar, better taste, and quicker preparation. And it seems that whenever the food processor removes something ("reduced sugar," "low fat"), the price increases.

Few Americans realize that we spend less of our income on food than do the citizens of any other country in the world—only about 9.5 percent. It would be even less if we did not choose to eat out so often.[57] Thirty-nine percent of our food bill is spent in restaurants.[58] People in the United Kingdom spend 10 percent of their paychecks to eat at home; the Japanese spend 16 percent, Italians 19 percent, Indians 48 percent, and Africans 90 percent. In fact, the cost of food for Americans is decreasing. The cost of feeding an average family (adjusted for inflation) is 5 to 10 percent less than thirty years ago.[59]

What food products should you buy with the $90 you spend each week for food?[60] There is a profusion of ads for sugary breakfast cereal flakes, Pop-Tarts, and Coca-Cola. Food processors and advertising agencies spend large amounts of money promoting processed foods. Kraft and Procter & Gamble are the leading producers, with Frito-Lay and General Mills trailing behind.[61] These companies and the others in the food production industry turn out an immense and ever-expanding variety of products. Kraft has more than 200 in North America alone, including such consumer favorites as 23 different Post cereals, Velveeta cheese, Oscar Mayer meats, Jell-O, and the ever-popular children's favorite, Kool-Aid.[62] Obviously no store can stock every food product produced by the food processors, so the competition among them for shelf space in supermarkets is ferocious. The number of products available in the average size supermarket is approximately 45,000, and the size of an average supermarket is now 44,000 square feet, (about twenty times the size of an average American house), but as large as this number seems, it is not large enough. The number of new products introduced by food processors in 2003 topped 20,000.[63]

Natural foods, as contrasted with food products, are rarely advertised. How often do you see or hear an ad expounding the virtues of fresh broccoli, tomatoes, or peaches? These "old," "original," or natural foods are the ones humans evolved with over hundreds of thousands of years, so it is safe to assume that our digestive system is well adapted to their nutritive value. In contrast, the products of food processors are likely to be loaded with substances that are completely alien to our evolutionary heritage. Technology has given us an impressive number of new foods, but there is no evidence that it has improved the quality of natural food: fruit, vegetables, and unmodified natural meat. The food we eat today is no healthier than the food humans ate a hundred or a thousand years ago, and in many cases it is clear that modern processed food is not healthful. Whole grain bread is better than white bread. Water is better than soft drinks. Beef raised naturally on forage is more healthful than grain-fed beef from a feedlot. The list is endless. How much of your diet is based on food your parents or grandparents would not recognize?

Parents with small children commonly go food shopping with them. Advertisers know this and know that children tend to believe TV commercials touting certain food products. Hence, children are increasingly the target of advertisers of processed food products. Junk food marketers dominate the ads in cartoon programs. Ninety percent of food commercials aired on Saturday morning children's TV shows are for products of low nutritional value such as sugary cereals, candy, and fast food.[64] According to the National Institute on Media and the Family, advertisements target children as young as three years old.[65] Consider how hard it is for an adult to resist using a money-off coupon to try a new processed food product or how hard it is to resist the free samples of processed food given out by the salespeople stationed at the end of the aisles in Costco or Sam's Wholesale. Is a five year old going to resist a box of sugary cereal flakes with a picture of Tony the Tiger or Spiderman on it? How often can a harried parent in a supermarket say no when every aisle is full of unhealthy but heavily promoted items for children? The Institute of Medicine has called on companies to make more healthy products and promote them more aggressively. The panel said that if the companies failed to do so within two years, Congress should mandate changes.[66] Children need legal protection from food pornography.

In January 2006, two consumer activist groups announced they would use consumer protection laws in Massachusetts to file a class-action lawsuit against the Kellogg Company and Nickelodeon, the nation's most popular children's channel.[67] The groups say the two companies harm children's health because they primarily advertise food high in sugar, fat, and salt and low in nutrients.

Difficult as it is to believe, Kellogg responded to the activist groups that it "is proud of its products and the contributions they make to a healthy diet."[68] Nickelodeon said it is "an acknowledged leader and positive force in educating and encouraging kids to live healthier lifestyles."

The consumer groups are not seeking monetary damages but want companies to stop advertising unhealthy products on shows where at least 15 percent of the audience is under age eight. The also want the network to stop allowing its popular human characters from being used to promote unhealthful foods. In June 2007, the Kellogg Company announced it would phase out advertising its products to children under age 12 unless the foods met specific nutrition guidelines for calories, sugar, fat, and sodium. The two activist groups responded by dropping the lawsuit against the company. Nickelodeon has not yet announced changes in their advertising to children. The activist groups have indicated they may sue General Mills, McDonald's, Burger King, and other companies in the future.

Both adults and children succumb to the power of advertisers. If you want your family to be as healthy as possible, natural food, preferably organically produced from local farms, is best. This seems obvious when you consider how industrial crop and animal farms produce their products: genetically modified plants dowsed with synthetic chemicals, plants laden with heavy doses of pesticides, farm animals raised on a diet rich in antibiotics and growth enhancers. Then consider the additional things that food processors add to the unnaturally raised plants and animals they are given. Can anyone claim that Americans eat well? As we will see in the next chapter, they eat a lot, but certainly not well.

No one quarrels with the need of corporations to make a profit from their food production activities, but this does not mean that they should be free of social obligations. It should be, and is, possible to combine the need of the corporation to seek its own best self-interest with fulfilling a social obligation that clearly is desirable. That obligation is to deliver the healthiest food possible to the public.

Corporations maximize profits for their shareholders primarily by externalizing as many of their social and environmental costs as possible.[69] They are not financially responsible for the damage their refined sugar and saturated fat products do to people or for the higher health maintenance costs caused by the weight gain they cause. As noted in earlier chapters on grain and meat production, the corporate producers of basic foodstuffs externalize these costs as well. They are not financially responsible for the water pollution or other environmental degradation they cause.

However, pressure is building for corporations to change their ways. Activist groups are targeting some of the most powerful and best-known firms with demands that they adopt more socially and environmentally responsible practices. Corporations that ignore these ultimatums are often subjected to organized protests, boycotts, and other forms of embarrassing publicity. Corporations spend $500,000 each year[70] to create positive images through advertising, and a sudden storm of negative publicity from the activities of large numbers of activists can quickly raise environmental issues to the top of the action list of boards of directors. The technique has been successful in recent years in changing the lending policies of Citibank and Bank of America. It also changed the wood-buying practices of Home Depot and Lowes. There is no reason that the producers of the food we eat cannot also be convinced that changes in their production practices are necessary. The changes recently announced by McDonald's (chapter 9) are a small example of what can be done. It should not be difficult to create television advertisements that entice children to eat food that is good for them instead of the unhealthy products that dominate TV ads today.

In April 2005, the USDA unveiled a new food pyramid as a dietary guide, replacing the one it touted in 1992, and recommending which foods to base a diet on and which foods to eat sparingly.[71] The recommendations emphasize fruits, vegetables, whole grains, and skim or low-fat milk and milk products. They include lesser amounts of lean meats, poultry, fish, beans, eggs, and nuts, and recommend staying away from foods high in saturated fats, transfats, cholesterol, salt, and added sugars. The guide to the pyramid asks for your age, gender, and level of physical activity and will tailor an eating plan for you based on the information you provide. The goal is to balance calories in with calories out by eating less and exercising more.

The recommendations of the 1992 food pyramid were not widely adopted by the American public, judging by the increasing girth of Americans since then (chapter 11). It remains to be seen whether the new eating guidelines will be equally ignored. If you were asked to go to Las Vegas and place a bet on whether industrial agriculture, food additives, bacon double cheeseburgers, and pizza will be replaced by whole wheat bread, broccoli, and apples, on which side would you place your money?

11

Eating Poorly and Too Much: Poor Health and Body Bloat

The wise eat to satisfy hunger, the foolish eat to satisfy appetite.
—Michio Kushi, macrobiotic teacher and author

Food is abundant in the United States. And by and large, Americans take advantage of this fact, some of them in a sporting way. To be aware of this does not require searching out numerical data on food production or the number of people who need to be fed. One need only glance around at the size of our fellow citizens. They are large and getting larger.

Most Americans have heard of the professional golf circuit, tennis circuit, NAS-CAR, and the professional rodeo circuit. But how many are familiar with the professional eating circuit? To a devoted small minority of Americans, the records set by the 3,000 hearty eaters at about 100 annual events are as important as those set in other sports. For example, according to the 300-member International Federation of Competitive Eating (IFOCE), the governing body of stomach-centric sports, the current record for devouring hot dogs was set in 2007 by a 230-pound 24-year-old veteran Joey "Jaws" Chestnut, who devoured 66 wieners (with buns) in 12 minutes or 1 hot dog with bun each 10.9 seconds.[1] This broke the record he set the previous month. His main competitor, Takeru Kobayashi (a 145 pounder who devoured only 63 wieners), still holds several records, having scarfed down 57 cow brains (17.7 pounds) in 15 minutes and 69 hamburgers in 8 minutes.[2] Not to be made second rate, Joey ate 32.5 grilled cheese sandwiches in 10 minutes and 18.5 eight-ounce waffles at Waffle House in 10 minutes.[3] On the heavyweight side, Cookie Jarvis (weight 409 pounds) devoured 4.5 pounds of fries in 6 minutes, 91 Chinese dumplings in 8 minutes, and 33.5 ears of corn in 12 minutes.[4] Eric Booker (weight 420 pounds) consumed 9.5 pounds of peas in 12 minutes and 4 pounds of corned beef hash in less than 2 minutes.[5]

Then there is the forty-year-old Sonya Thomas, the Korean American former manager of a Burger King restaurant who grew up hungry in South Korea. She is 5 feet 5 inches tall and weighs in at 105 pounds (formerly 99 pounds). She is known as "The Black Widow" on the food event circuit because of the way she devours her male opponents. She has also devoured 167 chicken wings in 32 minutes, 48 tacos in 11 minutes, 4 pounds of toasted ravioli in 12 minutes, 80 chicken nuggets in 5 minutes, 552 oysters in 10 minutes, 65 hard-boiled eggs in 6 minutes 40 seconds (6 seconds per egg), 35 bratwursts in 10 minutes, 11 pounds of cheesecake in 9 minutes, and 44 Maine lobsters in 12 minutes (11.3 pounds of meat), among a host of other comestibles.[6] Her lobster-mania was halted only by the time clock. She insisted she could have eaten more. She won the lobster-eating prize only one day after earning a prize for eating 21.5 grilled cheese sandwiches in 10 minutes and only a week after copping the bratwurst prize.[7] Sonya owns 22 of the seventy-seven records recognized by the IFOCE. In the first seven months of 2005, she took home $50,000 in prize money and appearance fees.[8]

Understandably, Kobayashi, Thomas, and Chestnut are consistently ranked among the top eaters by the IFOCE. Jarvis and Booker, the 400 pounders, are also among the IFOCE champions. Other big winners in the world of competitive eating are Jason "Crazy Legs" Conti at 202 pounds, Mark "The Human Vacuum" Lyle at 291 pounds, and Chris "The Juggernaut" Patton at 285 pounds. Imagine the fear in the eyes of the owner of an all-you-can-eat for $8.99 buffet restaurant when Takeru, Sonya, Joey, and a few of their competitive buddies enter the eatery.

The IFOCE recognizes only human contestants, possibly because even Takeru, Joey, and Sonya would lose out to the star-nosed mole.[9] This subterranean competitor can eat ten mouthful-sized chunks of earthworm, one at a time, in 2.3 seconds, or 0.23 seconds a chunk. That is over twenty-six times faster than Sonya Thomas's record-shattering egg-eating performance.

Of course, few, if any, people are able to compete with the IFOCE's distinguished group of rapid-fire eaters, and probably few would want to. But the prowess of these IFOCE champions makes one thing clear: both thin people and fat people can be big eaters, but the appearance of some people reflects their eating habits while the appearance of others gives no clue of their gluttony. How is this possible, and why can't the rest of us be as fortunate as the remarkable 105-pound Sonya Thomas?

Body Mass Index

We find no sense in talking about something unless we specify how we measure it.
—Herman Bondi, *Relativity and Common Sense*, 1964

Who decides who is overweight, how did they get the authority to do so, and why should we listen to them? The answer to all of these questions is that acceptable weight limits for each height were originally established by American insurance companies and were based on changing rates of illness and death in the United States with increasing weight.[10] Death is a subject near and dear to the hearts of companies that issue life insurance policies. However, a better method for deciding who should be considered overweight was developed by Adolphe Quetelet in Belgium between 1830 and 1850[11] and finally came into general use in the United States about twenty years ago. Like the table developed by the insurance companies, it correlates body fat with the metabolic complications of excess poundage and is called the body mass index (BMI). In the metric units used in Belgium, it is calculated as follows:

BMI = weight in kilograms/(height in meters)(height in meters) or, in the units
       used in the United States,

BMI = [weight in pounds/(height in inches)(height in inches)] × 703.

The values of BMI for adults more than twenty years of age over a range of weights and heights are shown in table 11.1. According to the table (and the World Health Organization), the best BMI to have from a health standpoint lies between 18.5 and 24.9,[12] although recent evidence suggests that the upper limit should be increased by a small amount.[13] A BMI less than 18.5 is underweight and may indicate malnutrition, an eating disorder, or other health problem. A BMI between 25.0 and 29.9 suggests an excess of girth, which may mean an excess of serious illnesses as the years roll on. Numbers of 30 or more are probably a forerunner of many life-threatening illnesses in adulthood, and a BMI of 40 or more is an almost certain indicator of premature death.

Body mass index calculations are not only for adults. BMIs for children and young adults two to twenty years old are calculated the same way as for adults but are classified differently. For children, a BMI that is less than the fifth percentile is considered dangerously skinny (95 percent of children weigh more) and above the ninety-fifth percentile (only 5 percent of children weigh more) is considered obese. The interval between eighty-fifth and ninety-fifth percentiles is considered the underweight or overweight zone.[14] The BMI is not an infallible predictor of future health because it does not distinguish between fat and muscle. Individuals with a significant amount of lean muscle, such as body builders, have high BMIs, which does not indicate an unhealthy level of fat.[15] Nevertheless, few Americans can use this excuse.

The index is primarily a statistical tool designed for public health studies, which enables the investigation and comparison of any medical data set in which the height and weight of subjects were recorded, to determine whether overweight or obesity was a factor in health outcomes. The use of the BMI allows the assessment of changes over time within a community and can be used to evaluate the impact of intervention strategies and economic development on nutrition status.

Table 11.1
Body mass index by height and weight

	Appropriate weight						Overweight					Obese					
BMI	19	20	21	22	23	24	25	26	27	28	29	30	31	32	33	34	35
Height (inches)	Body Weight (pounds)																
58	91	96	100	105	110	115	119	124	129	134	138	143	148	153	158	162	167
59	94	99	104	109	114	119	124	128	133	138	143	148	153	158	163	168	173
60	97	102	107	112	118	123	128	133	138	143	148	153	158	163	168	174	179
61	100	106	111	116	122	127	132	137	143	148	153	158	164	169	174	180	185
62	104	109	115	120	126	131	136	142	147	153	158	164	169	175	180	186	191
63	107	113	118	124	130	135	141	146	152	158	163	•169	175	180	186	191	197
64	110	116	122	128	134	140	145	151	157	163	169	174	180	186	192	197	204
65	114	120	126	132	138	144	150	156	162	168	174	180	186	192	198	204	210
66	118	124	130	136	142	148	155	161	167	173	179	186	192	198	204	210	216
67	121	127	134	140	146	153	159	166	172	178	185	191	198	204	211	217	223
68	125	131	138	144	151	158	164	171	177	184	190	197	203	210	216	223	230
69	128	135	142	149	155	162	169	176	182	189	196	203	209	216	223	230	236
70	132	139	146	153	160	167	174	181	188	195	202	209	216	222	229	236	243
71	136	143	150	157	165	172	179	186	193	200	208	215	222	229	236	243	250
72	140	147	154	162	169	177	184	191	199	206	213	221	228	235	242	250	258
73	144	151	159	166	174	182	189	197	204	212	219	227	235	242	250	257	265
74	148	155	163	171	179	186	194	202	210	218	225	233	241	249	256	264	272
75	152	160	168	176	184	192	200	208	216	224	232	240	248	256	264	272	279
76	156	164	172	180	189	197	205	213	221	230	238	246	254	263	271	279	287

W is at height 62/63, BMI 29. M • at height 69, BMI 27.

Note: The locations of the average American woman and man in 2004 are shown by dots.
Source: Adapted from *Clinical Guidelines on the Identification, Evaluation, and Treatment of Overweight and Obesity in Adults: The Evidence Report.*

Obese				Morbid obesity														
36	37	38	39	40	41	42	43	44	45	46	47	48	49	50	51	52	53	54
172	177	181	186	191	196	201	205	210	215	220	224	229	234	239	244	248	253	258
178	183	188	193	198	203	208	212	217	222	227	232	237	242	247	252	257	262	267
184	189	194	199	204	209	215	220	225	230	235	240	245	250	255	261	266	271	276
190	195	201	206	211	217	222	227	232	238	243	248	254	259	264	269	275	280	285
196	202	207	213	218	224	229	235	240	246	251	256	262	267	273	278	284	289	295
203	208	214	220	225	231	237	242	248	254	259	265	270	278	282	287	293	299	304
209	215	221	227	232	238	244	250	256	262	267	273	279	285	291	296	302	308	314
216	222	228	234	240	246	252	258	264	270	276	282	288	294	300	306	312	318	324
223	229	235	241	247	253	260	266	272	278	284	291	297	303	309	315	322	328	334
230	236	242	249	255	261	268	274	280	287	293	299	306	312	319	325	331	338	344
236	243	249	256	262	269	276	282	289	295	302	308	315	322	328	335	341	348	354
243	250	257	263	270	277	284	291	297	304	311	318	324	331	338	345	351	358	365
250	257	264	271	278	285	292	299	306	313	320	327	334	341	348	355	362	369	376
257	265	272	279	286	293	301	308	315	322	329	338	343	351	358	365	372	379	386
265	272	279	287	294	302	309	316	324	331	338	346	353	361	368	375	383	390	397
272	280	288	295	302	310	318	325	333	340	348	355	363	371	378	386	393	401	408
280	287	295	303	311	319	326	334	342	350	358	365	373	381	389	396	404	412	420
287	295	303	311	319	327	335	343	351	359	367	375	383	391	399	407	415	423	431
295	304	312	320	328	336	344	353	361	369	377	385	394	402	410	418	426	435	443

BMI is only one indicator of potential health risks associated with excess weight. It has gained prominence because weight and height are recorded regularly by physicians, so an enormous number of measurements are available for analysis. However, other predictors of future health problems associated with excess weight are also useful. One favorite is the waist-to-hip ratio. Measure the waist at its narrowest point and the hip at its widest point. When the ratio exceeds 0.8 for women or 1.0 for men, the person is considered obese. Other useful overweight indicators are the person's waist circumference and skin-fold thickness.[16] In each of these measurements, smaller is better.

Trends in the Weight of Americans

If you looked at any epidemic—whether it's influenza or plague from the Middle Ages—they are not as serious as the epidemic of obesity in terms of health impact on our country and our society.

—Dr. Julie Gerberding, director of the Centers for Disease Control, 2004

A person's basal metabolic rate (BMR) is the amount of energy the body needs to function at rest. This accounts for 60 to 70 percent of calories burned in a day and includes the energy required to keep the heart beating, the lungs breathing, the kidneys functioning, and the body temperature stabilized. In general, men have a higher BMR than women.

The laws of physics dictate that you cannot gain weight unless you eat more calories (typically in heavily processed foods) than you burn. Overweight and obesity result from an energy imbalance over time: eating too many calories (a measure of the energy content of food) and engaging in too little physical activity to keep the excess calories from increasing the size or number of fat cells. Excess weight results when the size or number of fat cells in a person's body increases greatly. A mature fat cell contains a huge, clear droplet of fat that takes up nearly the entire cell and appears on a microscope stage as a shining sea of fat stored as molecules of triglyceride.[17] A normal-sized person has between 30 billion and 40 billion of these fat cells; when a person gains weight, the cells first increase up to six times in size and later in number, up to 100 billion.[18] Each fat cell weighs a very small amount; it would take about 5 million fat cells to accumulate just 1 ounce of fat, 80 million cells to the pound.

An astounding 127 million Americans, 43 percent of adults, are overweight; 60 million, or 20 percent, are obese, and 9 million, or 3 percent, are extremely or morbidly obese (they weigh more than 400 pounds).[19] The average man is 5 feet, 9 1/2 inches tall and weighs 191 pounds today, 1 1/2 inches taller than in 1960 and 25 pounds heavier.[20] Hence, his BMI has increased from 25.2 to 27.8. The average woman increased in height by 1 inch to 5 feet 4 inches, in weight by 24 pounds to 164 pounds.[21] Hence, her BMI increased from 24.8 to 28.1. Obesity rates have increased consistently in recent decades, from 15 percent in 1980 to 32 percent in 2004.[22] The percentage of Americans with severe or morbid obesity also increased significantly.

In 1991 six states had obesity rates below 10 percent, and in no state did obesity exceed 19 percent. In 2006, the percentages of obese Americans was higher in every state, with Colorado the leanest state at 17.6 percent, and Mississippi the most obese at 30.6 percent. West Virginia, Alabama, and Louisiana and Kentucky rounded out the top five. Mississippi, the fattest state, also has America's worst rates of cardiovascular disease and hypertension and second worst for diabetes. According to the American Obesity Association's survey of cities, the highest percentage of overweight citizens in 2005 lived in Houston, Texas, which also won this dubious championship in 2003, 2002, and 2001.[23] They were outgrown by Detroit in 2004 but were able to regain the title the next year. San Antonio holds the title for the city with the highest percentage of obese citizens.

Clearly excess girth has become an epidemic in America and is poised to become the nation's leading health problem and number 1 killer. According to the Centers for Disease Control, it is already the cause of 365,000 deaths annually, and if present trends continue, which seems likely, it will soon overtake tobacco use as the leading cause of preventable death.[24] The percentage of overweight children aged six to eleven nearly tripled between 1980 and 2004, from 6 percent to 17.5 percent.[25] Since 1963 both the average ten-year-old male child and the average ten-year-old female child are 11 pounds heavier. The childhood obesity rate more than tripled from 1980 to 2004, from 5 percent to 17 percent. The probability that overweight preschool children will become obese adults is over 30 percent; for overweight adolescents, the likelihood is 80 percent.[26]

Education is a factor in the occurrence of obesity. Although obesity has increased in the past two decades across all educational levels, those with less than a high school diploma are consistently nearly twice as likely to be obese as college graduates. Those with better educations are more likely to control their weight, but nearly one-sixth of college graduates are obese. Because educational attainment correlates with income, we would expect those lower on the economic ladder to have higher obesity rates, and this is indeed the case. Obesity rates are approximately twice as high in low-income groups as they are in higher-income groups, probably because of different eating habits; foods rich in calories (cookies, candy bars, and processed foods in general) are less expensive than the lower-calorie, more nutritious ones such as fresh vegetables. And those with less education are less likely to understand the differences in caloric content among different foods. However, the difference in obesity rate between rich and poor is narrowing. The percentage of obese people among those with incomes of more than $60,000 nearly tripled, increasing from 9.7 percent to 26.8 percent between the early 1970s and early 2000s. Among those making less than $25,000, the increase was much smaller (but still substantial), from 22.5 percent to 32.5 percent.[27]

Women tend to be overrepresented in lower-income groups, and more women than men are obese.[28] Thirty-four percent of women are obese compared to 28 percent of men. Over 38 percent of African American women are obese compared with 27.5 percent of Latino women and 21.1 percent of Caucasian women. Obese women tend to give birth to babies who are bigger than normal and at an increased risk of becoming overweight during childhood. Babies with obese mothers are more likely to become obese themselves.

Many of us have heard the claim that people tend to have pets that look like them in some way. One way is in their size. Many of our pets are apparently being overfed in keeping with the eating habits of their owners. Sixty percent of Americans own a pet, with 44 percent having a dog and 29 percent a cat, and 40 to 60 percent of U.S. dogs and cats are too heavy.[29] The problem is apparently not restricted to the United States. In Australia, 40 to 44 percent of dogs and more than one in three household cats are overweight. In Great Britain, three-quarters of vets now run fat clinics to combat the problem. In 2005 they established the first national fat-camp-style clinic to help pampered pets shed a few pounds. The Small Animal Hospital at the University of Liverpool established the Royal Canin Weight Management Clinic for overfed and underexercised dogs and cats.

Many businesses that are concerned with the size of their employees or customers are adapting to their increased girth. Disneyland's Magic Kingdom has redesigned some of its costumes for ride operators, shop clerks, waitresses, and other employees. A couple of decades ago, the park's wardrobe department stocked a narrow range of sizes. Today the uniforms for women extend to size 30; men's trousers have stretched to 58-inch waists.[30] A woman who wears a size 30 would weigh about 350 pounds; a man with a 58-inch waist would weigh about 380 pounds.

The Puget Sound ferries in Washington State have increased the width of their seats from 18 inches to 20 inches to allow room for adults with bigger bottoms.[31] Nearly 300,000 of America's children aged one to six are too heavy for standard car seats; more than 10 percent of those ages two to five are obese. Car seat manufacturers now offer the "Husky" car seat, which is 10 pounds heavier and 4 inches wider than the standard size, and seat belt extenders for adults are now available. Airlines are considering similar adaptations, such as charging more for bulky passengers.[32] Wheelchairs are being built that can handle a 600-pound occupant.[33] The Big John toilet seat has an outside rim that is 5 inches wider than a standard 14-inch seat and can hold up to 1,200 pounds. In Colorado an ambulance company has retrofit its vehicles with a winch and a compartment to handle patients weighing up to 1,000 pounds. The American Medical Response company in Las Vegas has an extrawide ambulance with a larger gurney and a winch and ramps capable of loading up to 1,600 pounds. A company official said that between October 2005 and the end of March 2006, they handled seventy-five calls involving patients who weighed more than 600 pounds.[34]

Perhaps nowhere is the issue of obesity better illustrated than at Goliath Casket Company of Lynn, Indiana, specialty manufacturers of oversize coffins.[35] It produces a triple-wide coffin, 44 inches across compared with 24 inches for a standard model. It is designed to handle corpses of 700 pounds. Sales have been expanding by about 20 percent annually. Currently, 17 percent of Americans over 70 are obese, and so the future for Goliath seems bright.[36]

All segments of the funeral industry have recognized the problem of caring for the obese dead. Woodlawn Cemetery in New York City recently increased the width of its standard burial plot from 2 feet to 4 feet, and the cemetery's newest mausoleum has four crypts designed to hold oversize coffins.[37] The Cremation Association of North America trains its members in the handling of obese bodies, and hearse manufacturers are making their vehicles wider and with bigger rear doors.[38] Capitalism responds quickly to the needs of its citizens, dead or alive.

Body Mass Index and Illness

Now learn what and how great benefits a temperate diet will bring along with it. In the first place you will enjoy good health.

—Horace, *Satires*, Book II, circa 30 B.C.

There is an old medical saying that we dig our grave with our spoon, a truism verified by life insurance companies many decades ago when they determined that overweight and obese people are more likely to live shorter-than-average lives. Those who are overweight have a 50 to 100 percent greater risk of premature death from all causes than do people at a healthy weight. Being obese may reduce life expectancy by nearly thirteen years, depending on the degree of obesity.[39] Obese men and women are 62 percent more likely to die of cancer, 100 percent more likely to develop heart disease, and more than 1,000 percent more likely to develop diabetes. There are more than thirty diseases and physical and mental conditions associated with obesity in addition to some types of cancer, heart disease, and diabetes. Among them are hypertension, congestive heart failure, stroke, blindness, osteoarthritis, bone fracture, gall bladder disease, insomnia, indigestion, asthma, depression and suicidal tendencies, hip and-knee-deterioration, decreased male fertility, and impotence.[40] Furthermore, overweight and obese people are significantly more likely to suffer dementia and Alzheimer's disease.[41] The reason obesity increases the likelihood of these diseases is not always known, but it is obvious for some, such as coronary heart disease and hip and knee deterioration. The heart must work harder to pump the blood in those who are overweight, and hips and knees that bear excessive weight for many years will not last as long.

One of the best documented correlations between being overweight and having a higher incidence of disease is for diabetes.[42] A new word has been coined for this—*diabesity*. Obesity interferes with the body's ability to use the hormone insulin and control blood sugar levels because "fat" fat cells, unlike "thin" fat cells, produce chemicals that interfere with the effectiveness of insulin. Among the states and the District of Columbia, Alabama, Mississippi, and West Virginia rank 1, 2, and 3 in percentage of obese citizens and 6, 2, and 1, respectively, in the percentage of diabetic adults. Colorado has the lowest percentages of obese citizens and those with diabetes. Diabetes is the leading cause of kidney failure, adult blindness, and limb amputations.

Women who are overweight have a risk five times higher than those of appropriate weight of developing type 2 diabetes, and women who are severely obese (BMI of 35 or more) have a risk more than fifty times higher.[43] It is estimated that 80 percent of diabetes cases worldwide result from obesity.[44] Between 1994 and 2004, the prevalence of diabetes increased more than 50 percent.[45] The rise is projected by health authorities to continue because an increasing percentage of children are being diagnosed with "adult-onset" diabetes, a disease that normally can be treated by changes in diet and lifestyle, as contrasted to type 1 diabetes, which requires treatment with insulin. Ninety-five percent of diabetics have the type 2 variety. Newly diagnosed cases of diabetes in children have increased from less than 5 percent a year before 1994 to 30 to 50 percent since then.[46] In New York City, 21 percent of the kindergartners are obese, a sign that the epidemic will not end. The health commissioner of New York commented, "I will go out on a limb and say 20 years from now people will look back and say 'What were they thinking? They're in the middle of an epidemic and kids are watching 20,000 hours of commercials for junk food.'"[47]

As of 2004 the average American male could expect to live 75.2 years and the average female 80.4 years.[48] Presumably the 5.2-year difference would be even greater were it not for the higher rate of obesity among females. Life spans have been increasing by about two-tenths of a year annually since 1993, but some researchers believe that the rising rate of obesity will reverse this consistent increase in longevity over the coming decades.[49] Today's children may become the first generation in American history whose life expectancy will be shorter than that of their parents. Other researchers argue that increased public health education and continuing medical advances will offset the declines resulting from increasing obesity, but no one regards the increasing girth of Americans as unimportant.

Why Are Americans So Fat?

Fast food corporations, like the tobacco companies before them, are largely aware of the ill-effects of their products, but as long as the customers keep lining up they continue to sell their toxic wares.

—Yusef Al-Khabbaz, an academic working in the Middle East, 2004

A 2005 survey by the International Food Information Council found that at least 89 percent of American adults believe that diet, exercise, and physical activity influence health, a belief reflected in the astonishing sales of books, magazines, and weight-loss programs offering dietary and health advice. Nevertheless, the percentage of overweight and obese individuals is growing in the United States (62 percent), as well as in an increasing number of relatively rich Western countries. In the United Kingdom, about two-thirds of the English are overweight or obese, with obesity up almost 400 percent since 1980. In France, 32 percent of the population is over-weight and 11 percent are obese, and four times as many children need to lose weight as in 1960.[50] However, the percentage of overweight citizens is only 25 per-cent in Japan, a nation comparable in wealth to the Western democracies. This sug-gests that more than one factor is responsible for the epidemic of expanded girth. Suggested factors include evolution, individual genetic inheritance, availability of food, the way food is produced, amount and type of food consumption, lack of nutritional information on food packages, environmental pollution, and a reduction in physical activity. What is the evidence for the involvement of these factors?

Evolution

There is reason to believe that evolutionary factors may be partly to blame for the current epidemic of overweight and obesity. There may be a genetic human ten-dency to overeat whenever possible. For the first few hundred thousand years of its history and until perhaps 1850, the human race adapted to living in a world of food scarcity. Starvation was always just around the corner for nearly everyone. It was only sensible to eat every good-tasting thing in sight when you could find it. Genet-icists have found that famine survivors preferentially pass on a gene that helps the body store fat.[51] But today people in the Western world live in an artificial techno-logical environment where food is readily available all the time and in whatever amounts they want. We need to fight our genes to keep our desire to eat under con-trol. Our bodies are good at converting food into fat and then hanging on to it. We are storing fat for the famine that never comes. We also automatically lower our metabolic rate when food is not available. Clearly, in the Western world, the genes that favor preserving life and gaining weight have outlived their usefulness, but they are not ours to willfully change. Evolution has betrayed us. We have a built-in predisposition toward binge eating and a built-in preference for fatty and sweet food. Fat is the densest source of calories.

Scientists report that just looking at tempting foods activates a part of the brain that controls drive and pleasure.[52] Parts of the brain responsible for sensation in the mouth, lips, and tongue are more active in obese people than in people of nor-mal weight,[53] and that correlates closely with the desire for food. The areas of the brain that are involved in the desire for food are the same brain circuits involved in drug addiction, and in the Western world there is food all around us, making us want to eat all the time: full color photos in magazines, chefs preparing tempting dishes on television, food courts in most shopping malls to complement an array of candy and soda vending machines, and stimulating aromas from your neighbor's kitchen wafting into your yard. And your own well-stocked kitchen is always nearby.

Individual Genetic Inheritance

Some people can eat amounts of food beyond imagining without gaining weight, and some need only to smell food to put on a few pounds. If we understood genetics well enough, we could fingerprint people when they are born and say, "Ah, good genes. Lucky you. You can eat whatever you want." Or: "Uh-oh. Poor kid. You better never eat a doughnut. Your genetic type is Bill Clinton, not Mahatma Gandhi." Recent advances in genetics are moving in this direction. Researchers have found that those who have an altered copy of a gene called FTO have an increased risk of obesity of 30 percent; two altered copies raises the risk to 70 percent.[54] Sixteen percent of white adults have two altered copies, and these people in a study of 39,000 Finns and Britons were 6 pounds heavier than those with normal copies of FTO.[55]

Increasingly, researchers are demonstrating that the urge to eat may be controlled mostly by a powerful biological system of hormones, proteins, neurotransmitters, and genes that regulate fat storage and body weight and tell the brain when, what, and how much to eat. According to the director of George Washington University's Weight Management Program, "Eating is largely driven by signals from fat tissue, from the gastrointestinal tract, from the liver. All those organs are sending information to the brain to eat or not to eat. So, saying to an obese person who wants to lose weight, 'All you have to do is eat less,' is like saying to a person suffering from asthma, 'All you have to do is breathe better.'"[56] In effect, if you are overweight, you are not at fault. You just did not get the most desirable set of genes at birth.

There are about 340 genes involved in weight control, so that genetic manipulation is not a likely path to controlling weight.[57] However, in 1999, researchers discovered a hormone that seems to be a leader in appetite regulation. Cycles of this powerful hormone, called ghrelin, reflect a complex interplay of chemical signals in the body, telling you how often and how much to eat. Reasonably enough, most ghrelin is produced in the stomach. This "hunger hormone" has lots of accomplices. At least two dozen chemicals, many of them hormones, stimulate food intake, and a similar number suppress appetite. A few of these substances seem to be star players in the dinner theater, while the rest serve as understudies.[58] Featured players include insulin, which is made in the pancreas, and leptin, manufactured by fat cells. The appetite-regulating hormones travel through the body carrying their messages to eat or not to eat. They trigger nerve signals running from the gut to the brain and are influenced in turn by messages returning from the brain.

The complexity of chemical signals that regulate eating can be illustrated by considering leptin, one of the featured players in the dinner theater.[59] About ten years ago, geneticists identified a gene in mice that tells the body how to make leptin, a hormone that decreases appetite. Leptin is produced in fat cells and is part of a thermostat-like system that maintains weight at a constant level. Lose weight and leptin levels fall, prompting you to eat more and gain back the weight. Put on some extra pounds and leptin increases: you eat less. Leptin tells the brain, "I'm full; stop eating." In addition, an increase in the amount of leptin sets off hormones that speed metabolism, burning calories faster. Perhaps regulation of leptin in some people is out of whack, and they simply have a stronger biological urge to eat. Perhaps there are fundamental differences in wiring between the obese and the lean. Brain imaging studies have revealed that obese people have fewer brain receptors for dopamine, a neurotransmitter that helps produce feelings of satisfaction and pleasure.[60] Eating increases the level of dopamine in the brain. Perhaps obese people tend to eat more because it makes them feel happy, and this may be satisfying their underserved reward circuits. Granting that there are strong genetic factors that determine susceptibility to obesity, biology is not necessarily destiny where weight is concerned. In studies of adult identical twins who share the same genes, measures of fatness appear to be 20 to 70 percent inherited. This suggests that perhaps half of the fatness of an overweight person is genetic and half is not.[61]

Perhaps most disturbing in the obesity saga was the discovery in 2000 of a human virus that causes obesity in animals. And in 2005 researchers found that the fat virus is contagious.[62] The results of investigations with animals so far give cause for concern. When a chicken infected with the virus is placed in a cage with an uninfected chicken, it takes only 12 hours for the viral DNA for fat to appear in the blood of the formerly uninfected chicken. Both the infected donor and the receiver become obese. Is it possible that close association with obese people can cause an alteration in a thin people's appetite-controlling hormones so they become fat? Should weight be a factor to consider in choosing friends? This may sound strange, but it has been recognized for some time that when female college students share a room, their menstrual cycles merge.[63] Apparently a close association is all that is needed for one person's bodily rhythms to influence another's.

Job insecurity, unpleasant surroundings, or marital unhappiness may also promote obesity.[64] It is known that acute stress can make some people lose weight, but in 2007, researchers found that chronic stress can induce weight gain by promoting the presence of particular amino acids in the body.

It has also recently been found that the microbes in the gut may cause people to get fat.[65] Two people may eat the same food containing, say, 300 calories. But one person may extract 250 of the calories during digestion, while another may extract only 150, depending on the particular combination of microbes in the gut. The complexities in turning food into flesh seem to become more complex as research increases.

Availability and Consumption of Food

Clearly evolutionary factors and individual genetic inheritance are important controls of obesity, but this is only part of the story. As two investigators in the field stated, "Despite obesity having strong genetic determinants, the genetic composition of the population does not change rapidly. Therefore, the large increase in ... [obesity in recent years] must reflect major changes in non-genetic factors."[66] This inference is supported by a Roper Poll in 2000, in which 70 percent of American adults said they were eating "pretty much whatever they want," an all-time high and up from 58 percent in 1997.

Surprisingly, despite eating whatever and whenever we want, most of us are not enjoying it. Only 39 percent of adults in a Pew Research Center survey say they enjoy eating "a great deal," down from the 48 percent who said the same in 1989.[67] Overweight people showed a bigger decline in enjoyment than those of lesser weight, suggesting that the increased publicity in recent years about the dangers of overweight and obesity may be having an effect.

All kinds of fresh, unprocessed foods are available in Western societies, and no one gets fat from eating too many low-calorie, highly nutritious fresh fruits and vegetables. They gain weight by eating too much processed food, an addiction supported by American farm policy (chapter 3). As we saw in the previous chapter, vegetarians are, by and large, healthier than the rest of American society, but vegetarians are a small percentage of all Americans. As we noted earlier in this chapter, the calories you eat are a measure of energy intake. More calories mean more energy, and if you do not burn it off, your clothing size will need to increase. According to the American Dietetic Association, 3,500 calories is 1 pound of body weight.[68] An increase in calorie consumption or reduction in caloric expenditure of that amount in a day or a week or a month will increase weight by 1 pound in that length of time. What do we know about the caloric intake of Americans?

The number of calories needed to maintain a healthy weight, as defined by the BMI, varies with age, level of physical activity, and other factors, but for most people it lies between 1500 and 2500 calories. Men tend toward the higher numbers, women the lower numbers. According to the USDA, the average American in 2000 ate 2,700 calories per day, about 530 calories more than in 1970, an increase of 24 percent.[69] We each ate 2,200 pounds of food in 2006, up from 1,775 pounds in 2000 and 1,497 pounds in 1970.[70] Note that the increase in caloric intake is significantly larger than the increase in food intake. This suggests that we are eating a higher percentage of high-calorie foods, that is, heavily processed foods.

A study in 2004 found that nearly one-third of the daily calories consumed by American adults came from junk food, sugary drinks, and beer. One of every five calories in the American diet is liquid. The nation's single biggest "food" is soda.[71] The top ten foods on our plates are soft drinks, cakes and other sweets, burgers, pizza, chips, rice, bread, cheese, beer, and french fries. Not a single fruit or green vegetable makes the list. People overindulge in these foods because most of them have a high glycemic index, meaning they produce a satisfying rapid increase in blood sugar.[72] However, the surge in blood sugar and insulin produced by such foods soon crashes and stimulates hunger within a short period. Most foods with low nutritional value, such as snacks and soft drinks, have a high glycemic index. Also, they are made from cheap ingredients such as sugar (or the less expensive and more easily blendable liquid high-fructose corn syrup) and potatoes and are therefore highly profitable for the companies that produce them. This is why they are pushed so aggressively by companies that spend billion of dollars a year marketing them. (National Junk Food Day is July 22.) Foods with a low glycemic index such as legumes, fruits, and vegetables release nutrients more slowly, which leads to less erratic eating habits.

We are eating more food and particularly high-calorie food largely because much of the food we eat has been made richer in calories and tastier than ever by increasing the amount of fat, sugar, and salt in it, all products that are pleasing to the palate but lead to weight gain. Fats add taste, texture, and palatability to foods, which is one of the reasons fast foods are so popular, but they contain more than twice the calories of proteins or carbohydrates. An ounce of fat contains 255 calories, an ounce of protein or carbohydrate only 113 calories.[73] On average, fast foods contain 70 percent more calories per unit weight than an average home-cooked meal.[74] People tend to eat similarly sized portions of fast food and ordinary food and get nearly twice as many calories from fast food as from a food that is low in fat.

Consider sugar (now often replaced by the less expensive and artificial corn-based sugar). During the past fifty years, the average American increased sugar intake dramatically. Nearly all of this increase in natural sugar intake is refined sugar, which has been depleted of all its vitamins and minerals. What remains is pure, refined carbohydrates. Lacking the natural minerals present in sugar beet or sugar cane, it drains the body of vitamins and minerals because of the demand its digestion, detoxification, and elimination make on the system. In the 1950s, one doctor called refined sugar a poison, and he was not far wrong.[75] Each of us now consumes the equivalent of thirty-one teaspoonfuls of sugar per day.[76] There are about 15 calories in a level teaspoon of sugar, so that sugar alone puts 465 calories into our bodies each day, about 25 percent of our total daily caloric need. Refined sugar gives empty calories, that is, calories with no other nutrients—no vitamins, no minerals, no fiber, and no protein.

Two-thirds of the sugar we consume is hidden inside processed foods,[77] so those who always use artificial sweeteners in their coffee or on fruit cannot escape the increase in their sugar consumption.[78] Food manufacturers know that humans are attracted to sweet tastes, so they lard their processed foods with sugars, commonly concealing the amount in the wording on the package. They may list all the different types of sugars separately on ingredient lists and omit a figure for total sugar content.[79] Or because sugar is a carbohydrate, the manufacturer may lump the sugar content with the carbohydrate content rather than listing it separately. Hidden sugar turns up in some unlikely places: pizza, bread, hot dogs, boxed mixed rice, soup, crackers, spaghetti sauce, lunch meat, canned vegetables, fruit drinks, flavored yogurt, ketchup, salad dressing, mayonnaise, and some peanut butter. Heinz tomato ketchup is 24 percent sugar, and its Original Sandwich Spread is 22 percent sugar. Safeway Pasta Pronto contains 21 percent sugar, and, somewhat shockingly, Weight Watchers' low-fat salad dressing is 11 percent sugar. Quaker Sugar Puffs is an accurate name; they are 49 percent sugar. Most brands of breakfast cereals are 30 to 50 percent sugar.[80] In addition, consumption of carbonated sodas has increased greatly in the past few decades, and a 12-ounce soft drink contains the equivalent of 10 teaspoonfuls of sugar (or HFCS), for 150 calories. Carbonated sodas provide more than a fifth (22 percent) of the refined and added sweeteners in the American food supply compared with 16 percent in 1970.[81] Chocolate bars are typically between 50 percent and 70 percent sugar.[82]

Not only is the food richer in calories and tastier, but portions and the plates they are served on are larger than ever before. Most American dinner plates are 12 inches in diameter, much larger than fifty years ago. Research indicates that large plates, bowls, and popcorn buckets lead to more eating. Dieticians have noted that portion sizes began to grow in the 1970s, rose sharply in the 1980s, and have continued to increase in parallel with expanded body weights.[83] In the 1970s, hamburgers were thin, and a candy bar was a bar and not a brick. In 1955 the company's largest soda was 7 ounces; today their Hugo cup holds 42 ounces, more than a quart. McDonald's gained prominence by selling reasonably sized hamburgers, french fries, and soft drinks cheaply. The company then discovered that most people would not order two hamburgers (280 calories each) or two small bags of french fries (210 calories each) but could be enticed to ask for a Big Mac (600 calories) or large-sized fries (540 calories). And many customers would, if offered the choice, choose supersize fries (610 calories).[84] It may not be coincidental that people who live in countries in which Big Macs are more expensive have fewer obese citizens than countries in which they cost less.[85]

McDonald's is not alone in catering to weighty Americans. In 2004, Hardee's devised the Monster Thickburger, a beef, bacon, cheese, mayonnaise, and butter delight on a sesame seed bun that contains 1,420 calories and a monstrous 107 grams of fat.[86] Dietitians recommend a maximum of 60 grams of fat per day. The company admits that this food "is not for tree huggers." Jay Leno joked that it was being served in little cardboard boxes shaped like coffins. Hardee's ad slogan for this artery-clogging feast is, "Be afraid. Be very afraid," a small triumph for truth in advertising. Not to be outdone, Subway offers a Double Meat Meatball Sub that weighs in at 1,560 calories.[87] Wash that down with a McDonald's Triple Thick Shake at 1,150 calories, and you may be able to waddle downtown to the emergency ward of your favorite hospital for a weight check before your arteries congeal. At least one in four adults now eats fast food on any given day.[88] Among children, one of every three meals is fast food. Granted, most people are not going to gorge on these monsterburgers and artery cloggers, but they do illustrate the trend in restaurants' attempts to please their customers.

Because the nearly 200,000 fast food establishments in the United States[89] are notably unhealthy places to eat, some obese people have sued McDonald's, the largest fast food chain, for making them obese.[90] McDonald's successfully defended itself, but because of impending lawsuits and a rising chorus of complaints from the public, the company has stopped "super-sizing" its products and has started promoting salads and low-fat sandwiches on its menu. Number two Burger King has followed the lead of McDonald's. But the wide variety of fat- and calorie-laden items on the menu of the two chains remains, and sales of double cheeseburgers and fried chicken sandwiches have increased. An average McDonald's store sells roughly 50 salads a day compared with 300 to 400 double cheeseburgers.[91]

A Seattle restaurant, 5 Spot, introduced an unhealthy dessert treat in 2003 but hopes to avoid obesity lawsuits when patrons order its fried banana fat fest called The Bulge.[92] Customers must sign a release statement that states: "The 5 Spot will not be held liable in any way if the result of eating this dessert leads to a Spare Tire, Love Handles, Saddle Bags, or Junk in My Trunk." Clearly we live in a toxic environment as far as food is concerned. Getting people to eat less is like trying to treat an alcoholic in a part of town where every third store is a bar. Bad food is cheap, convenient, heavily promoted, and engineered to taste good.

Perhaps the ultimate in catering to the voracious appetites that Americans have developed is symbolized by the Heart Attack Grill in Arizona. It serves "Double- Triple- and Quadruple-Bypass Burgers," the last weighing in at 8,000 calories, about four days worth of calories for a normal person. The burgers can be accompanied by "Flatliner Fries" (advertised as being fried in pure lard and smothered in mozzarella cheese and beef gravy). The restaurant advertises its food as having "Taste Worth Dying For."

Research has shown that the sizes of food portions in general in restaurants have increased. Forty-seven percent of Americans' meals are now consumed outside the home, and restaurant portion sizes have increased by between 20 percent and 60 percent over the past twenty years.[93] The more food served to patrons, the more they eat. After all, isn't it bordering on sin to waste food, particularly when you have paid for it? And the 80 percent of the spaghetti portion you have eaten so far certainly was tasty and enjoyable, so you might as well finish it. Even the size of hamburgers cooked at home has increased by almost 50 percent in the past twenty years.[94]

The Way Food Is Produced

Since the Great Depression of the 1930s, farmers have been the beneficiaries of a variety of subsidies and support programs meant to provide Americans with an affordable and reliable supply of food. Agricultural efficiency was emphasized. And what farmers are most proficient at producing are just a few highly subsidized crops: wheat, soybeans, and especially corn. Support for these crops has compelled farmers to ignore other crops, such as fruits, vegetables, and other grains. The market is flooded with products made from the highly subsidized crops, including sweeteners in the form of high-fructose corn syrup, fats in the form of hydrogenated fats made from soybeans, and feed for fat cattle and pigs. This flood drives down the prices of fattening fare such as prepackaged snacks, ready-to-eat meals, fast food, corn-fed beef and pork, and soft drinks. The price gap between the nutritionally valuable fruits and vegetables and the high-fat, high-sugar foods is made artificially large. For example, the cost of fresh fruits and vegetables is up 40 percent since 1985; meat, oil, and soft drinks are down 5 to 25 percent. The federal government subsidizes the growing obesity epidemic.

The very poorest Americans are undernourished and thinner than the general population, but if the poorest of the poor are excluded, obesity is associated with poverty.[95] One reason is that the fattening foods found at convenience stores and fast food restaurants are the cheapest and sometimes the only foods available in poor neighborhoods. And because poverty is inversely related to education, poor people are less likely to be aware of sound nutritional practices.

As was noted in chapter 1, the farmer's costs are only a small part of the price of a food product, and the elimination of subsidies would have a small to negligible effect on the price of the fattening processed products made from the subsidized crops. It would be essentially a political statement by the government. We should support healthy eating. There are better reasons to eliminate farm subsidies.

Nutritional Information on Food Packages

In 1990, the federal Nutrition and Labeling Act mandated certain information on packaging. Restaurants were specifically excluded from the provisions of the act, although public opinion polls reveal that 83 percent of Americans want restaurants to provide nutritional information. The act took effect in 1994 and required labels on food packages to have larger, more readable information about saturated fat, trans-fat, unsaturated fat, cholesterol, fiber, calories from fat, sugar, sodium content, and other important information. Most people do not object to having more nutritional information on packages, but it is not clear that labeling causes them to make better food choices. Since the labeling law came into force, the percentage of overweight and obese people has continued to increase, possibly because people ignore the labels when they purchase food, possibly because they are too ill informed about nutrition to make use of the information, possibly because reading the labels would double or triple the amount of time they would need to spend shopping, or possibly because they consider price and other factors to be of more importance. Whatever the reason, labeling seems to have been more of an academic exercise than anything else, a good idea but with little demonstrated effect.

Children, of course, are even less likely to read or care about nutritional labels than adults. In meals eaten at home and in restaurants, parents should supervise what their children eat, but this is not easy because restaurant menus do not give nutritional information. They are not required to and they have not volunteered it. The USDA has warned that children's appetites for sugary drinks, salty snacks, and other junk food are established before they reach school age.[96] The advertising divisions of the companies that produce these products are well aware of this, as evidenced by the enormous number of television food commercials aimed at children. In 2005 children in the two to seven age range saw twelve food ads a day, or about 4,400 a year. Those in the eight to twelve age group viewed twenty-one food ads a day, or more than 7,600 a year. Teenagers in the thirteen to eighteen age group saw seventeen food commercials a day, or about 6,000 a year.[97]

Food is by far the dominant product advertised to children, and 72 percent are for unhealthy products: 34 percent are for candy and snacks, 28 percent for sugary breakfast cereals, and 10 percent for fast food. Of the remaining 28 percent, only 4 percent are for dairy products and 1 percent for fruit juices. There are no ads for fruits or vegetables. In response to a rising torrent of complaints from parents and environmental organizations, in 2006 ten of the largest U.S. food and beverage companies, fearing government regulation and lawsuits, agreed that at least half their advertising for children under the age of twelve would promote healthier foods or encourage active lifestyles.[98] There are loopholes in these agreements, but they are a start. Public pressure must continue.

Advertisers know that children under eight years old are not able to differentiate factual information from advertising.[99] Research indicates that children consume an extra 167 calories for every hour of TV they watch per day, most of it in snacks featured in TV ads.[100] The average child watches three hours of television per day and has an in-between-meals calorie intake of 500 calories, perhaps 30 percent of their recommended daily intake. Parents often must struggle against "pester power," also known as the "power of nagging." In Israel, it has been estimated that about $20 per week is added to the family grocery bill by this universal children's technique. But there is no escaping the fact that parents are ultimately responsible for what their children eat.

Children of school age spend perhaps six hours a day without their parents in school buildings, most of which have vending machines dispensing unhealthy processed snack foods, sugary soft drinks, and candy bars. Nearly every high school and more than half of elementary schools have soft drinks and sugared juices for sale on campus.[101] Ninety-two percent of teachers surveyed and 91 percent of parents with at least one child in public school believe school vending machines should instead sell healthful foods and beverages.[102] In response to this concern, government recommendations, and threatened lawsuits, the beverage industry companies in May 2006 announced a voluntary modification of policy regarding the contents of school vending machines. The modification prohibits sales of sodas, sports drinks, juice drinks, and apple or grape juice in elementary schools, with lesser restrictions in middle schools and high schools. Purveyors of other junk food products such as candy and potato chips agreed in October 2006 to voluntarily start providing more nutritious food to schools. Critics have denounced these changes as inadequate and unenforceable.

Why have schools installed the vending machines without requiring approval of their contents? The answer is money. Citizens do not want to pay more taxes to maintain or upgrade school programs, and the vending machine companies arrange contracts with the schools that bring greater proceeds to the schools if children consume larger amounts of the products being sold.[103] Sugar and processed junk foods are the children's favorites. It is always worthwhile to work with the food industry to make improvements, but it is naive to expect it to voluntarily stop its highly profitable promotion of unhealthy foods and drinks to children. Voluntary approaches are politically attractive but by themselves are ultimately ineffective. The need to increase profits is nearly always more appealing to companies than being too concerned about the long-term healthfulness of their products. If it meets federal requirements for safety, that is, it is not poisonous in the short term, they are satisfied.

Although the beverage industry's new policy should be applauded, John Sicher, editor and publisher of *Beverage Digest*, noted that the policy would have no impact on the industry's bottom line. The sale of sugar-carbonated sodas in schools is less than 1 percent of the firms' U.S. sales.[104]

Although thirty-two states have no legislative or regulatory policy regulating the sale of drinks in schools, there has been a reaction in some states to the unhealthy products in school vending machines. As of mid-2006, forty-three states had enacted or introduced legislation to improve school nutrition, and thirty percent of school districts have banned junk food in vending machines.[105] In 2003, Arkansas passed a law eliminating all vending machines from its schools and requiring that the body mass index of every schoolchild be measured and reported to parents.[106] The American Beverage Association reports that 45 percent of all school vending sales are sweetened soda, so Connecticut has banned the sale of all sodas in its schools. California has banned sales of soda and junk food in all its public schools, and New York City has overhauled its school lunches, reducing the fat content of every meal by 30 percent.[107] Whole milk is now banned in their public schools.[108] In Texas public schools, deep-fat frying is banned, as are pizza fundraisers. There is a limit on how often french fries may be served and how much fat and sugar a meal can contain. This restriction is particularly important for girls, as researchers have found that for each serving of french fries that a preschool girl consumes per week, her adult risk of breast cancer increases by 27 percent.[109]

Environmental Pollution

Recently evidence has appeared indicating that increased chemical pollution in the environment may be partly responsible for the growing obesity epidemic.[110] In 2004 the National Institutes of Health made an urgent request for more research on the link between hormone-disrupting chemicals and obesity, noting that exposure during adulthood and, crucially, in the womb can permanently disrupt the body's weight control mechanisms. Industrial chemicals that act like hormone disrupters profoundly alter several aspects of human metabolism and appetite control. We might call these industrial compounds chemical calories. Their importance compared to the other factors involved in weight gain is not yet known.

Exercise

Here we come to perhaps the most important controllable cause of overweight and poor health: physical exercise. It may be more important than weight as a factor in a person's health. Neither adults nor children get the amount of physical exertion they need to burn off the calories they take in. Because we live in a technological environment, machines have replaced human exertion, with horrendous effects on physical well-being. According to the Centers for Disease Control and Prevention, almost one-fifth of people age eighteen and older exercise for less than ten minutes a week. Only 46 percent of adults perform the recommended thirty minutes or more of brisk walking or other moderate exercise five days a week.[111] Driving a car does not qualify as exercise.

As evidence of the importance of exercise, we can look at the Old Order Amish of Ontario, Canada,[112] who live much as their ancestors did 150 years ago. Insofar as possible, they live divorced from the world around them, without electricity (no television or radio), cars, or wide-circulation newspapers. They see no advertising and know little or nothing about current health controversies or fads. Food is raised and grown, preserved, and prepared the old-fashioned way. Processed dietetic foods are unavailable. The Old Order Amish diet consists of high-fat, high-sugar foods: meat, potatoes, gravy, eggs, vegetables, pancakes, bread, milk, pies, and cakes. They have a high caloric intake, yet have an obesity rate of only 4 percent, less than one-seventh the norm in the United States.

The explanation of the healthy weights of the Amish lies in their low-technology lifestyle, which entails vastly more physical exertion than is usual in the wider world. David R. Bassett, a professor of exercise science, gave pedometers to ninety-eight of these Amish adults[113] and found that the men averaged 18,425 steps per day, the women 14,196—about 9 miles and 7 miles, respectively (walking 30 minutes a day provides about 10,000 steps). The Amish men averaged 10 hours a week of vigorous activities like shoveling or tossing bales of hay (women, 3.5 hours) and 43 hours of moderate exertion like gardening or doing laundry (women, 39 hours). They also engaged in 43 hours of moderate intensity activity. For example, laundry was done by hand rather than using electrically powered washers and dryers. On average, these Amish participated in roughly six times the amount of weekly physical activity performed by other Americans. The lifestyle of the Old Order Amish suggests that physical activity played a critical role in keeping our ancestors fit and healthy. "The Amish are not freaks," says professor of anthropology Daniel Lieberman, a skeletal biologist.[114] "They are just anachronisms. Human beings are adapted for endurance exercise. We evolved to be long-distance runners. Running a marathon is not a freak activity. We can outrun just about any other creature."

The way we do our work has changed, and so has the way we spend our leisure time. The average number of television hours watched per week is close to a full-time job. People used to go for walks and visit their neighbors, commonly some distance away. Not only do we now spend much of our work lives sitting in front of a computer screen, but the design of public spaces outside work eliminates physical activity. In tall buildings, it is often hard to find the stairs, but no one cares because the bank of elevators is nearby. Electronic sensors in public restrooms are eliminating even the most minimal actions of flushing toilets or turning faucets on and off. The world is being redesigned so that we need never get up from overstuffed cushions other than to walk a few steps to the bathroom or the bed. We are engineering physical activity out of our daily lives: no more cranking a handle to get fresh air in the car (we have power windows); no more lifting of garage doors to store the car (we have remote control clickers); no more fanning ourselves to get cool (we have air conditioners). The list of devices invented to decrease physical activity is endless.

Few Americans want to give up modern conveniences and live in Old Order Amish communities. Fewer still could. But that is not a reason to give up or resign to the trend of overweight and obesity. We need to realize that the modern environment has changed for the worst in terms of promoting activity and good health. It is up to each of us to adapt to this reality by finding new opportunities to become and stay active. One fitness expert who spends lots of time at his computer has devised an ingenious way to exercise.[115] He says that "you have a natural tendency to want to move your legs." So he stationed a treadmill under his computer. He says he "works at 0.7 miles an hour." At this speed, he reports, "You don't get sweaty. It's about one step a second. It's very comfortable. Most people seem to like it around 0.7."

Whether we exercise or not depends on the value we put on our health and that of our children. They have the same problem with exercising that their parents do, and because of the chronic shortage of funds in American schools, required physical education classes are unknown to most of today's children. Only 4 percent of elementary schools, eight percent of middle schools, and 2 percent of high schools offer daily physical education.[116] Only Illinois requires physical education in all grades in school. Even getting to school is less physically demanding than it used to be. Only 18 percent of children walk or bicycle to school; 70 percent of their parents did.[117] One-quarter take buses, and about 60 percent are transported in the family car, often driven by a teenager. October 3 is National Walk to School Day, a holiday not celebrated by many young people.

Cities are compatible with automobiles, not for healthier ways of getting around like walking or bicycling. Today 90 percent of all trips made by adults and 70 percent of all trips made by children are in cars.[118] Only 6 percent are by foot or bike. American cars consume nearly 1 billion gallons more gas annually than they did in 1960 because drivers and passengers weigh two dozen more pounds than they did then.[119] Urban sprawl, defined as low-density development that outpaces population growth, increases yearly and has aggravated the lack-of-exercise problem. Sprawl development is associated with both increased time spent in cars and increases in body weight.[120] People living in the least dense, most sprawling communities are likely to weigh 6 pounds more than those living in the most dense, compact communities.[121] For every 30 extra minutes of commuting time per day, there is a 3 percent greater likelihood of obesity than for those who drive less.[122] People who live within walking distance (defined as a half-mile) of shops are 7 percent less likely to be obese than those who live farther away.[123]

For people who simply cannot bear the thought of exercise, take heart. Dr. Claude Bouchard, director of the Pennington Biomedical Research Center in Louisiana, points out that "there are large segments of the population who seem to have an inherited tendency to be real couch potatoes. We have several studies to support this contention and we have candidate genes."[124] Nevertheless, studies indicate that lack of exercise is a greater risk factor for premature death than obesity.[125] Your parents' slothful genes may be killing you. And before they kill you, they may be decreasing your IQ. Recent research suggests that physical exercise encourages healthy brains to function at their optimum levels. Fitness prompts nerve cells to multiply, strengthens their connections, and protects them from harm.[126]

Dieting

One should eat to live, and not live to eat.
—Molière, 1668

Most Americans want to lose weight. In 1985 the percentage of Americans who said that they would like to lose at least 20 pounds was 54 percent. Today that percentage stands at 61 percent.[127] On any given day, about 29 percent of men and 44 percent of women are trying to lose weight.[128] Clearly most of us weigh more than we want to. The reason for this desire to lose weight is not clear. Perhaps it results from idolization of the possibly anorexic or bulimic young women who work as clothing fashion models (average height 5 feet, 9 inches, weight 123 pounds, BMI 18.3, in the underweight and unhealthy category).[129] Perhaps it results from the well-known fact that most men and women are not sexually attracted to seriously overweight people. Possibly the desire is motivated by health concerns. But whatever the reason, it seems that most Americans want to lose weight, but it is difficult to do and even more difficult to maintain.[130] And we recognize this. During the past twenty years, we have become much more accepting of overweight peers. In 1985, 55 percent of respondents to a survey said they completely agree with the statement, "People who are not overweight are a lot more attractive." In 2005, only 24 percent completely agree with the statement.[131]

There is no evidence that commercial weight-loss programs are effective in helping people drop excess pounds, although there is always the exceptionally successful case that is widely trumpeted by the companies. Companies are generally unwilling to conduct studies of the success of their programs, perhaps because they know what the results will be.[132] The reason it is very difficult to lose weight may be that the biological control of body weight tends to be more powerful than the will-power a person has. Sluggish metabolisms and out-of-control hormones, rather than gluttony and sloth, may drive the tendency to gain weight and keep it on. Some researchers believe that many obese people have a resistance to leptin, which is produced by fat cells, with weight gain as the result. When the sensitivity to leptin is dulled, people with too much body fat do not receive the signal to stop eating. More leptin is required to quell hunger, but the only way to increase the level of leptin is to increase the amount of fat you have, so you end up regulating your body at an elevated leptin level. Researchers have also found that ghrelin, a hormone that sends hunger signals to the brain, is increased in dieters who tried to lose weight by eating less.

Fighting body chemistry can be tough. Eat too little, and the body ratchets down its metabolism so that it does not need as much energy and you regain weight more easily. Most obesity experts now believe that obesity is triggered by a food-rich, activity-scarce environment in people who are genetically susceptible to weight gain.

It is not clear how the many genes involved in regulating body weight interact, and genes are estimated to be up to 70 percent responsible for body weight, controlling a variety of unconscious drives, including when you feel full and a taste for fatty, sweet foods. The biochemistry of human weight is an extremely complex topic that will not soon or easily give up its secrets. But those who eat less will certainly lose weight. If you doubt this, examine some of the many photos of victims of Nazi or Serbian concentration camps. It is virtually certain that some of the prisoners were overweight or obese when they were cast into the camps, but none of them was when rescued. No one would advocate starvation as a healthy way to lose weight, but the effectiveness of reduced food intake on weight cannot be questioned.

Cuba provides a clear example of the effectiveness of reducing calorie intake.[133] One result of the economic crisis in the 1990s was a reduction in the amount people ate, from an average of 2,899 calories in 1989 to 1,863 in 2002. Obesity and diabetes rates were halved, heart disease declined 35 percent, and death rates from all causes dropped by 20 percent.

We have often heard that necessity is the mother of invention, and the help for those who need to lose weight is on the way in the form of a recently developed mouth device.[134] A company in Georgia has developed a device that fits in the mouth and forces the wearer to take smaller bites of food to help fend off overweight. The system costs about $500 and needs to be custom-designed by a dentist to fit the roof of a weighty person's mouth. The principle of the device is that, based on normal bites, the brain needs 15 to 20 minutes to tell the stomach that it is full and to stop eating. If you take smaller bites and hence ingest less food during this period, you will eat less while the brain-to-stomach message is being transferred.

Nir Barzilai at the Albert Einstein College of Medicine studied 213 centenarians, trying to determine the factors that sustained them to such an advanced age.[135] He reported that not a single one was a yogurt eater, no one was a vegetarian, 30 percent of them smoked, and none of them was concerned with exercise. However, 80 percent of them had a lot of big, fluffy, fatty protein molecules. So did many of their offspring, suggesting that good genes are important. Fortunately, for those unfortunates who have undersized lipoprotein molecules, research indicates that the molecules are increased in size by exercise.

Surgery

With the help of a surgeon, he might yet recover.
—William Shakespeare, *A Midsummer-Night's Dream*, 1595

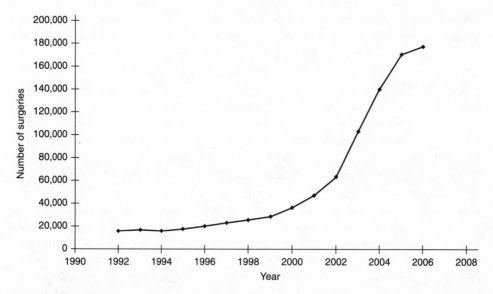

Figure 11.1
Number of bariatric surgeries in the United States, 1990–2006. *Source*: American Society for Bariatric Surgery.

An increasing number of Americans have given up on trying to lose weight by changing the amount and type of food they eat and have turned to surgery for quick fixes. Liposuction, the vacuum removal of fat, typically 5 to 10 pounds from the middle part of the human frame, is one of the leading forms of cosmetic surgery in the United States, rivaling breast implants and nose jobs. There were 455,000 of them in the United States in 2005.[136] The surgery is becoming so common that it has been speculated that liposuction may become like a gym membership, where you pay a doctor $10,000 for the year and can have as much liposuction surgery as you want.

But there is no evidence that decreasing weight by liposuction has a beneficial effect on health, as contrasted with losing the same amount of weight through diet and exercise.[137] Liposuction without accompanying diet and exercise is largely a waste of money. However, the cost of this procedure may be recoverable if you are a prominent personality. A bar of soap made from fat said to have been sucked from Italian Prime Minister Silvio Berlusconi during liposuction sold for about $18,000.[138]

In 2006, researchers discovered that cellulite could be zapped away using heat from an infrared laser.[139] The laser breaks down fatty tissue without harming the overlying skin.

In recent years, an increasing number of severely obese people have turned to an even more drastic measure to shed weight, bariatric surgery. There are two main types. The less drastic type is called the Lap Band adjustable gastric banding system, in which a synthetic ring is attached to the upper end of the stomach. The more common and drastic procedure is stomach bypass surgery. Also called gastric bypass surgery, the operation shrinks the stomach's capacity from wine bottle to shot-glass size and reconfigures the small intestine. Most patients lose about two-thirds of their excess weight within a year of surgery. The number of stomach bypass operations for severely obese Americans has ballooned in recent years (figure 11.1). Children as young as 12 have had the surgery.[140]

Bariatric procedures should be limited to people who have been unsuccessful with other weight-loss methods over an extended period of time and are at least 100 pounds overweight (5 percent of Americans).[141] The surgery and six months of follow-up care costs about $30,000 and is risky, with many possible complications that can triple the cost. Studies reveal that 10 to 20 percent of patients need additional surgery for complications such as bleeding, blood clots, bowel obstructions, hernias, and severe infections, which boosts the cost to $65,000.[142] Bariatric surgery results in death in about 1 percent of the cases.[143] However, if you survive, you will certainly live longer.[144]

12

Conclusion

It is crazy that the study of food has been reduced to the fields of nutrition and food technology, which are now completely in the grasp of an inhuman industry.
—Carlo Petrini, 2004, founder, University of Gastronomic Science

Farming in the United States has many serious problems, usually unnoticed by the public because urban dwellers live divorced from food production. Hardly anyone is a farmer anymore, and most Americans live in and near cities, where food is an abstraction that becomes real only in the supermarket or at the kitchen table. But a host of unsustainable practices in the agricultural economy needs to be changed or modified for reasons of both ecology and human health.

The popular image of the landscape in farming country is a visual metaphor for human life in harmony with nature. As described by Joan Nassauer, "A mix of crops weaves a varied field pattern. Livestock graze on the land. Woodlands and streams make sinuous borders along the fields. Tiny farmsteads dot the landscape. There are fish in the ponds, birds in the sky, and wildlife in the woods. The air is clean. There is a small town nearby with a school, stores, and churches. You might not live in this landscape but you would like to visit it, and when you did, you could stop and enjoy a friendly talk with the farmer and buy fresh produce you couldn't find in the city."[1] This nostalgic image of the American farm stems from the outdated belief that our agriculture is ecologically and socially healthful. But this image is far from reality.

Although most farms are small- to medium-size family farms, they produce only a tiny percentage of the nation's farm products. Monstrously sized farms that practice industrial agriculture (factory farms) are the scale of farming that grows nearly all grain and livestock today. These farms are quite divorced from Nassauer's idyllic visage. Today's most productive farms grow only one crop, there are no livestock, the woodlands are no longer there, a stream may still exist but it is polluted by fertilizer chemicals and pesticides, and it contains no fish. The friendly small town is now deserted and boarded up, the gas station pump still recording gasoline at 30 cents a gallon. No farmer can be seen, except perhaps in the distance riding a massive farm machine guided by a satellite tracking system.

The Small Farm Today

As a work of art, I know few things more pleasing to the eye, or more capable of affording scope and gratification to a taste for the beautiful, than a well-situated, well-cultivated farm.
—Edward Everett, 1857

The operators of small- to medium-sized farms today are probably the oldest and lowest-paid group of workers in American society, except possibly crossing guards near elementary schools. They enjoy no pensions, no paid vacations, no overtime pay, no unemployment compensation, no job security, and no specified hours of work; they have no time off they can call their own. The average farmer is male and fifty-five years old, fifteen years older that the average nonfarm worker. His children probably work elsewhere and have no interest in the family farm. The farmer himself earns most of his income in the city because his small- to moderately sized farm cannot compete with the monstrous factory farms in the area. His farm loses money in most years. When the farmer dies, hundreds of acres that have been producing food for generations, perhaps for the same family, are likely to be sold to a land developer, who will convert the fertile agricultural soil into a paved urban sprawl suburb. With the current price of farmland $1,900 an acre, the sale of the 300 acres he bought in the 1960s for $150 an acre will provide economic security for his children. He is sad to think that the farm will be lost, but he understands the economic needs of his children and is resigned to witnessing the end of an era.

Operators of small- to medium-sized farms have many handicaps when competing with industrial or factory farms. First, an ecologically sound farm containing both grains and animals is less likely to be financially successful than a farm that harbors only plants or animals. Not only must the small farm choose between plants and animals, but it also must concentrate on one type of plant or animal. If a plant is chosen, the farmer knows that the best chance for financial success in a difficult agricultural financial environment is to grow a government-subsidized crop such as corn, soybeans, or wheat, and the crop must have limited species variation to maximize output. The crop farmer on a small holding must buy large amounts of artificial commercial fertilizer because the source of natural fertilizer, manure, is not available on the farm and trucking it in from elsewhere is too costly. In addition to the cost of the fertilizer, the bags of commercial NPK supplements sold by the local feed-and-seed store do not contain the organic matter and micronutrients present in manure. Perhaps the soil on the farm contains enough of them; perhaps not. Costly chemical tests may be necessary to determine this.

This producer of food will need to purchase large amounts of artificial pesticides, because the use of natural methods of pest control, such as integrated pest management, crop rotation, and fallowing, will put him at a disadvantage against his larger neighbors. They can use massive amounts of pesticides because they are not financially responsible for the groundwater, stream, and soil pollution they cause (externalities) or for any medical costs to workers and ordinary citizens that may result. These medical costs and the cost of the cleanup of the environment are borne by taxpayers as a hidden subsidy to the industrial farmers.

If the farmer chooses to operate an animal farm, specialization will be as necessary as it would be on a grain farm. Probably the animals will be chickens, cattle, or swine, and the animal farmer must develop a way to store and dispose of the mountains of manure the animals will produce. Although the small farmer will not have the thousands to hundreds of thousands (or even millions) of animals contained on neighboring factory farms, it is no secret that even limited numbers of farm animals are prolific manure producers that create waste disposal problems, because there are no grain fields on the farm that can use the manure as natural fertilizer. When it comes to feeding the animals, there may be enough grassy acreage to permit the cattle to forage for their food, but to produce economically competitive beef, the farmer will have to send them to finishing feedlots to eat prescribed grain mixtures so their flesh will be uniform, marbled with unhealthy saturated fat, and pleasing to the ultimate consumers. If an animal farmer chooses to raise chickens, they will need to be raised in conditions that are physically and ecologically unsound to maximize the production of meat and eggs. As on a grain farm, the most ecologically unsound practices bring the greatest financial rewards.

The operator of a small- to medium-sized farm must also compete for federal dollars against neighbors with much larger farms. And the small farmer knows the chances of winning this battle are small to nonexistent. The "bigger is better" philosophy of the federal government ensures that the small farmer will always be at a disadvantage in the competition for federal money.

In summary, the small farmer must choose between adopting the methods of the neighboring farm that practices industrial farming or fold up his tent, or work part-time in the city to support the family. The time-honored, ecologically sound, and highly productive farming methods that served forebears well for generations and indeed, for thousands of years, are nearly extinct, preserved only in organic farming.

The Large Farm Today

It seems anomalous that institutions bent on private greed need not apologize for their polluting activities, whereas agencies that are devoted to the protection of public health under due process of law, must apologize in terms that it's good for the economy.

—Ralph Nader, 1978

The methods of production, processing, distribution, and consumption of food in the United States are ecologically and environmentally unsound and unhealthy.[2] They need to be changed. If continued in the way they have developed over the past sixty years, they are bound to collapse. Ecologically unsound ways of working cannot persist indefinitely, but because of a combination of the seemingly low cost of food and governmental politics the food industry will not be easily altered.

Norman Wirzba, professor of philosophy at Georgetown College in Kentucky, stated, "The fact of the matter is that agribusiness practices cannot be sustained over the long term, because they depend on techniques that either produce a crop that often is equaled, even doubled, in weight by the soil lost to erosion, or kill the microbial, life-promoting soil structure with heavy pesticide use and soil compaction, or deplete and contaminate groundwater systems—all predicated on a cheap and limitless supply of fossil fuel."[3] The goal of the American food culture has not been fidelity to the land. Our culture measures human progress in terms of our ability to combat and control the land, to establish human flourishing in opposition to the well-being of the earth.

Fertilizer, Pesticides, and Productivity

The abundance of food in America in 1981 is being achieved at a very high cost—not so much in dollars, but in soil, in irreplaceable underground water, in squandered oil, wasted gas and environmental degradation that nature will be able to repair only many decades after the current American food machine grinds to a halt.
—Robert Rodale and Thomas Dybdahl, *Cry California*, 1981

Farm productivity in the United States, the ratio of input to output, has more than doubled since 1950 because of the increased use of fertilizer and advances in technology. However, since 1980, fertilizer applications have averaged about 19 million tons with little variation from year to year. The reason fertilizer use has been unchanged on America's farms for more than twenty-five years is that the saturation level has been reached. Additional fertilizer has been found to have little effect on production. Most crops on farms are physiologically incapable of absorbing more nutrients. American farmers are now getting the maximum possible crop production per acre, barring new developments in genetic engineering and technology.

Serious pollution problems have been caused by fertilizer runoff from industrial megafarms. Because the artificial fertilizers are wastefully applied and inefficiently used by the crops, much of it ends up in surface runoff or in groundwater. During the past fifty years, runoff, mostly from the corn belt in the Midwest, has caused a marked increase in the lower Mississippi River in the concentration of nitrate from fertilizer. This has caused a dead zone around the mouth of the river that averages more than 5,000 square miles in area. It reached 7,900 square miles in 2007, an area roughly the size of New Jersey. Brown shrimp harvests have routinely been 25 percent of their historic catch size.[4] The effect of nitrate pollution on near-shore areas is worldwide. The number and size of oxygen-deprived dead zones in the world's oceans has risen every decade since 1970, and now stands at 200, with the largest in American waters covering an area the size of Ohio. Excessive use of fertilizer on the land is killing the life in the near-shore areas of the ocean.

A high and unsafe level of fertilizer nitrate in well water has been reported in some farming areas, and this has caused a few cases of "blue baby" syndrome, a potentially fatal condition that occurs when the concentration of nitrate in the water exceeds 100 ppm. The nitrate reacts with the hemoglobin that carries oxygen to human cells and produces metheglobin, which cannot carry oxygen. The baby develops a blue coloration of its mucous membranes and possibly digestive and respiratory problems. But the condition is reversible if caught in time, and no deaths have yet been reported. Families getting their water from wells should have the water analyzed for nitrate concentration and also for pesticide pollution. Farmers using well water are marginally poisoning themselves.

Pesticide pollution of America's groundwaters and surface waters is widespread. In 1992 the EPA published the results of a twenty-year study of pesticides in groundwater. The study covered 68,824 wells in forty-five states, and pesticides were detected in 16,606 wells in forty-two states.[5] Pesticides were, as expected, more likely to be found in groundwater in agricultural areas. Surface waters are in even worse condition. According to a report from the environmental group American Rivers, 60 percent of the nation's impaired rivers are degraded by agricultural pollution.[6] Stream samples examined by the U.S. Geological Survey and reported in 2004 found at least one pesticide in 95 percent of the streams and pesticide mixtures of at least ten chemicals in 25 percent of them.[7] The amounts of each pesticide in the streams were below regulatory thresholds in all cases, but there are no regulations concerning pesticide mixtures. Mixtures are known to have harmful effects not predictable from the individual chemicals alone.

It is important to remember that although scientific research is the best path to knowledge, there is an element of uncertainty in all research findings. This is a particular concern in studies of pesticide toxicity because the levels allowed in food are so small that harmful effects may not be immediately obvious. There can be a long time lag between exposure to a carcinogen or other substance, for example, and someone developing cancer or another debilitating illness. Well-known examples include asbestos, cigarette smoke, and coal dust.

It is known that pesticides can have harmful effects on babies and adults, including low birth weights and mental impairment. Pesticide exposure is also linked to damaged immune systems, Parkinson's disease, and several types of cancer. None of this should come as a surprise to pesticide producers or the general public.

Pesticides are poisons designed to kill living organisms, and common sense should tell us that they are dangerous and may cause harm to humans. This commonsense expectation is confirmed by the many cases of pesticide poisoning of farm workers.

In fact, the use of pesticides is largely unnecessary, as organic farmers have shown. Crop pests can be controlled without the massive use of artificial pesticides and with little or no loss of crop productivity. As Rachel Carson said in 1962, the war against agricultural pests cannot be won. Pesticides poison humans and the environment and, over the long run, have increased crop loss to pests rather than decreasing it. Why do we continue these clearly harmful applications? The answer is advertising and brainwashing by pesticide manufacturers. We must learn to coexist with the other hungry small creatures on the planet, controlling them only to the extent necessary to maintain our food supply and using natural methods whenever possible. If there was ever a case where a Pogo cartoon hit the nail on the head, this is it. We have indeed met the enemy, and it is us.

The massive use of pesticides in agriculture is unnecessary and self-defeating. Current understanding of plant growth and pest biology enables pest control without poisonous pesticides and with little, if any, loss of productivity. The use of modern pest control methods, as in organic farming, does require additional human labor, but this is readily available from America's legal labor pool if higher salaries were paid and better working conditions instituted. This would have little effect on the cost of food because three-quarters of the cost of food in the supermarket results from processing, marketing, and distribution costs, not from money paid to the farmer.

Our ingenuity with genetic engineering of crops cannot save us from the small creatures that share our bounty. They are already developing resistance to our gene manipulations. Humans insist on fighting a "war of the worlds," us against every other creature. Our efforts only seem to make things worse.

One technological development that has the potential to increase crop yields without harmful side effects is the use of satellites. Space technology is creating new tools to increase productivity. For example, farmers who can afford it now have a satellite guidance system. Radio signals from a network of satellites are picked up by receivers mounted on the cabs of combines and tractors and help farmers pinpoint hot spots in their fields that need to be monitored more carefully for fertility and yield. A map of the data is created that displays various nutrient levels, soil types, fertility, and yield potential. Small farmers are, of course, less likely to be able to afford such high-tech gadgetry, and there is no sign the federal government will subsidize their purchase.

Food Production

In simplest terms, agriculture is an effort by man to move beyond the limits set by nature.
—Lester Brown and Gail Finsterbusch, *Man and His Environment: Food*, 1972

Because of the use of fertilizers (natural and artificial), increased understanding of plant biology, and increased sophistication of farming equipment, each American farmer can now feed about 212 people, a far cry from 8 people a century ago. The United States grows more food now than it did fifty years ago, on about 25 percent less acreage, and with a small fraction of the workers it once used. This increased production has created huge surpluses. As a result, farmers and ranchers now rely heavily on export markets to sell the part of their production that is not needed for domestic consumption. Fifty percent of our wheat and 19 percent of our corn are exported. The United States is the world's major food exporter, providing more than half the food aid that feeds hungry people around the world. The dominant role of U.S. agriculture in the global economy has been likened to OPEC's in the field of petroleum.

But the increase in crop yields has had negative effects on the nourishment they provide. Plants cultivated to produce higher yields tend to have less energy for other activities, such as growing deep roots and generating phytochemicals humans need to remain healthy.

The global corn harvest continues to increase, but the increased yields of corn are not having a commensurate effect on the amount of food available for the world's people. The reason is that as economies improve, people eat more meat and less grain, and farm animals are fed a diet rich in corn. The conversion from corn into meat is an inefficient process. It takes 2 to 4 pounds of grain to produce 1 pound of chicken, and 7 pounds of grain to make 1 pound of beef. Most of the world's corn is fed to animals, not people. The desire to increase meat consumption seems to be universal and is likely to grow in the foreseeable future as populous nations such as China improve their economies.

Organic farming is the only sustainable farming method. It has been successfully practiced for 10,000 years, and only in the past sixty years has it been regarded as exotic or suitable only for a fringe of agriculturalists. The large numbers of humans who were malnourished and died of starvation over the centuries have been taken as evidence that the ecologically sound farming methods of the past were inadequate to feed the human population. In fact, the starvation was due largely to the same causes as starvation today: undemocratic governments, poverty, and inequitable land distribution.

The view that ecology can be ignored in the march to a brave new agricultural world has been championed by industries that benefit from chemically oriented agricultural technologies, such as mass-produced fertilizers and pesticides. They have used sophisticated advertising and promotional techniques to convince farmers and the general public that we are courting starvation if we do not use copious amounts of their products (table 12.1). This is false, as has been amply demonstrated by organic farmers. Although the number of organic farmers is growing, they still form only a minuscule percentage of all farmers, and it is not clear that they will be a major part of the farming community over the next few decades.

Table 12.1
Pesticide use in U.S. row crops, fruits, and vegetables

Crop	Proportion of area treated		
	Herbicide	Insecticide	Fungicide
Row crops[a]			
Maize	97%	30%	1% or less
Cotton	92	79	6
Soybean	97	1	1 or less
Winter wheat	56	12	1
Spring wheat	88	3	1 or less
Tobacco	75	96	49
Potato	87	83	89
Fruits and vegetables[b]			
Apple	63	98	93
Oranges	97	94	69
Peaches	66	97	97
Grapes	74	67	90
Tomato, fresh	52	94	91
Lettuce, head	60	100	77

Note: Data for row crops are for 1996 and for fruits and vegetables, 1997. Fungicide amounts do not include seed treatments.
Sources: For row crops: Agricultural Chemical Usage 1996 Field Crop Summary, USDA, September 1997. For fruits and vegetables: Agricultural Statistics, 1997, NASS Crop Branch.

A growing problem facing organic farmers is the increasing abundance of genetically engineered crops. Seeds from these crops are blown by the wind and carried by birds and other flying creatures into organic farms, and they have the potential to destroy the character of these farms. There is no obvious way to stop this from happening, as the law seems to be on the side of the genetic engineering companies. If a non-GE farm is found to have GE crops growing on it, the farmer is judged to be growing patented GE crops without a license, even though he did not plant them and does not want them.

There is as yet no evidence that genetic engineering of crops will have a beneficial effect on food production. Hopeful scenarios have been presented but have yet to materialize. So far, the products of agricultural genetic engineering have concentrated on curing some of the problems created by the change from organic, sustainable agricultural methods to industrial agriculture: engineering plant genetics to control weeds so the use of herbicides can be reduced, and altering plant genetics to control the increased loss of crops to insects that has occurred in the sixty years since organic methods were relegated to a minor part of crop growing. The process smacks of insanity. Insects and weeds can be controlled by natural methods that do not use artificial chemicals, and they should be. At some point, we humans will have to recognize our place in nature: we are only a strand in the web of life, not the web itself. We have no choice but to accommodate to the world in which we evolved. We cannot change it and can only harm ourselves by attempting to change it.

Meat

The industrial production of meat is difficult to justify from ecological and health perspectives. Of all the water used in the United States, more than half goes to the production of livestock (including the water used to grow the grain they eat in the feedlot).[8] It takes more than 100 times more water to produce 1 pound of beef than 1 pound of wheat.[9] For corn the ratio is 70; for soya beans, 50.

Chicken is not as thirsty a crop as beef but still requires almost twice as much water to produce 1 pound of meat as it does to produce any grain. The conversions from pounds of grain to pounds of meat are inefficient in terms of the human food supply. As of 2007, the world's production of grain is still sufficient to tolerate the ecologically wasteful and unnecessary raising of animals for food.

The animals whose flesh we find tasty produce more than a hundred times more feces and urine than humans do and, because the factory farms that produce most farm products do not raise both animals and grain, there is an enormous waste disposal problem on animal farms. Imagine having to deal with the generation of 86,600 pounds of excrement pouring into a group of small areas every second! No one has found an adequate solution for this problem, and there is none on the horizon. Meanwhile, tens of thousands of miles of rivers and untold volumes of groundwater have become polluted by farm animal waste.

Nearly everyone in the United States today is aware that eating meat and dairy products raises the amount of cholesterol and saturated fats in their blood and that these substances are a major cause of clogged blood vessels and resulting heart attacks. The development of clogged arteries is a slow process resulting from heavy consumption of meat and other animal products over several decades, although early stages of the process are increasingly detected in America's children. Perhaps children at fast-food restaurants should be told by Ronald McDonald that hamburgers are ground up cows that were killed by a metal rod shot into their heads before their throats were slit by a machete so their blood could be drained out. At least they should be informed that the first Ronald McDonald is now a vegetarian.

Seafood

We are running out of large fish in the ocean and shellfish near the shore. We are close to totally exterminating many of the vertebrates and invertebrates we are used to eating. Tuna may go the way of the wooly mammoth. This is not only sad for our palates but is also having noticeable negative effects on ocean ecology.[10] Remedying the disaster we have created will require less fishing, catching only the right kinds of fish, only the right sizes of fish, and only a certain amount of each kind of fish, while avoiding nontarget fish, marine mammals, seabirds, turtles, and damage to the seabed. This is a tall order, and it is unclear whether the United States and the other fishing nations of the world are politically capable of doing it.

The structure of a fisheries salvation plan is discussed in some detail by Pauly and Maclean and the Environmental Defense organization[11] and involves not only reducing our fishing effort, but also buying back and destroying fishing vessels, establishing no-fishing zones, reducing subsidies to fishing fleets, imposing carbon taxes and more efficient engines on ships, educating consumers by ecolabeling, and an increase in international cooperation to save the commons that is the ocean. Implementation of any one of these suggestions is fraught with political difficulties, and it is unclear whether many of them can be implemented.

With regard to shellfish, the problem involves not only reducing the harvest. A drastic reduction in the input of pollutants into near-shore areas is at least as important. The situations in Chesapeake Bay and the Gulf of Mexico offshore of Louisiana are clear examples of the devastation human greed and ineptness have wrought.

Food for All

Considering the people of the earth as a whole, experts agree that there is no food shortage and that if there were an equitable distribution, the earth currently provides human beings with enough food to make everyone overweight, if not obese.[12] The existing starvation and malnutrition in the world are not due to inadequate food production. Given present methods of food production in the industrialized countries, feeding the 6.7 billion people on the planet now and the peak of 9 billion projected for 2050 should not be a problem. Despite the eternal political and social problems in the world and the feeding of most corn to animals, increases in world food production and international food aid to poor countries has decreased the number of undernourished and starving people since the 1960s, and according to the UN, the number of undernourished people dropped by 13 percent between 1980 and 2000, even though the world's population increased by 1.6 billion. We are making consistent progress in alleviating world hunger. Unfortunately, this information does not sell newspapers or raise TV ratings as well as photos of starving people. The problem of hunger in the world cannot be solved overnight, but it is not politically acceptable to emphasize this reality.

The many articles written about America's need to produce more food to feed the 850 million or so hungry people of the earth are commercially and politically motivated. In the United States, the amount of fertile soil, the amount of water available, and the climate are adequate to feed not only everyone in the United States but much of the rest of the world as well. With the addition of food-producing areas in Europe and Asia, more than enough food is being produced today. This does not mean we should decrease food production, because the situation can change fairly rapidly if there are prolonged droughts in grain-producing areas resulting from climate change, political turmoil in agriculturally productive nations, terrorism, or other factors. An adequate world food reserve needs to be maintained. But no one has demonstrated that massive amounts of artificial chemicals are necessary to do this.

People are starving and malnourished in many parts of the world because of international and intranational political problems, not because of inadequate understanding of the biology of crop production, soil erosion, loss of soil fertility, or lack of water or sunshine. While it is true that soil erosion, excessive mining of soil nutrients, and drought are problems in some areas of the world, the central causes of hunger and malnutrition are political. People are hungry today primarily because they own no agricultural land, they are too poor to purchase food, and their governments are insufficiently motivated to help them, not because of an outright world shortage of food. It is a disgraceful situation.

Food Consumption

Down where I fear there's a terrible lot o' me'
down where some people are hippopotami,
in the department of laparotomy,
that's where the vest begins.
—Arthur Guiterman, "Vulgar Lines for a Distinguished Surgeon," 1917

Americans have poor eating habits, and the other developed nations of the world are copying them. We eat too few unprocessed foods such as fresh fruits and vegetables and too many processed foods. Three-quarters of us fail to meet the minimum recommended five servings each day of fruits and vegetables.[13] Technology is used irresponsibly to create taste-tempting, nearly irresistible edible products that are beneficial only to the bottom lines of company balance sheets. The numbers on their bottom lines increase in tandem with the size of our bodies. Corporate responsibility seems to extend no further than ensuring that their food products are not lethal in the short run. They are well aware that most processing removes nutrients and introduces unhealthy substitutes. The sugar, salt, and fats they add to the grains and meats that came from the farm are known by all to be bad for humans but, hey, they sure taste good. The food companies are not doing anything illegal, only immoral, and there are few laws against that.

Government Intervention

More than half of all Americans favor taxing fast foods and junk foods as a way of reducing their consumption. Should the federal government become involved in our eating habits? Eating is widely perceived to be a matter of individual responsibility and none of the government's business. In general, government intervention is economically justified when the cost of someone's actions is borne by others (social costs or externalities), as in the case of industrial pollution, drunk driving, or secondhand smoke. Costs borne by individuals themselves are not a matter for governmental involvement. In the case of overeating and unhealthy weight gain, those who favor federal involvement point out that medical costs in the United States are skyrocketing and that a growing part of the reason is the increase in disease caused by or aggravated by overweight and obesity. Everyone pays for this through their taxes for Medicare and Medicaid, and people of normal weight pay for it through higher private insurance costs. But the jury is still out on the amount of these costs and whether they justify government intervention.

Another factor that some groups believe justify government action in the area of food consumption is social inequity. There are documented disparities in obesity across racial/ethnic groups. Low-income and black and Hispanic groups are disproportionately obese, at least in part because of the higher cost of fruits and vegetables and the lesser access of disadvantaged groups to them (fewer supermarkets in lower-income areas), and the accompanying forced increased consumption of processed, calorie-dense foods.

Even if we assume that government intervention in the food consumption area is justifiable, what form should the intervention take? It has been suggested that clearly unhealthy snack foods such as soft drinks, potato chips, pork rinds, and cheese puffs might be taxed to increase their price and reduce consumption—in other words, a "fat tax."[14] Revenues raised from the tax might be mandated for use in expanded federal educational programs in nutrition. But the tax would be regressive, falling most heavily on low-income consumers, who spend more of their income on food than do higher-income consumers. And there is no guarantee that the money saved by not buying the snacks would result in an increase in fruit and vegetable consumption. Then there is the question of whether an increase of, say, five cents on a $1.99 item would dissuade consumers from buying it. These items are a minuscule part of a food shopping trip and are impulse purchases probably not on the list of food needs the shopper made out at home. Studies by the Department of Agriculture's Economic Research Service indicate that the purchase of salty snack foods is not very responsive to price. If it pleases our taste buds, we do not pay much attention to its cost.

Another possible federal intervention might be in advertising for candy, soft drinks, fast foods, and sugared cereal aimed at children.[15] The principle would be the same as that for the restriction on cigarette advertising: detrimental to human health and a drain on worker productivity and medical costs. But some studies have found that aggregate cigarette consumption actually increased for a while after the government banned broadcast cigarette advertising. Cigarette companies were forced to compete on price rather than through broadcast commercials and were able to do so from savings in their advertising budgets. Restrictions on advertising of unhealthy foods might have a similar effect on price and consumption. American adults and children could end up fatter, not fitter.

Some European nations have led the way in imposing a greater degree of social responsibility on corporate advertisers.[16] In 2007, Britain banned ads for foods high in fat, salt, and sugar around children's television programs. In 2006 Ireland imposed a similar ban. Wrappers must now carry warnings like, "Fast food should be eaten in moderation as part of a balanced diet," or, "Snacking on sugary foods and drinks can damage teeth." The new code also prohibits using celebrities or sports stars to promote junk food to children. France prohibited vending machines selling soft drinks and chocolate bars from schools in 2005. Sweden and Greece have also enacted legal limitations on food advertising to children. Similar laws are needed in the United States.

There probably is no substitute for self-control. If individuals do not care enough about their health to eat more responsibly, no one can force them to care. There is little outside intervention can do. An alcohol, saturated fat, or sugar addict must want to change for a cure to be possible. There is no other way that harmful behavior can be stopped. Education is the most promising route to improved eating habits, but unless the producers of processed foods act more responsibly to keep unhealthy foods off the market, it is doubtful that food education will have the desired effect. Most people are ruled by their taste buds, as the epidemic of overweight and obesity in America amply demonstrates.

Slow Food

In response to globalization and the dominant position and sameness of fast-food in Western culture, the Slow Food movement was started in Italy in 1986. It has since spread to more than a hundred countries, with chapters in Switzerland, Germany, United States, France, Japan, and the United Kingdom and more than 80,000 members. In the United States, the organization's quarterly magazine is called the *Snail*. In 2004 Slow Food opened the University of Gastronomic Sciences in Italy, whose goal is to promote awareness of good food and nutrition. Because of the grassroots nature of Slow Food, few people in the United States and Europe are aware of it.

The aims of the Slow Food movement include the preservation of native seed varieties; promotion of local and traditional food product know-how; educating consumers about the hidden risks of fast food, agribusiness, factory farms, and monoculture; the importance of biodiversity; lobbying against artificial pesticides; support of government support of organic farms; opposition to genetic modification of food; and teaching gardening to students. In short, the Slow Food movement wants to promote knowledge of healthy food and tasteful eating. It is too early to gauge the effectiveness of the movement, but its objectives are commendable.

Compassion and Ethics

The basis of all animal rights should be the Golden Rule: we should treat them as we would wish them to treat us, were any other species in our dominant position.
—Christine G. Stevens, 1980, Founder, Animal Welfare Institute

The question is not, can they reason? nor, can they talk? but can they suffer?
—Jeremy Bentham, *An Introduction to the Principles of Morals and Legislation*, 1789

There are two questions involved in killing other animals. The first is whether we have the ethical right to kill them at all when it is unnecessary for a healthy human diet and, as a practical matter, when eating them is demonstrably harmful to our bodies. It shortens our lives just as surely as smoking tobacco does. Many Americans think there is something lacking in the psyche of those who kill animals simply for sport. Shooting Bambi is regarded by most people as a heartless act. Sixty-three percent of us "strongly agree" with the statement that humans have moral duties and obligations to other animal species; another 31 percent "somewhat agree." Only 6 percent either disagree or have no opinion.[17]

But hunting and fishing are significant national pastimes, particularly with the half of the American population whose bodies are rich in testosterone. This is the same 49.1 percent of the population that is responsible for 87 percent of the murders in the United States.[18] The presence of testosterone is a major factor in the desire to kill. In 2001, 13 million Americans hunted and 34 million fished. These numbers have been decreasing over the past two decades; the percentage who hunt or fish declined from 26 percent to 18 percent.[19] How much of the decline has resulted from increasing ethical concerns is uncertain. It may result mostly from a ripple effect of suburbanization and the decline of rural communities. Today more than two-thirds of these "sportsmen" live in metropolitan areas, where their children grow up less familiar with firearms, removed from daily contact with blood and dirt, and often less comfortable with the pursuit of animals as a sport. Just as successive generations of immigrant families lose touch with the language and customs of the old country, the descendants of rural America do not have the same strong cultural attachment to the land and to hunting as in the past.

The second question involved in the human obsession with meat consumption is the treatment of the animals we raise for food. Our society is showered with images of happy animals living on farms where cows graze in lush green fields and chickens have the run of the barnyard. This vision of free-roaming animals living out their days in sunny fields is far from reality. Most of the animals that are raised for food live miserable lives in intensive confinement in dark, overcrowded facilities, commonly called factory farms. There is no doubt that they are thought of only as parts in a well-oiled food manufacturing machine rather than as sentient beings.

A few organizations are actively concerned about the treatment of animals existing below our radar screen of "cute." Above the radar screen for most Americans are dogs, cats, baby seals, bunny rabbits, and some types of birds. Apparently below the screen are cattle, fowl, hogs, and many types of sea creatures. There seems to be no defining characteristic that separates the two groups other than how we regard their physical appearance. Certainly the distinction is not based on intelligence or brain development, because swine are as smart as dogs and most of us eat pork but are repulsed at the thought of eating dog, although many East Asians do.

As noted in chapter 9, only 2 percent of Americans refuse to eat meat, and for many of them the decision is based on health concerns rather than on an objection to killing animals. But as noted in chapter 7, most Americans favor strict laws concerning the treatment of farm animals, indicating that we believe that it is okay to kill them but not to torture them beforehand. Most Americans seem to agree with the writer who said, "I believe that animals should be treated with dignity and respect. We should not be cruel and we should be as 'humane' as is possible. But at the end of the day, I want to see them on my dinner table as well. I do not believe these two thoughts are incompatible."[20]

The treatment of animals has been viewed by many humans as one measure of humankind's humanity ever since the Enlightenment in the eighteenth century. It is probably no coincidence that two of the leaders of the movement to abolish the slave trade helped found the Royal Society for the Prevention of Cruelty to Animals in the 1820s.[21] Animal activists believe that we have an obligation not to cause pain to animals that are sentient, that is, have the capacity to suffer. We may doubt that an ant or a cockroach has the neural complexity that allows it to feel pain, but there is no question that cattle, swine, and fowl can sense pain. As noted in chapter 6, chickens with feet that hurt from standing on wire in cages all the time choose feed that contains pain-killing drugs over feed that does not. And having their beaks cut off is not a pleasant experience. Why don't we care about their pain? What rights accrue to animals simply by their virtue of being alive? It is not a simple question and is complicated by religious beliefs, making discussion doubly difficult. People cannot agree on a system of rights for each other after hundreds or thousands of years of debate, so the ground is bound to get even shakier when other animals are included.

Presumably the most fundamental right humans have is the right to life, thought to be an unalienable right in the language of the American Declaration of Independence in 1776. It is a right that, according to the dictionary definition, cannot be taken away or transferred. It clearly is not unalienable, however. Seventy-four percent of Americans believe in capital punishment, and a large part of the American public believes in denying this supposedly unalienable right to humans who have been conceived but not yet born. The validity of these two exceptions is hotly debated, and opinions about them have changed over time, but despite these exceptions, Americans agree that ending the life of innocent persons is not ethically acceptable.[22]

Why do we not grant the right to life to the animals with which we share the planet? Apparently the reason is that they have bodies shaped differently from ours, and over the millennia humans have discovered that the flesh of many of them tastes good. We eat them regardless of their position on the evolutionary ladder. In various parts of the world, human carnivores eat a wide variety of vertebrates, including mammals (cows, sheep, whales, and many others), birds (chickens, turkeys, ostriches, and others), reptiles (turtles, lizards, snakes), amphibians (frogs), and fish (many kinds). We also eat many types of invertebrate animals on the evolutionary tree. Among the invertebrates judged particularly tasty in some cultures are sea animals such as octopus, squid, lobster, shrimp, clam, oyster, crab, sea urchin, and sea cucumber. Edible invertebrate land animals include snails, worms, and a wide variety of insects such as crickets, termites, and ants. Should we feel guilty about depriving them of their "unalienable" right to life?

Every animal is potential food for another. But because of a highly developed brain and sensibilities, the human animal has a choice about eating habits. In contrast, animals in the cat family (lions, tigers, house cats) have no choice. They do not have molar teeth with which to grind plant materials, so they must be carnivorous, and their digestive systems and biochemistry are adapted to this type of food consumption. Humans have several different types of teeth, and are able to eat both animal flesh and plants. They have a choice, although their digestive systems and biochemistry are not well adapted for handling meat. And they can easily live without it. As vegetarians know, their bodies are healthier when they do not eat other animals.

Is there a rational ethical basis on which to decide whether humans should terminate the life of an animal only because it is structured differently from us? On an ethical basis, the ability to feel pain seems a most useful criterion.[23] Most of us recoil from causing physical pain to another person, and it is a short step from there to the view that it is not ethical to cause pain to a dog, cat, or parrot. Only people in the United States who are generally considered to be mentally deranged deliberately hurt these animals. We all know that animals this high on the evolutionary tree can sense pain. However, termites and cockroaches are too neurologically undeveloped to do so, so it is okay to eat them if we choose. In fact, edible insects are on the menu for an estimated 80 percent of the world's population.

Characteristics in the natural world of evolution are gradational, so there will always be a middle ground where we are unsure whether *pain* is a meaningful word to apply to a particular type of organism, but certainly the vertebrate animals that most Americans eat know what pain is.[24] This has been recognized since at least the time of Charles Darwin. That being the case, the treatment they receive at human hands is unconscionable. It is bad enough that we deprive them of their unalienable right to life, but the lifetime of torture they receive on factory farms before they are slaughtered is reprehensible and should be stopped immediately.

What We Can and Should Do

There is, however, a limit at which forbearance ceases to be a virtue.
—Edmund Burke, 1769

Farm policy in the United States does not serve the interests of either the citizens or the irreplaceable soil. People are being poorly served by the agricultural interests that produce the food we eat. I suggest the ways to change this unhealthy situation are the following:

1. *Farm subsidies should be discontinued, but if they are not, they should be redirected so they serve real needs.* Government farm programs should provide public goods such as environmental benefits, a healthy and safe food supply, and support for farmers who have financial difficulties due to circumstances beyond their control, such as natural disasters or low prices resulting from national or international manipulations.

Federal aid should be directed to sustainable organic farms and away from those whose existence depends on the massive use of artificial fertilizers and pesticides. Such redirection will improve the health of Americans and have no significant effect on America's food production. Fruits and vegetables should be part of the government's program so that healthy foods can better compete in price with the clearly unhealthy processed foods. The money can come from reductions in corn subsidies.

Revamping of the American government's subsidy program will have the added benefit of helping to alleviate poverty and malnourishment in the Third World. We cannot overthrow the military dictatorships that stifle growth and progress in many countries in Africa and Asia, but we can encourage their citizens to produce their own food.

2. *More resources should be directed toward inspection of the way food is produced, processed, and packaged.* The small percentage of inspections is a national disgrace, and there is no sign that an increase is on the way.[25] The government's food safety system is severely underfinanced and perhaps more concerned with serving private interests than with protecting public health.

3. *The food processing industry needs to show more concern for the health of Americans and decrease the amount of salt, sugar, and fat in their products.* How do their executives sleep at night, knowing that they are feeding themselves and their friends nutritional garbage? They are certainly capable of giving us healthier products, but they seem to be saying that health does not translate into profits. It is hard to believe this. However, it is not obvious how the ethical standards of the processing industry can be improved. As the New Testament warns (1Timothy 6:10), love of money is the root of all evil.

4. *The American people should decrease their meat consumption and increase their consumption of fish, fruits, and vegetables.* We can choose to eat products produced locally rather than far-traveled ones. We can buy organic foods and tell local supermarket managers that we want them to carry more of these products. We can tell the managers that we want produce to be labeled with its place of origin. We can choose healthier processed products. We can petition our representatives at both the local and federal levels about the changes we want in agriculture. As stockholders in agriculturally related industries, we can tell boards of directors about our concern for environmentally friendly farming methods. As voting citizens, we are responsible for the current state of agriculture. We can change it if we wish.

5. *We as citizens need to inform our elected officials that we want major changes to be made in the agricultural sector.* America is a nation of elected officials who respond to both the voters who elected them and big-money agribusiness contributors, whose agenda may differ from that of the voters. Our representatives need the money to finance their reelection campaigns, but they also need our votes. No one likes to feel controlled by others, and businesses are no exception. They want a free hand to run their activities, which is usually termed voluntary cooperation with governmental authorities. The record reveals that their cooperation commonly does not go far enough when major changes are required. And there is no question that major changes are required in the agricultural sector of the American economy. But they will only happen if enough people are concerned and act. Are you, and will you?

Notes

References to pages in *Time*, *Newsweek*, and *U.S. News & World Report* refer to the International Edition. Page numbers may be different in the edition for readers in the United States.

Introduction

1. *Ecologist*, September 2005, p. 10.

2. *Ecologist*, July–August 2007, p. 11.

3. Live Cattle, Lean Hogs, & Lumber. dailyfutures.com, 2005, available at dailyfutures.com/livestock/.

4. L. R. Brown, Grain Harvest Falls, *Vital Signs 2000* (New York: Norton, 2000), p. 34.

5. R. Marks, Cesspools of Shame, *Natural Resources Defense Council*, 2001, available at nrdc.org/water/pollution/cesspools/cesspools.pdf.

6. Facts about Pollution from Livestock Farms. *Natural Resources Defense Council*, Fact Sheet, 2001, available at nrdc.org/water/pollution/ffarms.asp.

7. R. A. Myers and B. Worm, "Rapid Worldwide Depletion of Predatory Fish Communities." *Nature*, May 15, 2003, pp. 280–283.

Chapter One

1. Ahearn, M. C., Farm Operators No Longer Have Lower Levels of Educational Attainment. Amber Waves, February 2008. available at ers.usda.gov/AmberWaves/February08/Indicators/inthelongrun.htm.

2. A. Jerardo, Estimating Export Share of U.S. Agricultural Production, *Amber Waves*, November 2003, available at ers.usda.gov/AmberWaves/November03/Indicators/behinddata.htm.

3. In Brief, *Environment*, January–February 2003, p. 8; H. Norberg-Hodge, Think Global … Eat Local, *Ecologist*, September 2002, p. 29; B. Halweil and D. Nierenberg, Watching What We Eat, in *State of the World 2004*, ed. L. Starke (New York: Norton), p. 82; J. E. McWilliams, Food That Travels Well, *New York Times*, August 6, 2007, p. A19.

4. Trends in U.S. Agriculture: Farm Numbers and Land in Farms, USDA National Agricultural Statistics Service, n.d., available at usda.gov/nass/pubs/trends/farmnumbers.htm.

5. Trends in U.S. Agriculture: A History of American Agriculture, n.d., available at agclassroom.org/teacher/history/farm_tech.htm.

6. Look No Driver, *New Scientist*, March 7, 1998, p. 11.

7. A History of American Agriculture.

8. S. Moore, History of Horse-Powered Farming in America, Rural Heritage, 2000, available at ruralheritage.com/equip_shed/history.htm.

9. Trends in U.S. Agriculture.

10. A. Trewaves, Malthus Foiled Again and Again, *Nature*, August 8, 2002, p. 668.

11. A History of American Agriculture.

12. Land Owned by Ownership Characteristics, Agriculture Economics and Land Ownership Survey, U.S. Department of Agriculture, 1999, Table 68, p. 247.

13. Farms, Land in Farms, and Livestock Operations 2005 Summary, 2006, available at usda.mannlib.cornell.edu/usda/current/FarmLandIn-01-31-2006.txt; R. A. Hoppe and D. E. Banker, *Structure and Finances of U.S. Farms: 2005 Family Farm Report*, USDA Economic Research Service, Economic Information Bulletin no. EIB-12, 2006, available at ers.usda .gov/publications/EIB12/.

14. Trends in U.S. Agriculture.

15. Agricultural Productivity in the United States, USDA Economic Research Service, 2007, available at ers.usda.gov/Data/AgProductivity; K. O. Fuglie, J. M. MacDonald, and E. Ball, Productivity Growth in U.S. Agriculture, USDA Economic Research Service, Economic Brief Number 9, 2007.

16. K. Casa, The Changing Face of Farming in America, *National Catholic Reporter*, February 12, 1999, pp. 13–16.

17. Hoppe and Banker, Structure and Finances of U.S. Farms.

18. Ibid.; Federal Spending on the Budget, National Center for Policy Analysis, 1999, available at ncpa.org/pd/budget/pd081099d.html.

19. C. Dimitri, A. Effland, and N. Conklin, *The 20th Century Transformation of U.S. Agriculture and Farm Policy*, USDA Economic Information Bulletin no. 3, 2005, p. 2.

20. U.S. Census Bureau, *Statistical Abstract of the United States 2003, Section 23, Agriculture* (Washington, D.C.: U.S. Government Printing Office, 2004), p. 665.

21. K. M. Dudley, *Debt and Dispossession: Farm Loss in America's Heartland* (Chicago: University of Chicago Press, 2000), p. 162.

22. H. Stewart, *How Low Has the Farm Share of Retail Food Prices Really Fallen?* USDA Economic Research Report Number 24, 2006; H. Elitzak, Calculating the Food Marketing Bill, *Amber Waves*, February 2004, USDA Economic Research Service, available at ers.usda .gov/amberwaves/february04/indicators/behinddata.htm; R. Boehm, AgriNotes and News, Michigan Farm Bureau, 2005, available at michiganfarmbureau.com/press/2005/20050407 .php.

23. P. D. Johnson, The Future of Farming, 2003, available at ewg.org/farm/news/story .php?id=1780.

24. S. Offutt and C. Gundersen, Farm Poverty Lowest in U.S. History, *Amber Waves*, September 2005, USDA Economic Research Service, available at ers.usda.gov/AmberWaves/scripts/print.asp?page=/September05/Features/.

25. America's Last Family Farms? Turning Point Project, n.d.

26. U.S. Department of Agriculture, *Agriculture Fact Book 2001–2002* (Washington, D.C.: U.S. Government Printing Office, 2003), p. 35, available at usda.gov/factbook/2002factbook.pdf.

27. D. E. Banker and J. M. MacDonald, eds., *Structural and Financial Characteristics of U.S. Farms: 2004 Family Farm Report*, USDA Agriculture Information Bulletin 797, 2005; B. Gardner, The Little Guys Are O.K., *New York Times*, March 7, 2005, p. A17, available at nytimes.com/2005/03/07/opinion/07gardner.html?_r=1&oref=slogin.

28. R. Strickland, *Farm Income and Costs: Multiple Well-Being*, USDA Economic Research Service, 2002.

29. Hoppe and Banker, Structure and Finances of U.S. Farms, p. 8; Census of Agriculture 2002, U.S. Department of Agriculture, National Agricultural Statistics Service, 2004, available at usda.gov/Newsroom/0063.04.html.

30. T. Egan, Amid Dying Towns of Rural Plains, One Makes a Stand, *New York Times*, December 1, 2003, pp. A1, A18–A19.

31. R. Doyle, Leaving the Farm, *Scientific American*, August 2004, p. 18.

32. Egan, Amid Dying Towns of Rural Plains, One Makes a Stand, pp. 18–19.

33. C. Le Duff, A Farmer Fears His Way of Life Has Dwindled Down to a Final Generation, *New York Times*, October 2, 2006, p. A10; M. Davey, Agricultural Mainstay Get a New, Urban Face, *New York Times*, November 16, 2006, p. 30A.

34. Why Family Farmers Need Help, 2004, available at farmaid.org.

35. R. Doyle, Down on the Farm, *Scientific American*, August 2002, p. 27.

36. S. Cohen, D. Morgan, and L. Stanton, Farm Subsidies over Time, *Washington Post*, July 2, 2006, p. A1.

37. Banker and MacDonald, *Structural and Financial Characteristics of U.S. Farms*; Gardner, The Little Guys Are O.K.

38. B. Riedl, Still at the Federal Trough: Farm Subsidies for the Rich and Famous Shattered Records in 2001, Heritage Foundation, April 30, 2002, available at heritage.org/Research/Agriculture/BG1542.cfm.

39. J. Kelly, Corporations, Agencies Get Lion's Share of Farm Subsidies, *Kansas City Star*, September 9, 2001, available at 2001.kcstar.com/item/pages/home.pat,local/3accf5b7.909.html.

40. S. Cohen, Deceased Farmers Got USDA Payments, *Washington Post*, July 23, 2007, p. A1.

41. Farm Subsidy Database, Environmental Working Group, 2004, available at ewg.org/farm/region.php?fips=00000&yr=2002.

42. A. Barrionuevo, Imports Spurring Push to Subsidize Produce, *New York Times*, December 3, 2006, p. 1.

43. E. Becker, Western Farmers Fear Third-World Challenge to Subsidies, *New York Times*, September 10, 2003, pp. A1, A8; American Monoculture, *Scientific American*, May 2007, p. 34.

44. Farm Subsidy Database, Environmental Working Group, 2007, available at ewg.org/farm/top_recips.php?fips=00000&progcode=totalcons.

45. K. Casa, The Changing Face of Farming in America, *National Catholic Reporter*, February 12, 1999, pp. 13–16.

46. F. Pearce, Crops without Profit, *New Scientist*, December 18, 1999, p. 10.

47. Becker, Western Farmers Fear Third-World Challenge to Subsidies; American Monoculture.

48. A. Mittal, U.S. Farm Bill Turns Ploughshares Back into Weapons, Inter Press Service, September 6, 2002, available at foodfirst.org/archive/media/opeds/2002/mittalfarmbillips .html; United States Dumping on World Agricultural Markets, Institute for Agriculture and Trade Policy, Cancun Series Paper no. 1; Food and Environment: The Costs and Benefits of Industrial Agriculture, 2003, available at ucsusa.org/food_and_environment/biotechnology/page.cfm?pageID=350tradeobservatory.org/library/uploadedfiles/United_States_Dumping_on_World_Agricultural Ma.pdf.

49. U.S. Agricultural Trade Value, USDA Economic Research Service, 2005, available at ers.usda.gov/briefing/baseline/gallery/gallery2005/agtradebalance.gif.

50. L. Rohter, South America: World's New Breadbasket, *International Herald Tribune*, December 13, 2004, pp. 1, 7; J. Wilkins, Think Globally, Eat Locally, *New York Times*, December 18, 2004, p. A19, available at nytimes.com/2004/12/18/opinion/18wilkins.html?th =&pagewanted=print.

51. P. Gumbel, A Spoonful of Reform, *Time Magazine*, July 4, 2005, p. 26.

52. Becker, Western Farmers Fear Third-World Challenge to Subsidies, pp. A1, A8.

53. United States Dumping on World Agricultural Markets; A. K. Chowdhury, The Key to a Breakthrough on Trade and Poverty, *International Herald Tribune*, March 26, 2004, p. 7.

54. Public Citizen, *Down on the Farm: NAFTA's Seven-Years War on Farmers and Ranchers in the U.S., Canada, and Mexico* (Washington, D.C.: Public Citizen, 2001); *Effects of NAFTA on United States Agriculture*, available at tiger.towson.edu/users/mogle1/; S. Vaughn, How Green Is NAFTA? *Environment*, March 2004, pp. 26–42.

55. North American Free Trade Agreement, FASonline, 2001, available at ffas.usda.gov/info/factsheets/NAFTA.asp.

56. The Ten Year Track Record of the North American Free Trade Agreement: U.S., Mexican and Canadian Farmers and Agriculture, Public Citizen, 2004; Effects of NAFTA on United States Agriculture, 2000, available at tiger.towson.edu/users/mogle1/; J. E. Stiglitz, The Broken Promise of NAFTA, *New York Times*, January 6, 2004, p. A23.

57. Top Destinations for U.S. Agricultural Exports, 2006. Amber Waves, available at ers.usda.gov/AmberWaves/February08/Indicators/indicators.htm]].

58. Wilkins, Think Globally, Eat Locally, p. A19.

59. M. Carmichael, Produce: Wash Out, *Newsweek*, December 8, 2003, p. 58; Importing Health Hazards, *U.S. News & World Report*, December 8, 2003, p. 68.

60. L. Rohter, South America: World's New Breadbasket, *International Herald Tribune*, December 13, 2004, pp. 1, 7.

61. Ibid.; T. Petry, Market Advisor: Comparing U.S. and Canadian Cattle Numbers, North Dakota State University, October 16, 2003, available at ext.nodak.edu/extnews/newsrelease/2003/101603/06livstk.htm.

62. E. Becker, Two Farm Acres Lost per Minute, Study Says, *New York Times*, October 4, 2002, p. A22.

63. R. Dahl, Population Equation, *Environmental Health Perspectives*, September 2005, p. A604.

64. Sprawl, *New Scientist*, December 24–31, 2005, p. 72.

65. B. Harden, Anti-Sprawl Laws, Property Rights Collide in Oregon, *Washington Post*, February 28, 2005, p. A1.

66. Farmland Preservation Issues, Van Buren County Community Center [Michigan], n.d., available at vbco.org/planningeduc0006.asp; W. Yardley, Anger Drives Property Rights Measures, *New York Times*, October 8, 2006, p. 34.

67. C. Runyan, Studies Find Sprawl Exacerbates Drought, Threatens Farmland, *Worldwatch*, January–February 2003, p. 10; Farming on the Edge, American Farmland Trust, 2002.

68. C. D. Elvidge, C. Milesi, J. B. Dietz, B. T. Tuttle, P. C. Sutton, R. Nemani, and J. E. Vogelmann, U.S. Constructed Area Approaches the Size of Ohio, *EOS* (American Geophysical Union), June 15, 2004, p. 223.

69. Trends in U.S. Agriculture: Farm Numbers and Land in Farms.

70. D. Coleman, Farm and Ranch Lands Protection Program, USDA Natural Resources Conservation Service, 2004, available at nrcs.usda.gov/programs/frpp/.

71. L. A. Greene, Accounting for Lost Acreage, *Environmental Health Perspectives*, May 2000, p. A209.

72. Ibid.

73. Runyan, Studies Find Sprawl Exacerbates Drought; E. L. Nizeyimana, G. W. Petersen, and E. D. Warner, Tracking Farmland Loss, *Geotimes*, January 2002, pp. 18–20.

74. J. R., Sprawling over Croplands, *Science News*, March 4, 2000, p. 155.

75. The Nation's Road Capacity: How Fast Is It Growing? available at transact.org/library/roadandmiles.asp; F. Wooldridge, Food and Environment: Part 26—Next Added 100 Million Americans, 2002, available at americanchronicle.com/articles/viewArticle.asp?articleID=27686.

76. Population-Interaction Zones for Agriculture, USDA Economic Research Service, 2005, available at ers.usda.gov/Data/PopulationInteractionZones/.

77. Smog and California Crops, California Air Resources Board, 1991, pp. 2–3.

78. The Farmland Bubble, *New York Times*, December 26, 2003, p. A42; L. Belsie, Farm Subsidies Prop Up Midwest Land Values, *Christian Science Monitor*, January 4, 2002, available at csmonitor.com/2002/0104/p3s1–usec.html.

79. USDA Economic Research Service, Land Use, Value, and Management: Agricultural Land Values, 2005, available at ers.usda.gov/Briefing/LandUse/aglandvaluechapter.htm.

80. J. M. Reilly, ed., *The Potential Consequences of Climate Variability and Climate Change for the United States* (Cambridge: Cambridge University Press, 2001), p. 3.

81. P. Ehrlich, *The Population Bomb* (New York: Ballantine Books, 1968), and P. Ehrlich, *The End of Affluence* (New York: Ballantine Books, 1975).

82. Lester Brown, *Who Will Feed China?* (New York: Norton, 1995) and *Tough Choices: Facing the Challenge of Food Scarcity* (New York: Norton, 1996); L. Yingling, China Releases Latest Census Results, Worldwatch Institute, 2006, available at worldwatch.org/features/chinawatch/stories/20060321-1.

83. M. O. Sheehan, Population Growth Slows, in *Vital Signs 2003*, ed. L. Starke (New York: Norton, 2003), p. 66.

84. World Agriculture: Towards 2015/2030, UN Food and Agriculture Organization, 2002, available at fao.org/docrep/004/y3557e/y3557e00.htm.

85. Reilly, *The Potential Consequences of Climate Variability and Climate Change for the United States*, pp. 12, 110–111.

Chapter Two

1. T. Jones, The Scoop on Dirt, *E Magazine*, September–October 2006, pp. 27–39.

2. *Soil and Water Quality: An Agenda for Agriculture* (Washington, D.C.: National Academy Press, p. 1.

3. P. Blaikie and H. Brookfield, *Land Degradation and Society* (London: Methuen, 1987), p. 91; D. Pimental, Soil Erosion, *Environment*, December 1997, p. 4.

4. U.S. Fertilizer Use, Fertilizer Institute, 2003, available at tfi.org/factsandstats/statistics.cfm.

5. W. Cohen, United States Deep in Manure, *U.S. News & World Report*, January 12, 1998, p. 46.

6. Schiffman, S. S., E. A. S. Miller, and M. S. Suggs, The Effect of Environmental Odors Emanating from Commercial Swine Operations on the Mood of Nearby Residents. Brain Research Bulletin, v. 37, pp. 369–375, 1995; R. Meadows, Livestock Legacy, *Environmental Health Perspectives*, December 1995, pp. 1096–1100; J. Mason, Fowling the Waters, *E Magazine*, September–October 1995, p. 33.

7. F. R. Troeh and L. M. Thompson, *Soils and Soil Fertility*, 5th ed. (New York: Oxford University Press, 1993), p. 186.

8. Fruit and Veg Not What It Was, *Ecologist*, April 2004, p. 8; A. Herro, Crop Yields Expand, But Nutrition Is Left Behind, Worldwatch Institute, 2007, available at worldwatch.org/node/5339/print; A. Halweil, Still No Free Lunch: Nutrient Levels in U.S. Food Supply Eroded by Pursuit of High Yields, Organic Center, 2007.

9. N. Nosengo, Fertilized to Death, *Nature*, October 30, 2003, pp. 894–895.

10. L. R. Brown, Eradicating Hunger: A Growing Challenge, in *State of the World 2001*, ed. L. Starke (New York: Norton, 2001), p. 50; L. R. Brown, Struggling to Raise Cropland Productivity, in *State of the World 1998*, ed. L. Starke (New York: Norton, 1998), pp. 79–95.

11. P. Blaikie and H. Brookfield, *Land Degradation and Society* (London: Methuen, 1987); D. Pimental, Soil Erosion, *Environment*, December 1997, p. 4.

12. Soilhealth.com, 2001, available at agric.uwa.edu.au/soils/soilhealth/animals/.

13. M. B. Bouche and F. Al-Addan, Earthworms, Water Infiltration and Soil Stability: Some New Assessments, *Soil Biology and Biochemistry* 29:3–4 (1997): 441.

14. D. L. Hey, Nitrogen Farming: Harvesting a Different Crop, *Restoration Ecology*, March 2002, pp. 1–10; F. Pearce, Planet Earth Is Drowning in Nitrogen, *New Scientist*, April 12, 1997, p. 10.

15. Lifeless Oceans, *New Scientist*, April 3, 2004, p. 7; J. Raloff, Limiting Dead Zones, *Science News*, June 12, 2004, pp. 378–380, and Dead Waters, *Science News*, June 5, 2004, pp. 360–362; R. W. Howarth, Hypoxia, Fertilizer, and the Gulf of Mexico, *Science*, May 25, 2001, pp. 1485–1486; S. Joyce, The Dead Zones: Oxygen-Starved Coastal Waters, *Environmental Health Perspectives*, March 2000, p. A120; C. Furniss, Sea Change, *Geographical*, February 2007, p. 54.

16. S. Fields, Global Nitrogen Cycling out of Control, *Environmental Health Perspectives*, July 2004, p. A561; Ratloff, Dead Waters pp. 360–362.

17. J. N. Galloway et al., The Nitrogen Cascade, *Bioscience* 53 (2003): 341–356.

18. In Brief, *Environment*, June 2004, p. 6; M. Schrope, The Dead Zones, *New Scientist*, December 9, 2006, p. 39; N. Parks, Dawn of the Dead Zones, October 26, 2006, available at sciencenow.sciencemag.org/cgi/content/full/2006/1026/1.

19. S. Simpson, Shrinking the Dead Zone, *Scientific American*, July 2001, pp. 10–11.

20. Organic Trade Association newsletter, October–November 2001, available at ota.com/pics/documents/19.pdf.

21. D. Tenenbaum, The Beauty of Biosolids, *Environmental Health Perspectives*, January 1997, p. 33.

22. D. Kulling, F. X. Stadelmann, and U. Herter, 2002, Sewage Sludge: Fertilizer or Waste? *EAWAG* (Swiss Federal Institute for Science and Technology) *News 53* (2002): 9, available at eawag.ch/publications/eawagnews/www_en53/en53e_screen/en53e_stadelm_s.pdf.

23. Tenenbaum, The Beauty of Biosolids, p. 34; D. MacKenzie, Waste Not, *New Scientist*, August 29, 1998, p. 30; R. Renner, Is Sludge Safe? *Environmental Health Perspectives*, November 2002, p. A667.

24. M. Barnett, Making a Stink, *U.S. News & World Report*, August 5, 2002, pp. 48–50.

25. University of California IPM Online, 2002, available at ipm.ucdavis.edu/PMG/selectnewpest.small-grains.html.

26. G. Munkvold, Disease Resistance and Crop Rotation, Iowa State University, 1996, available at ipm.iastate.edu/ipm/icm/1996/1–26–1996/disres.html.

27. *The Quality of Our Nation's Waters: Nutrients and Pesticides*, U.S. Geological Survey Circular 1225, 1999, p. 28; D. A. Pfeiffer, Eating Fossil Fuels, Wilderness Publications, 2004, available at fromthewilderness.com/free/ww3/100_303_eating_oil.html; *2000–2001 Pesticide Market Estimates: Sales*, Environmental Protection Agency, available at epa.gov/oppbead1/pestsales/01pestsales/sales2001.html; A. McCauley, *Environmental Impacts of Industrial Farming*, Oxfam America, 2006, available at oxfamamerica.org/whatwedo/where_we_work/united_states/news_publications/food_farm/art2565.html.

28. K. Casa, The Changing Face of Farming in America, *National Catholic Reporter*, February 12, 1999, pp. 13–16.

29. M. Yudelman, A. Ratta, and D. Nygaard, *Pest Management and Food Production: Looking to the Future*, International Food Policy Research Institute: Food, Agriculture, and the Environment, Discussion Paper 25, 1998.

30. Pesticide Data Program, Annual Summary Calendar Year 2005, USDA, 2006, available at ams.usda.gov/science/pdp.

31. Pesticides in Produce, Environmental Working Group, 2006, available at foodnews.org/walletguide.php.

32. J. R. Barrett, More Concerns for Farmers, *Environmental Health Perspectives*, July 2005, p. A472; E. Hood, Bringing Home More Than a Paycheck, *Environmental Health Perspectives*, December 2002, p. A765; J. Ritter, Pesticide Battles on the Rise in USA, *USA Today*, April 12, 2005, p. 1A; J. Phelps, Neurobehavioral Deficits in Children from Agricultural Communities, *Environmental Health Perspectives*, March 2006, p. A159.

33. Understanding Cancer Series: Cancer and the Environment, National Cancer Institute, 2006, available at cancer.gov/cancertopics/understandingcancer/environment/Slide28/print?page.

34. W. E. Larson and V. B. Cardwell, Current and Emerging Issues [in corn production], n.d., available at deal.unl.edu/cornpro/html/current/current.html.

35. P. Thomas, Sex, Lies and Herbicides, *Ecologist*, February 2006, pp. 14–20.

36. T. Riley, Chemical Consequences, *E Magazine*, July–August 2006, p. 12.

37. In Brief, *Environment*, September 2001, p. 8; Pesticide Data Program, Annual Summary Calendar Year, p. 7; P. S. Muir, Agriculture: Pesticides, Oregon State University, 1998, available at oregonstate.edu/%7Emuirp/agpestic.htm.

38. Yudelman, Ratta, and Nygaard, *Pest Management and Food Production*, p. 7.

39. D. Marty, Getting on Our Nerves, *E Magazine*, January–February 2002, pp. 40–41; J. Raloff, Pesticides May Challenge Human Immunity, *Science News*, March 9, 1996, p. 149; L. Hardell and M. Eriksson, Is the Decline in the Incidence of Non-Hodgkin's Lymphoma in Sweden and Other Countries a Result of Cancer Preventative Measures? *Environmental Health Perspectives*, November 2003, p. 14; O. Izakson, Farming Infertility, *E Magazine*, January–February 2004, pp. 40–41; M. Bremner, Committing Pesticide, *Ecologist*, December 2002–January 2003, pp. 13–18; Garden Cancer, *New Scientist*, March 11, 1995, p. 13; New Studies Challenge Pesticide Safety, *Ecologist*, April 2006, p. 9; T. P. Brown et al., Pesticides and Parkinson's Disease—Is There a Link? *Environmental Health Perspectives*, February 2006, p. 156.

40. M. Day, Dipping into Danger, *New Scientist*, February 21, 1998, p. 5.

41. Environmental Protection Agency, Organophosphate Pesticides in Food—A Primer on Reassessment of Residue Limits, 2005, available at epa.gov/pesticides/op/primer.htm.

42. My Grandma, What Toxic Genes You Have, *Ecologist*, July–August 2005, p. 13.

43. P. S. Muir, Agriculture: Pesticides, 1998, available at oregonstate.edu/%7Emuirp/whycare.htm.

44. R. J. Gilliom et al., *Pesticides in the Nation's Streams and Groundwater, 1992–2001*, U.S. Geological Survey Circular 1291, 2005.

45. Yudelman, Ratta, and Nygaard, *Pest Management and Food Production*, p. 10.

46. L. Horrigan, R. S. Lawrence, and P. Walker, How Sustainable Agriculture Can Address the Environmental and Human Health Harms of Industrial Agriculture, *Environmental Health Perspectives*, May 2002, pp. 445–456.

47. Health Effects of Air and Water Pollution, Natural Resources Defense Council, 1998, available at nrdc.org/health/effects/qendoc.asp.

48. V. Howard, Synergistic Effects of Chemical Mixtures—Can We Rely on Traditional Toxicology? *Ecologist*, September–October 1997, pp. 192–195.

49. C. Cox and M. Surgan, Unidentified Inert Ingredients in Pesticides: Implications for Human and Environmental Health, *Environmental Health Perspectives*, December 2006, pp. 1803–1806; D. J. Epstein, Secret Ingredients, *Scientific American*, August 2003, p. 12; S. Marquardt, C. Cox, and H. Knight, Inert Ingredients in Pesticides, *Global Pesticide Campaigner* 8, September 1998, available at panna.org/resources/pestis/PESTIS980925.2.html.

50. G. Lyons, Endocrine Disrupting Pesticides, *Pesticides News*, no. 46, December 1999, pp. 16–19, available at pan-uk.org/pestnews/actives/endocrin.htm.

51. Z. Goldsmith, Legalized, Random Genocide, *Ecologist*, January–February 1998, p. 3.

52. B. Harder, Class Acts from New Pesticides, *Science News*, September 3, 2005, p. 149.

53. Yudelman, Ratta, and Nygaard, *Pest Management and Food Production*, p. 13.

54. In Brief, *Environment*, September 2001, p. 8.

55. Yudelman, Ratta, and Nygaard, *Pest Management and Food Production*, p. 13.

56. Ibid.

57. B. Halweil, Pesticide-Resistant Species Flourish, in *Vital Signs 1999*, ed. L. Starke (New York: Norton, 1999), pp. 124–125.

58. Yudelman, Ratta, and Nygaard, *Pest Management and Food Production*, p. 15.

59. O. Tickell, Against the Grain, *New Scientist*, November 28, 1998, p. 20.

60. Bugs and Beetles: Coming to a Plate Near You? Independent Online, available at iol.co.za/general/news/newsprint.php?art_id=qw1099914302936B255&sf=; T. Abell, The Six-Legged World of Protein, *Richmond Review*, 2004, available at richmondreview.com/portals-c.../list.cgi?paper=45&cat=23&id=243921&more.

61. P. Muir, Genetic Resistance to Pesticides, Oregon State University, n.d., available at oregonstate.edu/~muirp/resistan.htm.

62. P. Muir, Why Not?, Oregon State University, n.d., available at oregonstate.edu/~muirp/whynot.htm.

63. E. Samuel, How Insect Pest Stays Ahead of the Game, *New Scientist*, October 19, 2002, p. 17.

64. Calling All Wasps, *Smithsonian*, March 2006, p. 18.

65. G. T. Miller, Jr., *Living in the Environment*, 11th ed. (Pacific Grove, Calif.: Brooks/Cole, 2000).

66. E. Stokstad, The Case of the Empty Hives, *Science*, May 18, 2007, pp. 970–972; J. Eilperin, Pollinators' Decline Called Threat to Crops, *Washington Post*, October 19, 2006, p. A10; Re-Bugging Our Environment Equals Big Bucks for the Economy, available at mymotherlode.com/News/article/kvml/1161796469; A. Barrionuevo, Honeybees Vanish,

Leaving Keepers in Peril, *New York Times*, February 27, 2007, p. A1; P. Thomas, Give Bees a Chance, *Ecologist*, June 2007, pp. 30–35.

67. In a Nitrogen Fix, Environmental Science and Technology online news, June 13, 2007, available at pubs.acs.org/subscribe/journals/esthag-w/2007/june/science/rr_nitrate.html.

68. F. Wooldridge, Food and Environment: Part 26—Next Added 100 Million Americans, available at americanchronicle.com/articles/viewArticle.asp?articleID=27686; D. A. Pfeiffer, Eating Fossil Fuels, 2004, available at fromthewilderness.com/free/ww3/100303_eating_oil.html.

69. D. Murray, Oil and Food: A Rising Security Challenge, Earth Policy Institute, 2005, available at earth-policy.org/Updates/2005/Update48_printable.htm.

70. D. Pimental and M. Giampietro, Food, Land, Population and the U.S. Economy, 1994, available at dieoff.com/page40.htm; Pfeiffer, Eating Fossil Fuels.

71. J. Woodward and I. Foster, Erosion and Suspended Sediment Transfer in River Catchments, *Geography* 82:4 (1997): 357.

72. D. Pimental et al., Environmental and Economic Costs of Soil Erosion and Conservation Benefits, *Science*, February 24, 1995, p. 1117.

73. Ibid., pp. 1117–1123; J. Glanz, Erosion Study Finds High Price for Forgotten Menace, *Science*, February 24, 1995, p. 1088; P. Crosson, Soil Erosion Estimates and Costs, *Science*, July 28, 1995, pp. 461–464, and reply by Pimental et al., Environmental and Economic Costs of Soil Erosion and Conservation Benefits, pp. 464–465.

74. C. den Biggelaar, R. Lal, K. Wiebe, and V. Breneman, Impact of Soil Erosion on Crop Yields in North America, in *Advances in Agronomy*, ed. D. L. Sparks, vol. 72 (Orlando, Fla.: Academic Press, 2001), p. 6.

75. Seafriends—Soil Erosion and Conservation, 2000, available at seafriends.org.nz/enviro/soil/erosion.htm.

76. J. C. Knox, Agricultural Influence on Landscape Sensitivity in the Upper Mississippi Valley, *Catena* 42 (2001): 193–224.

77. H. Blatt, *Our Geologic Environment* (Upper Saddle River, N.J.: Prentice Hall, 1997), p. 67.

78. W. Berry, *The Unsettling of America* (San Francisco: Sierra Club Books, 1997), p. 11.

79. J. Kaiser, Wounding Earth's Fragile Skin, *Science*, June 11, 2004, p. 1618.

80. Pimental et al., Environmental and Economic Costs of Soil Erosion and Conservation Benefits, p. 1117.

81. R. P. Stone and N. Moore, Control of Soil Erosion, Ontario Ministry of Agriculture and Food, 1996, available at omafra.gov.on.ca/english/engineer/facts/95–089.htm.

82. E. C. Dickey, D. B. Shelton, and P. J. Jasa, NebGuide, Residue Management for Soil Erosion Control, 1997, Institute of Agriculture and Natural Resources, available at ianr.unl.edu/pubs/fieldcrops/g544.htm.

83. Ibid.; Coombs, A. The Dirty Truth About Plowing. Science NOW Daily News, August 7, 2007. sciencenow.sciencemag.org/cgi/content/full/2007/807/2.

84. Soil Erosion in Agricultural Systems, n.d., available at msu.edu/user/dunnjef1/rd491/soile.htm.

85. P. Prada, Use of No-Till Farming Spreads as Crops Thrive, *International Herald Tribune*, September 30, 2005, p. 20.

86. C. Karasov, Spare the Plow, Save the Soil, *Environmental Health Perspectives*, February 2002, p. A75.

87. *Global Warming and Agriculture: Fossil Fuel*, Soil Conservation Council of Canada, Factsheet vol. 1, no. 3, 2001, available at soilcc.ca/fact/pdfs/Factsheet%203%20–fossil %20fuel.pdf.

88. E. G. Gregorich et al., Changes in Soil Organic Matter, Agriculture and Agri-Food Canada, available at agr.gc.ca/nlwis-snite/index_e.cfm?s1=pub&s2=hs_ss&page=11; University of Minnesota Extension, Organic Matter Management, available at extension.umn.edu/distribution/cropsystems/DC7402.html.

89. J. Kaiser, Wounding Earth's Fragile Skin, *Science*, June 11, 2004, p. 1118.

90. P. Muir, Conservation Tillage Systems, Oregon State University, 2002, available at oregonstate.edu/%7Emuirp/constill.htm.

91. Pimental et al., Environmental and Economic Costs of Soil Erosion and Conservation Benefits, p. 1121.

92. R. Fee, No-Till Rebounds, Successful Farming, December 2000, available at agriculture .com/sfonline/sf/2000/december/0013notill.html.

Chapter Three

1. T. Hayden, Could the Grass Be Greener? *U.S. News and World Report*, May 16, 2005, p. 57.

2. M. Rajiv, The Soy Health Guide, *Hindu*, July 17, 2002, available at hindu.com/thehindu/mp/2002/07/17/stories/2002071700340400.htm; K. Beardsmore, Lose Weight with Soy, *eMaxHealth*, June 14, 2004, available at emaxhealth.com/11/414.html.

3. *United States Wheat Production*, Western Organization of Resource Councils fact sheet, October 2002, available at worc.org/pdfs/WORCproductionfactsheet.pdf.

4. U.S. Soybean Total Usage, 2004, available at cbot.com/cbot/pub/static/files/s_uscons.gif; A. Barrionuevo, China's Appetite Stirs a Shift in Global Trade, *International Herald Tribune*, April 6, 2007, pp. 9, 11.

5. Soybean: World Supply and Distribution, 2006, available at fas.usda.gov/psd/complete _tables/OIL-table2–24.htm; World Exports of Soybeans, ASA International Marketing, 2005, available at Asa-europe.org.

6. Soybeans and Oil Crops: Background, USDA Economic Research Service, 2002, available at ers.usda.gov/Briefing/SoybeansOilCrops/background.htm; U.S. Fertilizer Use and Price, USDA Economic Research Service, 2005, Table 21, available at ers.usda.gov/Data/FertilizerUse/.

7. McCay's Miracle Loaf, *Mother Earth News*, September–October 1981, p. 10.

8. Lisa Dennis, personal communication, November 29, 2004.

9. U.S. Soybean Use by Livestock, 2004, Syngenta, available at soystats.com/2004/page_20 .htm.

10. Statistical Conversions, ASA International Marketing, 2005, available at asasoya.org.

11. Welcome to Soy Stats 2005, ASA International Marketing, 2005, available at soystats .com/2005/page_02.htm.

12. J. Hill et al., Environmental, Economic, and Energetic Costs and Benefits of Biodiesel and Ethanol Biofuels, in *Proceedings of the National Academy of Sciences*, July 25, 2006, pp. 11206–11210.

13. R. Zelesky, Integrating Biofuels into the Fuel Supply, *Geotimes*, March 2007, p. 31.

14. A. Barrionuevo, Crop Rotation in the Grain Belt, *New York Times*, September 16, 2006, pp. C1, C9.

15. World of Corn, 2007, available at ncga.com/WorldOfCorn/main/production1.asp.

16. Corn Subsidies in the United States, Wheat Subsidies in the United States, Soybean Subsidies in the United States, Rice Subsidies in the United States, Environmental Working Group, 2006.

17. H. Schoonover and M. Muller, Food without Thought: How U.S. Farm Policy Contributes to Obesity, Institute for Agriculture and Trade Policy, 2006.

18. Agriculture of America, fundus.org, n.d., available at fundus.org/pdf.asp?ID=3686.

19. Maize, 2001, available at nationalimporters.com/history/history.asp?articleid=16.

20. Corn Trivia, n.d. CyberSpace Farm, available at cyberspaceag.com/kansascrops/corn/ corntrivia.htm; H. Schoonover and M. Muller, Food without Thought: How U.S. Farm Policy Contributes to Obesity, Institute for Agriculture and Trade Policy, 2006, available at iatp.org/ iatp/publications.cfm?accountID=258&refID=80627.

21. Educational Information, 2002, available at iowacorn.org.

22. Iowa Corn, 2006, available at iowacorn.org/cornuse/cornuse_20.html.

23. M. MacLean, When Corn Is King, *Christian Science Monitor*, October 31, 2002, available at csmonitor.com/2002/1031/p17s01–lihc.html.

24. R. Schubert, The Loss of Corn Exports to Europe: Something to Chew On at the Commodity Classic, *Crop Choice*, March 5, 2004, available at mindfully.org/GE/2004/ Corn-Exports-Loss5mar04.htm.

25. L. R. Brown, Supermarkets and Service Stations Now Competing for Grain, Renewable Energy Access, 2006, available at renewableenergyaccess.com/rea/news/story?id=45441.

26. D. Morrison, Corn Ethanol Yields an Energy Dividend But Gains Are Higher with Soy Biodiesel, a New Study Shows, Renewable Energy Access, 2006, available at renewableenergy access.com/rea/news/story?id=45457; M. Lavelle and B. Schulte, Is Ethanol the Answer? *U.S. News & World Report*, February 12, 2007, p. 34; Westcott, P. C., U.S. Ethanol Expansion Driving Changes Throughout the Agricultural Sector. *Amber Waves*, September 2007.

27. Brown, Supermarkets and Service Stations Now Competing for Grain.

28. Morrison, Corn Ethanol Yields an Energy Dividend; Lavelle and Schulte, Is Ethanol the Answer? p. 34; *Amber Waves*.

29. A. Barrionuevo, Boom in Ethanol Reshapes Economy of Heartland, *New York Times*, June 25, 2006, pp. A1, A18.

30. F. Pearce, Fuels Gold, *New Scientist*, September 23, 2006, p. 38.

31. D. Lorenz and D. Morris, How Much Energy Does It Take to Make a Gallon of Ethanol? Institute for Local-Self Reliance, 1995.

32. T. Philpott, Archer Daniels Midland: The Exxon of Corn? 2006, available at gristmill .grist.org/story/2006/2/2/52324/18981.

33. R. Hasan, The Ethanol Investment Craze, Renewable Energy Access, 2006, available at renewableenergyaccess.com/rea/news/story?id=45231.

34. Lorenz and Morris, How Much Energy Does It Take to Make a Gallon of Ethanol?.

35. MacLean, When Corn Is King.

36. U.S. Fertilizer Use and Price, USDA Economic Research Service, Table 9, 2005, available at ers.usda.gov/Data/FertilizerUse/.

37. D. Pimental, Changing Genes to Feed the World, *Science*, October 29, 2004, p. 815; History of U.S. Corn Production, n.d., available at deal.unl.edu/cornpro/html/history/history .html.

38. H. Blatt, *America's Environmental Report Card: Are We Making the Grade?* (Cambridge, Mass.: MIT Press, 2005), pp. 22–23.

39. Utilization of Wheat Other Than Durum, 1997, Alberta Farm Business Initiative, available at members.shaw.ca/bethcandlish/util.htm.

40. U.S. Fertilizer Use and Price, USDA Economic Research Service, Table 27.

41. Wheat: Bread Trivia, n.d., CyberSpace Farm, available at cyberspaceag.com/kansascrops/ wheat/breadtrivia.htm.

42. C. Uauy et al., A NAC Gene Regulating Senescence Improves Grain Protein, Zinc, and Iron Content in Wheat, *Science*, November 24, 2006, pp. 1298–1301.

Chapter Four

1. C. Dimitri and C. Greene, Recent Growth Patterns in the U.S. Organic Foods Market, USDA Economic Research Service, Agriculture Information Bulletin 777, 2002, p. iii.

2. Organic Product Sales Show Strong Growth, According to Manufacturer Survey, Organic Trade Association, 2004, axcessnews.com/environmental_050304.shtml.

3. Three Percent of U.S. Grocery Food and Beverages Now Organic, Organic Consumers Association, 2007, available at organicconsumers.org/articles/article_5371.cfm.

4. L. Oberholtzer, C. Dimitri, and C. Greene, Price Premiums Hold on as U.S. Organic Produce Market Expands, USDA Economic Research Service, 2005, p. 4.

5. Dimitri and Greene, Recent Growth Patterns in the U.S. Organic Foods Market, p. 1.

6. B. Halweil, Organic Gold Rush, *Worldwatch*, May–June, 2001, p. 27; Wal-Mart Goes Organic, 2006, available at msnbc.msn.com/id/11977666/page/2/.

7. Three Percent of U.S. Grocery Food and Beverages Now Organic; OTA Survey: U.S. Organic Sales Reach $10.8 Billion, *Organic Trade Association Information Newsletter 28*, 2004, available at ota.com/pics/documents/WhatsNews28.pdf.

8. Dimitri and Greene, Recent Growth Patterns in the U.S. Organic Foods Market, p. 2.

9. Organic Farming and Marketing: Questions and Answers, USDA Briefing Room, 2003, available at ers.usda.gov/briefing/Organic/Questions/orgqa5.htm.

10. Survey: Organic Foods Gaining in Popularity, *Austin Business Journal*, 2004, available at austin.bizjournals.com/austin/stories/2004/10/18/daily39.html.

11. Three Percent of U.S. Grocery Food and Beverages Now Organic; OTA Survey.

12. A. Hopfensperger, Organic Foods Continue to Grow in Popularity According to Whole Foods Market Survey, 2004, Whole Foods Market, available at wholefoodsmarket.com/pressroom/pr_10-21-04.html; OTA Survey; Food-Safety Concerns Causing U.S. Organic Food Sales to Grow, National Association of Convenience Stores, August 20, 2004, available at nacsonline.com/NACS/News/Daily_News_Archives/August2004/nd0820045.htm.

13. U.S. Organic Sales Top $10.8 Billion, 2004, available at sustainablebusiness.com/features/feature_template.cfm?ID=1128.

14. Dimitri and Greene, Recent Growth Patterns in the U.S. Organic Foods Market, p. 1.

15. A. Hopfensperger, Organic Foods Continue to Grow in Popularity According to Whole Foods Market Survey. 2004, available at wholefoods.com/cgi-bin/print1Opt.cgi?url=company/pr 10-21-04.html.

16. K. M. Mangan, Perks for Pets, *E Magazine*, November–December 2004, p. 56.

17. American Veterinary Medical Association, *U.S. Pet Ownership and Demographics Sourcebook, 2002* (Schaumburg, Ill.: American Veterinary Medical Association, 2002); Small Is "Big," *The Old Farmer's Almanac* (Dublin, N.H., 2006), p. 11.

18. H. Hubscher and D. Hubscher, Ogopogo Worm Farm, n.d., available at members.shaw.ca/ogopogowormfarm/.

19. M. T. Moore, Busy Suburban Farmers Markets Helping Growers Stay in Business, *USA Today*, April 8, 2005, p. 4A; J. Wilkins, Think Globally, Eat Locally, *New York Times*, December 18, 2004, p. A19; Farmers Market Facts, USDA, 2007, available at ams.usda.gov/farmersmarkets/facts.htm.

20. Moore, Busy Suburban Markets Helping Growers Stay in Business, p. 4A.

21. Community-Supported Agriculture, 2004, available at en.wikipedia.org/wiki/Community-supported_agriculture; R. Van En, L. Manes, and C. Roth, What Is Community Supported Agriculture and How Does It Work? University of Massachusetts Extension, 2000, available at umassvegetable.org/food_farming_systems/csa/index.html.

22. Dimitri and Greene, Recent Growth Patterns in the U.S. Organic Foods Market, p. 3.

23. Organic Production, USDA Economic Research Service, 2006, available at ers.usda.gov/Data/Organic/.

24. U.S. Organic, Information—Consumption, USDA, 2003, available at usembassy.org.uk/fas/us_organic_consu.htm.

25. Ibid.; C. Greene and A. Kremen, *U.S. Organic Farming in 2000–2001: Adoption of Certified Systems*, Agriculture Information Bulletin 780, U.S. Department of Agriculture, Economic Research Service, 2003, p. 17; U.S. Organic Information—Consumption, USDA, 2003, p. 17.

26. L. Oberholtzer, C. Dimitri, and C. Greene, Price Premiums Hold on as U.S. Organic Produce Market Expands, USDA Economic Research Service, 2005, p. 9.

27. Ibid.

28. Iowa City Bans Non Organic and GMO Foods, 2003, available at organicconsumers .org/organic/iowa042903.cfm; no author given, Iowa City Bans Non Organic and GMO Foods Organic Consumers Association, 2003. available at organicconsumers.org/organic/ iowa042903.cfm

29. H. Willer and M. Yussefi, *The World of Organic Agriculture: Statistics and Emerging Trends* (Bonn, Germany: International Federation of Organic Agriculture Movements, 2005), p. 14; M. Warner, Wal-Mart Eyes Organic Food, *New York Times*, May 12, 2006, pp. A1, C4.

30. Willer and Yussefi, *The World of Organic Agriculture: Statistics and Emerging Trends 200*; H. Warwick, Cuba's Organic Revolution, *Ecologist*, November–December, 1999, pp. 457–460; E. E. Dooley, Organic Farming Flourishes in Cuba, *Environmental Health Perspectives*, December 2000, p. A551; Cuba: Running on (Almost) Empty, *Ecologist*, October 2005, p. 52.

31. Organic Centre Wales, 2003, available at organic.aber.ac.uk/statistics/index.shtml; V. Gewin, Organic: Is It the Future of Farming? *Nature*, April 22, 2004, pp. 792–798.

32. Willer and Yussefi, *The World of Organic Agriculture: Statistics and Emerging Trends*, p. 15.

33. C. Dimitri and L. Oberholtzer, Market-Led Versus Government-Facilitated Growth, Development of the U.S. and EU Organic Agricultural Sectors, USDA Economic Research Service, 2005.

34. What Is Organic Farming? Soil Association, n.d., available at soilassociation.org/web/sa/ saweb.nsf/Living/whatisorganic.html; Everything I Need to Know About Organic Foods, World's Healthiest Foods, 2004, available at whfoods.org/organics.php.

35. *A Better Road to Hoe: The Economic, Environmental, and Social Impact of Sustainable Agriculture* (St. Paul, Minn.: Northwest Area Foundation, 1994), pp. 14–15.

36. USA: Americans Hunger for Healthy Options as Organic Foods go Mainstream, Organic Consumers Association, 2002, available at organicconsumers.org/organic/121202_organic .cfm.

37. B. Hileman, After Long Struggle, Standards for Organic Food Are Finalized, *Chemical and Engineering News*, January 8, 2001, p. 24.

38. *Organic Certification Process*, Midwest Organic and Sustainable Education Service Organic Fact Sheet 405, n.d., available at mosesorganic.org/factsheets/certprocess.pdf; Organic Certification Process, Clemson University, n.d., available at fscs.clemson.edu/Organic/Process .pdf.

39. Survey: Organic Foods Gaining in Popularity, 2004, *Austin Business Journal*, available at austin.bizjournals.com/austin/stories/2004/10/18/daily39.html.

40. R. Kortbech-Olesen, The United States Market for Organic Food and Beverages, World Trade Organization, 2002, available at intracen.org/Organics/documents/us-market.pdf.

41. G. Kuepper and L. Gegner, Organic Crop Production Overview, National Sustainable Agriculture Information Service [ATTRA], 2004, p. 4, available at attra.ncat.org/attra-pub/ organiccrop.html.

42. M. Burros, Is Organic Food Provably Better? *New York Times*, July 16, 2003, pp. F1, F5; S. Squires, Advantages of Organic Food Still Unproven, *Houston Chronicle*, July 7,

2004, available at chron.com/CDA/archives/archive.mpl?id=2004_3779329; E. Nagourney, Nutrition: Another Benefit Is Seen in Buying Organic Produce, *New York Times*, July 17, 2007.

43. R. Edwards, The Natural Choice, *New Scientist*, March 16, 2002, p. 10.

44. S. Deneen, Food Fight, *E Magazine*, July–August 2003, p. 28; Everything I Need to Know about Organic Foods, Whole Foods Market, 2005, available at whfoods.com/genpage .php?tname=faq&dbid=17.

45. B. Halweil, Organic Produce Found to be Higher in Health-Promoting Compounds, *Worldwatch*, July–August, 2003, p. 9; Better for You, *Ecologist*, September, 2007, p. 9.

46. Deneen, Food Fight, p. 28; B. Halweil, Organic Produce Found to Be Higher in Health-promoting Compounds, *Worldwatch*, July–August, 2003, p. 9.

47. Why Certified Organic Food Is Better Food, n.d., available at mofga.org/food.html.

48. Report Card: Pesticides in Produce, Environmental Working Group, 2004, available at ewg.org/sites/foodnews/walletguide.php; Consumers Union in Action, Consumers Union, 2002, available at consumerreports.org/main/content/aboutus.jsp?FOLDER%3C %3Efolder_...; B. P. Baker, C. M. Benbrook, E. Groth III, and K. L. Benbrook, *Food Additives and Contaminants* 19, May, 2002, pp. 427–446, available at consumersunion.org/ food/organicsumm.htm; How Safe Is Our Produce? Consumers Union, 1999, available at consumerreports.org/main/detail.jsp?CONTENT%3C%3Ecnt_id=19227...; Available at Greener Greens? *Consumer Reports*, January 1998, pp. 12–16.

49. How Safe Is Our Produce?

50. Shoppers Guide to Pesticides in Produce, Environmental Working Group, 2004, available at foodnews.org/walletguide.php; Report Card: Pesticides in Produce, Environmental Working Group, n.d., available at ewg.org/sites/foodnews/walletguide.php.

51. Greener Greens? *Consumer Reports*, January 1998, p. 14.

52. C. M. Cropper, Does It Pay to Buy Organic? *Business Week Online*, September 6, 2004, available at businessweek.com/magazine/content/04_36/b3898129_mz070.htm.

53. C. M. Cropper, Does It Pay to Buy Organic? Business Week Online, September 6, 2004, available at businessweek.com/print/magazine/content/04_36/b3898129_mz070.htm?.

54. J. Fetto, Home on the Organic Range: Consumer Trends Help Save a Fading Piece of America, *American Demographics*, August 1999.

55. R. Kortbech-Olesen, The United States Market for Organic Food and Beverages, 2002, International Trade Center, World Trade Organization, p. 9; 100,000 Organic Farmers in US by 2013, Rodale Institute, 2003, available at strauscom.com/rodale/pkpress03.html.

56. C. MacIlwain, L. Nelson, J. Giles, and V. Gewin, Organic: Is It the Future of Farming? *Nature*, April 22, 2004, pp. 792–798; M. A. Altieri, Agroecology in Action, 2000, University of California at Berkeley, available at cnr.berkeley.edu/~agroeco3/modern_agriculture.html.

57. Why Certified Organic Food Is Better Food, Maine Organic Farmers and Gardeners Association, n.d., available at mofga.org/food.html.

58. M. Yudelman, A. Ratta, and D. Nygaard, *Pest Management and Food Production*, International Food Policy Research Institute, Food, Agriculture, and the Environment Discussion Paper 25, 1998, p. 23, available at ifpri.org/2020/dp/dp25.pdf; J. Raloff, Herbal Herbicides, *Science News*, March 17, 2007, pp. 167–168.

59. A. Coghlan, Breaking the Mould, *New Scientist*, December 5, 1998, p. 16.

60. C. Ohmart, What Is Integrated Pest Management (IPM)? Protected Harvest, 2002, available at protectedharvest.org/learnmore/ipm.htm; Integrated Pest Management, n.d., University of California at Davis, available at veghome.ucdavis.edu/classes/winter2003/plb12/integrated%20pest.html.

61. Coghlan, Breaking the Mould, p. 16.

62. Greener Greens, pp. 14–15.

63. E. Dooley, Protected Harvest, *Environmental Health Perspectives*, May 2002, p. A237; P. S. Muir, Agriculture: Pesticides, Oregon State University, 1998, available at oregonstate.edu/%7Emuirp/agpestic.htm.

64. S. S. Batie, Managing Pesticide Tradeoffs, *Environment*, October 2001, p. 41.

65. Ibid., pp. 40–44.

66. B. Halweil and D. Nierenberg, Watching What We Eat, in *State of the World 2004*, edited by L. Starke (New York: Norton), p. 80.

67. B. Halweil, Organic Gold Rush, *Worldwatch*, May–June, 2001, pp. 22–32.

68. Organic Farming Best for Birds (and Bats and Bees and Plants), *Ecologist*, October 2005, p. 11.

69. B. P. Baker, C. M. Benbrook, E. Groth III, and K. L. Benbrook, Pesticide Residues in Conventional, IPM-Grown, and Organic Foods: Insights from Three U.S. Data Sets, *Food Additives and Contaminants* 19 (May 2002): 427–446; A. Avery, The Deadly Chemicals in Organic Food, *New York Post*, June 2, 2001, p. 15.

70. An Overview of Organic Crop Production, n.d., Appropriate Technology Transfer for Rural Areas (ATTRA), available at attra.org/attra-pub/organiccrop/whatisorg.html.

71. P. Muir, Environmental (Cultural) Pest Controls, 2002, Oregon State University, available at oregonstate.edu/%7Emuirp/envlcont.htm.

72. P. Muir, Diminished Crop Diversity, 2001, Oregon State University, available at oregonstate.edu/%7Emuirp/cropdiv.htm; Out of Stock, *Ecologist*, December 2002–January 2003, pp. 39–41; The Costs and Benefits of Industrial Agriculture, 2001, Union of Concerned Scientists, available at ucsusa.org/food_and_environment/sustainable_food/costs-and-benefits-of-industrial-agriculture.html; R. Edwards, Tomorrow's Bitter Harvest, *New Scientist*, August 17, 1996, pp. 14–15.

73. Variety Selection [Potatoes], n.d., available at Oregonstate.edu/potatoes/variety.htm.

74. History and Origin of the Potato, Sun Spiced, n.d., available at Sunspiced.com/phistory.html.

75. A. Kimbrell, ed., *Fatal Harvest: The Tragedy of Industrial Agriculture* (Washington, D.C.: Island Press, 2002), p. 102.

76. D. Mackenzie, Billions at Risk from Wheat Super-Blight, *New Scientist*, April 7, 2007, pp. 6–7; E. Stokstad, Deadly Wheat Fungus Threatens World's Breadbaskets, *Science*, March 30, 2007, pp. 1786–1787.

77. National Academy of Sciences, *Genetic Vulnerability of Major Crops*.

78. Industrial Agriculture; Features and Policy, Union of Concerned Scientists, 2001, ucsusa.org/food_and_environment/sustainable_food/industrial-agriculture-features-and-policy.html.

79. Out of Stock, *Ecologist*, December 2002, p. 41.

80. Ibid.

81. Ibid.

82. Ibid.

83. Food and Environment, Industrial Agriculture: Features and Policy. Union of Concerned Scientists, n.d.

84. R. Edwards, Tomorrow's Bitter Harvest, *New Scientist*, August 17, 1996, pp. 14–15.

85. M. Sever, Planning for Food after "Doomsday," *Geotimes*, April, 2007, p. 32.

86. A. Graner and A. Borner, Quest for Seed Immortality Is Mission Impossible, *Nature*, July 27, 2006, p. 353.

87. Vive la Difference, *New Scientist*, April 10, 2004, p. 4.

88. L. L. Ching, Organic Agriculture Fights Back, Institute of Science in Society, 2003, available at i-sis.org.uk/OrganicAgriculture.php; B. Halweil, Can Organic Farming Feed Us All? *Worldwatch*, May–June 2006, pp. 18–24; Organic Holds Up, *Worldwatch*, September–October 2007, p. 7; no author given, Low Yields a Myth. Ecologist, September, 2007, p. 11.

89. S. Theil, Let Them Eat Organic, *Newsweek*, August 19, 2002, p. 47.

90. D. Pimental, P. Hepperly, J. Hanson, D. Douds, and R. Seidel, Environmental, Energetic, and Economic Comparisons of Organic and Conventional Farming Systems, *BioScience 55* (2005): 573–582.

91. Organic Farming Does Same for Less, *Ecologist*, September 2005, p. 11; P. Zurer, Organic Farming. *Chemical and Engineering News*, June 3, 2002, p. 8.

92. J. Moyer, Making the Transition to Organic Farming, *The New Farm*, 2002, available at newfarm.org/depts/midatlantic/FactSheets/transition.shtml.

93. O. Tickell, Ploughing for Profits, *New Scientist*, September 5, 1998, p. 13.

94. T. Clunies-Ross, H. Evans, T. Jewell, N. Lampkin, H. de Lange, J. D. van Mansvelt, M. Miller, and M. Redman, Green Fields, Grey Future, Greenpeace International, 1993.

Chapter Five

1. GM Food and Safety, *Ecologist*, April 2003, p. 11.

2. Compilation and Analysis of Public Opinion Polls on Genetically Engineered (GE) Foods, Center for Food Safety, 2002, available at centerforfoodsafety.org; J. Halloran and M. Hansen, Why We Need Labeling of Genetically Engineered Food, 1998, available at consumersunion.org/food/whywenny798.htm.

3. R. Cummins and B. Lilliston, *Genetically Engineered Food: A Self-Defense Guide for Consumers*, 2nd ed. (New York: Marlowe & Co., 2004), pp. 4–5.

4. F. Wu, Perceptions of Food That Are an Ocean Apart, Truth About Trade and Technology, 2004, available at truthabouttrade.org/print.asp?id=1815.

5. Immoral Europe, *Wall Street Journal*, January 13, 2003, p. A10; B. Halweil, The Continuing GMO Quagmire, *Worldwatch*, November–December 2002, p. 2; News Updates—July

Through December 2002, *Transgenic Crops: An Introduction and Resource Guide*, available at cls.casa.colostate.edu/TransgenicCrops/current.html.

6. Far Less Scary Than It Used to Be, *Economist*, July 26, 2003, pp. 23–25; In Brief, *Environment*, July–August 2005, p. 7; GMOs Gain Ground for the Tenth Consecutive Year, available at gmo-compass.org/features/printversion.php?id=194.

7. Global Status of Commercialized Biotech/GM Crops: 2006, International Service for the Acquisition of Agri-Biotech Applications, 2006.

8. J. Fernandez-Cornejo, Adoption of Genetically Engineered Crops in the U.S., USDA Economic Research Service, 2007.

9. M. K. Hansen, Genetic Engineering Is Not an Extension of Conventional Plant Breeding, Consumers Union, 1998, available at consumersunion.org/food/widecpi200.htm.

10. Crop Science: Better Genes without Splicing, Truth about Trade and Technology, 2003, available at truthabouttrade.org/print.asp?id=1962; A. Coghlan, Best of Both Worlds, *New Scientist*, September 20, 1997, p. 17; N. Russell, The Agricultural Impact of Global Climate Change, *Geotimes*, April 2007, p. 33.

11. G. F. Combs, Jr., Case Teaching Notes, Cornell University, available at ublib.buffalo.edu/libraries/projects/cases/rice/rice_notes.html.

12. C. E. Mann, Crop Scientists Seek a New Revolution, *Science*, January 15, 1999, pp. 310–314.

13. Overview of the Process of Plant Genetic Engineering, AgBiosafety, University of Nebraska, 2001, available at agbiosafety.unl.edu/education/summary.htm; Genetically Modified Plants for Food Use, Royal Society, 1998, available at royalsociety.org/page.asp?id=2144.

14. Future Food, *Bionet*, n.d., available at bionetonline.org/English/Content/ff_tool.htm.

15. Ibid.

16. J. Ackerman, Food: How Altered? *National Geographic*, May 2002, pp. 33–50.

17. B. Halweil, The Emperor's New Crops, *Worldwatch*, July–August 1999, p. 22.

18. Ackerman, Food: How Altered? p. 41.

19. C. K. Yoon, On the "Pharm," Silk from Goats, *International Herald Tribune*, May 2, 2000, p. 2.

20. S. Deneen, Food Fight, *E Magazine*, July–August 2003, p. 27.

21. R. Weiss, Seas Yield Surprising Catch of Unknown Genes, *Washington Post*, March 14, 2007, p. A1.

22. Overview of the Process of Plant Genetic Engineering, p. 44.

23. C. Holdrege, Should Genetically Modified Foods Be Labeled? 2006, available at Saynotogmos.org/regulatory_2.htm.

24. B. Hileman, Biotech Wheat's Intense Debate, *Chemical and Engineering News*, January 12, 2004, p. 33.

25. D. Glickman and V. Weber, Frankenfood Is Here to Stay. Let's Talk, *International Herald Tribune*, July 3, 2003, p. 9.

26. E. Ayres, Scientists of 74 Countries Call for Ban on Genetically Engineered Food, *Worldwatch*, November–December 2003, p. 8.

27. *EU Shelves Mostly GM-Free—Greenpeace,* Business Report 2005, available at busrep .co.za/general/print_article.php?fArticleId=2397766&fSectionId=565.

28. B. Hileman, Defeated Wheat, *Chemical and Engineering News,* May 17, 2004, p. 10; A. Pollack, Monsanto Shelves Plan for Modified Wheat, *New York Times,* May 11, 2004, p. C1.

29. Use and Abuse of the Precautionary Principle, Institute of Science in Society, 2000, available at i-sis.org.uk/prec.php.

30. R. F. Service, A Growing Threat Down on the Farm, *Science,* May 25, 2007, pp. 1114–1117; M. Ritter, Herbicide Resistant Weeds and GM Crops, *Green Nature,* January 1, 2004, available at greennature.com.

31. R. Daniels, C. Boffey, R. Mogg, J. Bond, and R. Clarke, The Potential for Dispersal of Herbicide Tolerance Genes from Genetically-Modified, Herbicide-Tolerant Oilseed Rape Crops to Wild Relatives, Winfrith Technology Centre, Dorchester, UK, 2005; Scientists Play Down "Superweed," BBC News, July 25, 2005, available at news.bbc.co.uk/1/hi/sci/tech/ 4715221.stm; How Widespread Are Glyphosate-Resistant Weeds in the United States? Weed Resistant Management, 2005, available at weedresistancemanagement.com/layout/faq/faq _01.asp.

32. Weeds Revenge, *Ecologist,* July–August 2007, p. 11.

33. Cummins and Lilliston, *Genetically Engineered Food,* p. 37.

34. A. Pollack, Gene Maps: The Key to Better Food? *International Herald Tribune,* March 8, 2001, p. 13.

35. R. Bessin, Bt-Corn: What It Is and How it Works, 2004, University of Kentucky, available at uky.edu/Agriculture/Entomology/entfacts/fldcrops/ef130.htm.

36. Taking Stock, *Ecologist,* July–August 2003, p. 33.

37. Ibid.

38. B. Halweil, The Emperor's New Crops, *Worldwatch,* July–August 1999, p. 24.

39. A. Coghlan, Will Low-Fat Foods Sway Biotech Skeptics? *New Scientist,* March 19, 2005, p. 9.

40. E. Stokstad, Monsanto Pulls the Plug on Genetically Modified Wheat, *Science,* May 21, 2004, p. 1089.

41. Ibid., pp. 1088–1089; Plant Tweaked to Make Omega-3s, *New Scientist,* May 22, 2004, p. 18.

42. Golden Spuds, *New Scientist,* May 12, 2007, p. 5.

43. C. Dugger, In Africa, Holy Grail for Hunger Is a New Rice, *International Herald Tribune,* October 10, 2007, p. 2.

44. R. L. Naylor, Agriculture and Global Change, in *Earth Systems: Processes and Issues* G. Ernst, ed. (Cambridge: Cambridge University Press, 2000), p. 465; K. S. Fischer, J. Barton, G. S. Khush, H. Leung, and R. Cantrell, Collaborations in Rice, *Science,* October 13, 2000, p. 279.

45. D. Ben Shaul, Taking the Scourge out of Rice, *Jerusalem Post,* November 21, 1999, p. 7.

46. F. Pearce, Protests Take the Shine Off Golden Rice, *New Scientist,* March 31, 2001, p. 15.

47. N. Schnapp and Q. Schiermeier, Critics Claim "Sight-Saving" Rice Is Over-Rated, *Nature,* March 29, 2001, p. 503.

48. J. Gillis, Rice Genome Fully Mapped, *Washington Post*, August 11, 2005, p. A1.

49. M. Roosevelt, Cures on the Cob, *Time Magazine*, May 26, 2003, p. 57.

50. A. Zitner, Field of Genes: Biotech Tries to Grow Its Future, *International Herald Tribune*, June 6, 2001, p. 13.

51. B. Hileman, Drugs from Plants Stir Debate, *Chemical and Engineering News*, August 12, 2002, p. 25.

52. P. Cohen, Fighting over Pharming, *New Scientist*, March 1, 2003, pp. 22–23.

53. Too Tempting, *New Scientist*, February 19, 2005, pp. 3, 19.

54. Hileman, Drugs from Plants Stir Debate, p. 22.

55. Pharming Ban, *New Scientist*, October 26, 2002, p. 9.

56. J. Giles, Transgenic Planting Approved Despite Skepticism of UK Public, *Nature*, March 11, 2004, p. 107; E. Stokstad, Europe Takes Tentative Steps Toward Approval of Commercial GM Crops, *Science*, January 23, 2004, pp. 448–449; German Parliament Approves Gene-Food Law, Deutsche Welle, 2004, available at dw-world.de/popups/popup _printcontent/0,,1204252,00.html; J. Gillis and P. Blustein, WTO Ruling Backs Biotech Crops, *Washington Post*, February 8, 2006, p. D1; E. Rosenthal, GM Farming in Europe: At What Cost? *International Herald Tribune*, May 25, 2006, pp. 1, 8.

57. Wu, Perceptions of Food That Are an Ocean Apart.

58. D. G. McNeil, Jr., A Divide as Big as Atlantic over Food, *International Herald Tribune*, March 15, 2000, pp. 1, 3.

59. Crop Protection: Regional and Country Detail, Syngenta, 2004.

60. I. Serageldin, From Green Revolution to Gene Revolution, n.d., available at usinfo.state .gov/journals/ites/1099/ijee/bio-serageldin.htm.

61. Genetically Engineered (GE) Crops Contaminate Fields and Food Around the World, Greenpeace International, 2005, available at connectotel.com/gmfood/gp290505.txt.

62. J. Halloran and M. Hansen, Why We Need Labeling of Genetically Engineered Food, *Consumers International*, April 1998, available at consumersunion.org/food/whywenny798 .htm.

63. Mutated Arguments, *Geographical Magazine*, June 2004, p. 37.

64. Pollen Furor, *New Scientist*, October 18, 2003, p. 7.

65. The Fastest Pollen Explosion on Earth, *New Scientist*, May 15, 2005, p. 19.

66. M. Mellon and J. Rissler, Gone to Seed: Transgenic Contaminants in the Traditional Seed Supply, Union of Concerned Scientists, 2004; Keeping Seeds Safe, *International Herald Tribune*, March 2, 2004, p. 8.

67. J. Thomas, Patents and Pollution, *Ecologist*, July–August 2004, p. 10.

68. J. Hepburn, Imagine . . . , *Ecologist*, September 2004, p. 15.

69. R. Schubert, Farming's New Feudalism, *Worldwatch*, May–June 2005, pp. 10–15; E. Goldsmith, Percy Schmeiser, the Man That Took on Monsanto, *Ecologist*, May 2004, pp. 21–24.

70. Deneen, Food Fight, p. 30.

71. K. Roseboro, Biotech, Organic Coexistence Research Paper Skews Facts to Support Dubious Conclusion, 2004, available at Cropchoice.com.

72. A. Pollack, Can Biotech Crops Be Good Neighbors? *New York Times*, September 26, 2004, p. 12.

73. Deneen, Food Fight, p. 31.

74. B. Lilliston, Farmers Fight to Save Organic Crops, *Progressive*, September 2001, pp. 26–29.

75. A. M. Wilborn, More Biotech Battles, *E Magazine*, September–October 2001, p. 22.

76. Taking Stock, *Ecologist*, July–August 2003, p. 33; G. Brookes and P. Barfoot, Study Concludes That Biotech and Non-Biotech Crops Can Exist Successfully, 2004, available at whybiotech.com/index.asp?id=4505; K. Roseboro, Biotech, Organic Coexistence Research Paper Skews Facts to Support Dubious Conclusion, 2004, available at cropchoice.com.

77. Seeds of Doubt, Soil Association, Bristol, UK, 2002, p. 27; B. Lilliston, Farmers Fight to Save Organic Crops, *Progressive*, September 2001, pp. 26–29.

78. Ibid.

79. Biotech Crop Tax, *New Scientist*, December 3, 2005, p. 5.

80. J. Huang, C. Pray, and S. Rozelle, Enhancing the Crops to Feed the Poor, *Nature*, August 8, 2002, pp. 680–681.

81. P. J. Griekspoor, Report: Biotech Helps Farmers, Environment, *Wichita [Kansas] Eagle*, November 12, 2004, available at kansas.com/mld/Kansas/business/10159370.htm?template =contentModule.

82. Biotech Crops Increase Use of Pesticides, *Chemical and Engineering News*, December 8, 2003, p. 27.

83. J. Randerson, Genetically-Modified Superweeds "Not Uncommon," New Scientist.com, February 5, 2002, available at newscientist.com/news/print.jsp?id=ns99991882; A. Coghlan, Enter the Superweed, *New Scientist*, August 27, 2005, p. 17.

84. Soil Association, *Seeds of Doubt: North American Farmers' Experiences of GM Crops* (Bristol, UK: Soil Association, 2002), p. 11; Cummins and Lilliston, *Genetically Engineered Food*, p. 26.

85. Soil Association, *Seeds of Doubt*, p. 13.

86. H. Schuster, Grim Report Card Emerges in My Final Thoughts, CNN.com, 2006, available at edition.cnn.com/2006/US/12/22/schuster.column/.

87. Food and Population: FAO Looks Ahead, 2000, fao.org/News/2000/000704–e.htm; V. Smil, *Feeding the World: A Challenge for the 21st Century* (Cambridge, Mass.: MIT Press, 2000); F. M. Lappé, J. Collins, and P. Rosset, *World Hunger: Twelve Myths*, 2nd ed. (New York: Grove Press, 1998); R. W. Kates, Ending Hunger: Current Status and Future Prospects, *Consequences* 2:2 (1996): 3–11; A. K. Sen, *Poverty and Famines: An Essay on Entitlement and Deprivation* (Oxford, U.K.: Clarendon Press, 1982).

88. P. Harrison and P. H. Raven, Foodcrops, in *AAAS Atlas of Population and Environment* (Berkeley: University of California Press, 2000), p. 55.

89. Ibid.

90. M. Hutchinson, The Bear's Lair: The Economics of Doom, *United Press International*, July 26, 2004, available at upi.com/NewsTrack/Business/2004/07/26/the_bears_lair_the _economics_of_doom/3217/; B. Halweil, Grain Harvest and Hunger Both Grow, in *Vital Signs, 2005*, ed. L. Starke (New York: =Norton, 2005), p. 22.

91. Filling the World's Belly, *Economist*, December 13, 2003, p. 4.

92. Ibid.

93. P. A. Sanchez and M. S. Swaminathan, Cutting World Hunger in Half, *Science*, January 21, 2005, pp. 357–359.

94. A. K. Sen, *Poverty and Famine* (Oxford: Oxford University Press, 1981).

95. Fatal Harvest: Deadly Myths of Industrial Agriculture, *Ecologist*, October 2002, p. 41.

96. Gardner and Halweil, Nourishing the Underfed and Overfed, in State of the World, ed. L. Starke (New York: Norton, 2000), p. 64.

97. F. M. Lappé, Biotechnology Isn't the Key to Feeding the World, *International Herald Tribune*, July 5, 2001, p. 7.

98. F. M. Lappé, J. Collins, and P. Rosset, *World Hunger*, p. 14.

99. M. Reardon, About ERS: Frequently Asked Questions, USDA Economic Research Service, 2005, available at ers.usda.gov/AboutERS/FAQs.htm.

100. S. Sengupta, Globally, More People Go Hungry, UN Says, *International Herald Tribune*, November 26, 2003, p. 5.

Chapter Six

1. D. Nierenberg, *Happier Meals: Rethinking the Global Meat Industry*, Worldwatch Paper 171, ed. L. Mastny, Worldwatch Institute, 2005, p. 9.

2. Statistics: Global Farmed Animal Slaughter, United Poultry Concerns, 2004, available at upc-online.org/slaughter/92704stats.htm.

3. FAOSTAT Statistical Database, United Nations Food and Agriculture Organization, 2003, available at faostat.fao.org; Turkeys Raised. USDA, National Agricultural Statistics Service, 2006.

4. Meat, Worldwatch Institute, 2004, available at worldwatch.org/pubs/goodstuff/meat/.

5. C. Tudge, It's a Meat Market, *New Scientist*, March 13, 2004, p. 19.

6. D. Nierenberg, Meat Production and Consumption Grow, in *Vital Signs 2003*, edited by L. Starke (New York: Norton), pp. 30–31.

7. Ibid.

8. Animals Killed for Food in the United States, Animal Liberation Front, n.d., available at animalliberationfront.com/Practical/Health/health.htm; T. Regan and M. Rowe, What the Nobel Committee Also Failed to Note, *International Herald Tribune*, December 19, 2003, p. 8.

9. N. D. Barnard, To Eat or Not to Eat, *E Magazine*, September–October 2004, p. 7.

10. B. Carnell, Farm on the (Lack of) Effectiveness of World Farm Animals Day, Animal Rights, 2003; Sheep and Lambs, 1987–2004, *USDA National Agricultural Statistics Service*, 2004, available at nass.usda.gov.

11. Agriculture Fact Book, *USDA*, 2003, p. 15, available at usda.gov/factbook/2002factbook.pdf.

12. Animals Killed for Food in the United States in 2000, n.d., available at upc-online.org/slaughter/2000slaughter_stats.html; U.S. Broiler Production 1952–2002, USDA National Agricultural Statistics Service, 2003, available at nass.usda.gov/Publications/Statistical_Highlights/2002/graphics/broiler.htm.

13. Trends in U.S. Agriculture: Broiler Industry, USDA, National Agricultural Statistics Service, n.d., available at usda.gov/nass/pubs/trends/broiler.htm; J. C. Buzby, Chicken Consumption Continues Longrun Rise, *Amber Waves*, April 2006, p. 5.

14. F. Lawrence, Chicken, *Ecologist*, September 2004, p. 3.

15. Organic Sellers Tout Mad-Cow-Free Beef, *International Herald Tribune*, January 12, 2004, p. 5; Cattle, Sheep, and Hogs on Farms Jan. 1, 1975–2006, *Cattle Feeders Annual 2006*.

16. J. J. Miller and D. P. Blaylock, Dairy Backgrounder, USDA Economic Research Service, 2006.

17. C. G. Davis and B.-H. Lin, Factors Affecting U.S. Pork Consumption, USDA Economic Research Service, 2005, p. 1.

18. Briefing Room, Sheep and Wool: Overview, USDA Economic Research Service, 2003, available at ers.usda.gov/Briefing/Sheep/.

19. M. P. Ryan, Who Goes First? Blue Honey, n.d. bluehoney.org/Poultry.htm.

20. Egg Production Information, n.d., available at aeb.org/eii/production.html.

21. Y. Dror, Hebrew University Shortens a Chick's Life, *Haaretz* [Israel English-language newspaper], July 5, 2004, p. 10.

22. Untitled, *National Geographic*, April 2005, unnumbered page.

23. U.S. Chicken and Turkey Slaughter, USDA Economic Research Service, p. 4, available at ers.usda.gov/publications/aer787/aer787b.pdf.

24. J. D'Silva, Faster, Cheaper, Sicker, *New Scientist*, November 15, 2003, p. 19.

25. C. Greene and A. Kremen, *U.S. Organic Farming in 2000–2001: Adoption of Certified Systems*, USDA, Economic Research Service, Agriculture Information Bulletin 780, p. 24.

26. Nierenberg, *Happier Meals*, p. 65.

27. Ibid., pp. 13–14.

28. L. Oberholtzer, C. Greene, and E. Lopez, *Organic Poultry and Eggs Capture High Price Premiums and Growing Share of Specialty Markets*, USDA Outlook Report, Economic Research Service, 2006; USDA, 2007.

29. Nierenberg, *Happier Meals*, p. 15.

30. C. Salvi and D. Hatz, This Little Piggy Went to the Global Market, Worldwatch Institute, 2004, available at worldwatch.org/ pubs/goodstuff/meat/.

31. A. R. Sapkota, L. Y. Lefferts, S. McKenzie, and P. Walker, What Do We Feed to Food-Production Animals? A Review of Animal Feed Ingredients and Their Potential Impacts on Human Health, *Environmental Health Perspectives*, May 2007, pp. 663–670.

32. Resistance to Antibiotics, Consumer Reports.org, January 2007, available at consumerreports.org/cro/food/food-safety/chicken-safety/chicken-safety-1–07/resistance-to-

antibiotics/0107_chick_anti.htm; Germ Count, ConsumerReports.org, January 2007, available at consumerreports.org/cro/food/food-safety/chicken-safety/chicken-safety-1–07/levels-of-contamination/0107_chick_level.htm; 8 out of 10 Chickens Tested Harbor Dangerous Bacteria, Consumer Reports.org, January 2007, available at consumerreports.org/cro/food/food-safety/chicken-safety/chicken-safety-1–07/overview/0107_chick_ov.htm.

33. Antibiotics Used for Growth in Food Animals Making Their Way into Waterways, Colorado State University, 2004, available at sciencedaily.com/releases/2004/10/041025120141.htm.

34. D. Nierenberg, Some Antibiotics with Your Vegetables? *Worldwatch*, March–April 2006, p. 10.

35. The Chicken Factory Farm, 2004, Humane Society of the United States, available at hsus.org/farm.

36. M. Owen, Manure Matters, PlanTea, 1996, available at plantea.com/manure.htm.

37. Gizmo, *New Scientist*, December 10, 2005, p. 29.

38. Watch That Bird's Rear, *Economist*, July 14, 2005, p. 44.

39. K. R. Nachman, J. P. Graham, L. B. Price, and E. K. Silbergeld, Arsenic: A Roadblock to Potential Animal Waste Management Solutions, *Environmental Health Perspectives*, September 2005, pp. 1123–1124.

40. D'Silva, Faster, Cheaper, Sicker, p. 19; Factory Chicken and Turkey Production, Farm Sanctuary, available at farmsanctuary.org/issues/factoryfarming/poultry/.

41. Animals Killed for Food in 2000, Animal Liberation Front, 2003, available at http://www.animalliberationfront.com/Practical/FactoryFarm/Animals%20Killed%20for%20Food.htm.

42. Ryan, Who Goes First?.

43. Animals Get Protections with Teeth, *International Herald Tribune*, May 28, 2004, p. 3.

44. Japanese Hold "Funeral" for Chickens, *China Daily*, April 28, 2004, available at chinadaily.com.cn/english/doc/2004–04/29/content_327232.htm.

45. Untitled, *Ecologist*, July–August 2005, p. 10.

46. B. Halford, Going Beyond Feather Dusters, *Chemical and Engineering News*, September 6, 2004, pp. 36–39.

47. S. Battersby, Down, But Not Out, *New Scientist*, December 25, 2004–January 1, 2005, p. 40.

48. Halford, Going Beyond Feather Dusters; L. Frazer, Chicken Electronics, *Environmental Health Perspectives*, July 2004, pp. A564–A567.

49. Frazer, Chicken Electronics, pp. A564–A567; Battersby, Down, But Not Out, pp. 40–41.

50. C. Leonard, Not a Tiger, But Maybe a Chicken in Your Tank, *Washington Post*, January 3, 2007, p. D3.

51. A. Ault, From the Head of a Rooster to a Smiling Face Near You, *New York Times*, December 23, 2003, pp. F1, F2.

52. Eggs: Indicators of Production, Supply, and Disappearance, USDA Economic Research Service, 2006, available at usda.mannlib.cornell.edu.

53. Ibid.

54. U.S. Egg Industry Fact Sheet, 2004, available at aeb.org/Assets/PDF/IndustryFacts/EggFactSheet.pdf.

55. U.S. Egg Industry, United Egg, 2004, available at unitedegg.org/useggindustry_generalstats.aspx.

56. U.S. Egg Industry Fact Sheet, 2004.

57. Ryan, Who Goes First?.

58. Egg Trivia, n.d., available at aeb.org/LearnMore/Trivia.htm; Egg Production Information, n.d., available at aeb.org/Industry/EggProduction.htm.

59. Egg Farming, n.d., Farm Sanctuary, available at farmsanctuary.org/issues/factoryfarming/eggs/.

60. Egg Woes Cracked, *New Scientist*, July 8, 2006, p. 4.

61. Egg Farming.

62. Ibid.

63. What Is a Chicken? n.d. University of Illinois Extension, available at urbanext.uiuc.edu/eggs/res08–whatis.html; Chicken Facts, n.d., available at homestead,com/shilala/chickenfacts.html.

64. Ryan, Who Goes First?.

65. Dror, Hebrew University Shortens a Chick's Life, p. 10; Egg Farming, n.d., Farm Sanctuary, available at farmsanctuary.org/issues/factoryfarming/eggs/.

66. P. Singer, *Animal Liberation* (New York: HarperCollins, 2002), pp. 101–102.

67. Animal Care, n.d., *United Egg*, available at unitedegg.org/animal_care.aspx.

68. G. Gregory, Animal Care, United Egg, 2004, available at unitedegg.org/animal_care.aspx; Animal Care? Compassion over Killing, 2004, available at eggindustry.com/cfi/report/.

69. Zogby Poll on Americans' Attitude Toward "Animal Care Certified" Eggs, Compassion over Killing, 2003, available at eggindustry.com.

70. McDonald's USA Laying Hens Guidelines, 2005, available at mcdonalds.com/usa/good/products/hen.html; Nierenberg, *Happier Meals: Rethinking the Global Meat Industry*, p. 61.

71. E. Weise, "Cage-Free" Is Closer to Ruling the Roost, *USA Today*, April 11, 2006, p. 7D.

72. 1981 Swiss Ban on Battery Cages: A Success Story for Hens and Farmers 1995, available at awionline.org/farm/hens.htm.

73. EU Bans Battery Hen Cages, BBC News, January 28, 1999, available at news.bbc.co.uk/1/hi/uk/264607.stm.

74. Animals Get Protection with Teeth, *International Herald Tribune*, May 28, 2004, p. 3.

75. European Commission, Treaty of Amsterdam, 1997, p. 110.

76. no author given, Heritage Turkeys 2004, available at heritageturkeyfoundation.org.

77. Thanksgiving Day, 2004, U.S. Census Bureau, 2004, available at census.gov/PressRelease/www/releases/archives/facts_for_features_special_editions/002938.html.

78. Ibid.

79. Turkeys Raised, USDA, National Agricultural Statistics Service, 2006.

80. A. Jackson, Let's Talk Turkey, Alaska Division of Environmental Health, n.d., available at dec.state.ak.us/eh/fss/consumers/allturk.htm.

81. The Turkey Factory Farm, Humane Society of the United States, 2004, available at hsus .org/farm; In the Christmas Spirit? *Ecologist*, December 2003–January 2004, p. 10.

82. Wild Turkey Facts, n.d., National Wild Turkey Federation, available at nwtf.org/all _about_turkeys/wild_turkey_facts.html.

83. Turkeys Raised, U.S. Census Bureau, 2006.

84. Stop Playing Chicken with Food Safety. Consumer Reports.org, January 2007, available at consumerreports.org/cro/aboutus/mission/viewpoint/food-safety-1–07/overview/0107 _viewpoint_ov_1.htm.

85. C. Greene, U.S. Certified Organic Farmland Acreage, Livestock Numbers, and Farm Operations, 1992–2003, USDA, Economic Research Service, 2005; L. Oberhottzer, C. Greene, and E. Lopez, *Organic Poultry and Eggs Capture High Price Premiums and Growing Share of Specialty Markets*, USDA OUTLOOK Report from the Economic Research Service, 2006; USDA, 2007.

86. Ducks out of Water, Viva, 2005, available at vivausa.org/campaigns/ducks/ duckbriefing.html; J. Gellatley, Ducks Out of Water: A Report on the Duck Industry in the USA, part 1, 1999 *Viva*, available at vivausa.org/campaigns/ducks/duckreport01.htm.

87. Ducks out of Water.

Chapter Seven

1. Briefing Room: Cattle: Background, USDA Economic Research Service, 2000, available at ers.usda.gov/briefing/cattle/Background.htm.

2. K. Lydersen, Chicago Nears End of Era in Stockyards, *Washington Post*, July 18, 2005, p. A3.

3. K. Casa, The Changing Face of Farming in America, *National Catholic Reporter*, February 12, 1999, pp. 13–16.

4. Cattle Feeders Annual, 2006. Beefacts: Cattle, Sheep and Hogs on Farms, January 1, 1975–2006, p. 117.

5. Factors Affecting U.S. Beef Consumption, USDA Economic Research Service, 2005, p. 7; Background Statistics: U.S. Beef and Cattle Industry, USDA Economic Research Service, 2005, available at ers.usda.gov/News/BSECoverage.htm.

6. Ibid.

7. Background Statistics: U.S. Beef and Cattle Industry.

8. Current State of the United States Live Cattle Industry, 2002, R-CALF United Stockgrowers of America, available at r-calfusa.com/Trade/current_state_of_the_u_s__live_cattle _industry.htm.

9. Background Statistics: U.S. Beef and Cattle Industry.

10. D. Tillman, K. G. Cassman, P. A. Matson, R. Naylor, and S. Polasky, Agricultural Sustainability and Intensive Production Practices, *Nature*, August 8, 2002, pp. 671–677; The

Editors, Now, It's Not Personal, *Worldwatch*, July–August 2004, pp. 12–19; L. Horrigan, R. S. Lawrence, and P. Walker, How Sustainable Agriculture Can Address the Environmental and Health Harms of Industrial Agriculture, *Environmental Health Perspectives*, May 2002, pp. 445–456; The Humble Hamburger, *Worldwatch*, July–August 2004, inside cover; P. Muir, Trophic Issues, 2001, available at oregonstate.edu/~muirp/trophic.htm.

11. S. P. Greiner, Beef Cattle Breeds and Biological Types, Virginia Cooperative Extension Publication Number 400–803, 2002, available at ext.vt.edu/pubs/beef/400–803/400–803 .html.

12. J. R. Connor, R. A. Dietrich, and G. W. Williams, *The U.S. Cattle and Beef Industry and the Environment*, TAMRC [Texas A&M University] Commodity Market Research Report no. CM-1–00, 2000, available at agrinet.tamu.edu/centers/tamrc/pubs/cm100.htm.

13. Animal Production and Marketing Issues: Questions and Answers, USDA Economic Research Service, 2004, available at ers.usda.gov/Briefing/AnimalProducts/questions.htm.

14. J. Raloff, Hormones: Here's the Beef, *Science News Online*, January 5, 2002, available at sciencenews.org/articles/20020105/bob13.asp.

15. D. Nierenberg, *Happier Meals: Rethinking the Global Meat Industry*, Worldwatch Paper 171, ed. L. Mastny, Worldwatch Institute, 2005, p. 6.

16. Ibid.

17. Animal Production and Marketing Issues.

18. Connor, Dietrich, and Williams, *The U.S. Cattle and Beef Industry and the Environment*.

19. Raloff, Hormones.

20. Ibid.

21. Ibid.

22. Nierenberg, *Happier Meals*, p. 56.

23. J. Mercola, Why Grassfed Animal Products Are Better for You, 2004, available at mercola.com/beef/health_benefits.htm; Back to Nature, Kerr Farms, n.d., available at backtonaturebeef.com/Test%20Results.htm; M. Muller, A Disservice to Environmentally Appropriate Livestock Producers, *Worldwatch*, September–October 2004, pp. 6–7; J. Robinson, *Pasture Perfect* (Vashon, Wash.: Vashon Island Press, 2004).

24. D. Rakestraw, Nutrients Declining in Food Supply, *Worldwatch*, May–June 2006, p. 11.

25. Pesticides, Sustainable Table, n.d., available at sustainabletable.org/issues/pesticides/.

26. Ibid.

27. Yampa Valley Almanac Local Lore, n.d., available at yampavalley.info/weather0020.asp.

28. Ibid.

29. Milk: A Cruel and Unhealthy Product, PETA Media Center, 2004, available at peta.org/ mc/factsheet_display.asp?ID=98; M. Burros, Veal to Love, without the Guilt, *New York Times*, April 18, 2007, p. F1, F4.

30. T. Hayden, Going to Waste? *U.S. News & World Report*, October 20, 2003, p. 50.

31. R. Marks, Cesspools of Shame, Natural Resources Defense Council and the Clean Water Network, 2001, available at nrdc.org/water/pollution/cesspools/cesspools.pdf.

32. H. Blatt, *America's Environmental Report Card* (Cambridge, Mass.: MIT Press, 2004), p. 27; R. Marks, Cesspools of Shame, Natural Resources Defense Council and the Clean Water Network, 2001, available at nrdc.org/water/pollution/cesspools/cesspools.pdf.

33. Clean Water and Factory Farms: Water Contamination from Factory Farms, Sierra Club, n.d., available at sierraclub.org/factoryfarms/factsheets/water.asp.

34. Untitled, *Time Magazine*, March 7, 2005, p. 14.

35. How to Poison a River, *New York Times*, August 19, 2005, p. A18.

36. Animal Waste, 2005; Keep Animal Waste Out of Our Waters: Stop Factory Farm Pollution, n.d., Sierra Club. Sierraclub.org/factoryfarms/, available at nrdc.org/water/pollution/cesspools/cesspools.pdf.

37. Cow Urine for Dazzling Teeth? *Times of India online*, February 27, 2005, available at timesofindia.indiatimes.com/articleshow/msid-1034529,prtpage-1.cms.

38. What We Learnt Last Month, *Ecologist*, May 2006, p. 10.

39. Farm Animals Bigger Threat to Environment Than Autos, Says UN, Earthtimes.org, available at earthtimes.org/articles/printstory.php?news=11021; Steionfeld, H. et al., Livestock's Long Shadow: Environmental Issues and Options. Food and Agriculture Organization, United Nations, 2006.

40. J. H. Holtz, Where Delhi's Cows Find Sanctuary, *International Herald Tribune*, February 25, 2005, p. 7.

41. Ibid.

42. Outlawed in Europe: Three Decades of Progress, Animal Rights International, n.d., available at ari-online.org.

43. D. P. Blayney, *The Changing Landscape of U.S. Milk Production*, U.S. Department of Agriculture, Statistical Bulletin Number 978, 2002, p. 7.

44. Ibid.

45. Ibid., p. 19.

46. B. Worthington-Roberts, Milk, Encarta Online Encyclopedia, n.d., available at incarta.msn.com/text_761562453_0/Milk.html.

47. J. J. Rehmeyer, Milk Therapy, *Science News*, December 9, 2006, pp. 376–377; Mothers Get Heart Risk off Their Chest, *New Scientist*, February 17, 2007, p. 17; P. Thomas, Suck on This, *Ecologist Online*, 2006, available at theecologist.org/archive_detail.asp?content_id=586.

48. Thomas, Suck on This, p. 26.

49. Ibid.

50. Ibid., p. 31; H. Miller and E. von Schaper, Nestlé Adds Gerber Baby Food to Its Family of Brands, *International Herald Tribune*, April 13, 2007, p. 16.

51. Lactose Intolerance, National Institute of Diabetes and Digestive and Kidney Disorders, n.d., available at digestive.niddk.nih.gov/ddiseases/pubs/lactoseintolerance/.

52. The Issues: rBGH, Sustainable Table, 2005, available at sustainabletable.org/issues/rbgh/; Posilac: General Information, Monsanto, 2005, available at monsantodairy.com/about/general_info/; H. I. Miller, Don't Cry over rBST Milk, *New York Times*, June 29, 2007, p. A29.

53. U.S. Certified Organic Farmland Acreage, Livestock Numbers, and Farm Operations, 1992–2005, USDA Economic Research Service, 2005.

54. Maddening World, *Scientific American*, March 2004, p. 16.

55. U.S. Certified Organic Farmland Acreage, Livestock Numbers, and Farm Operations.

56. K. Severson, An Organic Cash Cow, *New York Times*, November 9, 2005, pp. F1, F10; C. Dimitri and K. M. Venezia, Retail and Consumer Aspects of the Organic Milk Market, USDA Economic Research Service, 2007.

57. U.S. Certified Organic Farmland Acreage, Livestock Numbers, and Farm Operations.

58. R. Weiss, FDA Says Clones Are Safe to Eat, *Washington Post*, December 29, 2006, p. A1; Cloned Meat, *New Scientist*, April 16, 2005, p. 7.

59. E. Weise, Clones' Beef, Milk Same as Regular, *USA* Today, April 12, 2005, p. D7.

60. Weiss, FDA Says Clones Are Safe to Eat.

61. R. Weiss, FDA Is Set to Approve Milk, Meat from Clones, *Washington Post*, October 17, 2006, p. A1.

62. C. G. Davis and B.-H. Lin, Factors Affecting U.S. Pork Consumption, USDA Economic Research Service, 2005, p. 1.

63. Ibid.

64. Cattle Feeders Annual 2006; U.S. Certified Organic Farmland Acreage, Livestock Numbers, and Farm Operations.

65. W. D. McBride and N. Key, *Economic and Structural Relationships in U.S. Hog Production*, U.S. Department of Agriculture Agricultural Economic Report no. 818, 2003, available at ers.usda.gov/publications/aer818/aer818.pdf.

66. Ibid., p. 15.

67. Ibid.

68. Animal Welfare, Sustainable Table, n.d., available at sustainabletable.org/issues/animalwelfare/.

69. N. H. Niman, The Unkindest Cut, 2005, available at veganrepresent.com/forums/showthread.php?t=5516.

70. Weiss, FDA Says Clones Are Safe to Eat, p. 4; McBride and Key, *Economic and Structural Relationships in U.S. Hog Production*, p. 5.

71. McBride and Key, Economic and Structural Relationships in U.S. Hog Production, p. 5; U.S. Hog Operations: Number of Operations and Percent of Inventory, 2003, USDA National Agricultural Statistics Service, 2004, available at usda.mannlib.cornell.edu; U.S. Hog Operations: Number by Size Group, 2002–2003, USDA National Agricultural Service, 2004, available at usda.mannlib.cornell.edu/reports/nassr/livestock/php-bb/2004/hgpg0604.txt; P. D. Johnson, The Future of Farming, *Denver Post*, June 15, 2003, available at ewg.org/farm/news/story.php?id=1780.

72. F. Barringer, A Search for Pearls of Wisdom in the Matter of Swine, *New York Times*, July 7, 2004, p. A12.

73. R. Loehr, Pollution Implications of Animal Wastes—A Forward-Oriented Review, Environmental Protection Agency, Water Pollution Control Research Series, Office of Research and Monitoring, 1968, p. 26.

74. Farm Animal Waste and the Clean Water Dilemma, *Sanctuary News*, Spring 1998, available at vsc.org/0502–animal-waste-on-factor.htm; S. N. Cousins, Animal Waste on Factory Farms, Vegetarian Society of Colorado, n.d., available at vsc.org/0502–animal-waste-on-factor.htm.

75. Ibid.

76. R. L. Huffman and P. W. Westerman, Estimated Seepage Losses from Established Swine Waste Lagoons in the Lower Coastal Plain in North Carolina, *American Society of Agricultural and Biological Engineers Transactions* 38:2 (1995): 449–453.

77. Farm Animal Waste and the Clean Water Dilemma, *Farm Sanctuary News*, 1998, available at farmsanctuary.org/newsletter/news_cleanwater.htm; G. Vander Wal, 44 States Regulate Odors on Hog Farms, National Hog Farmer, 2001, available at nationalhogfarmer.com/mag/farming_states_regulate_odors/.

78. *Cattle Feeders Annual 2006*; U.S. Certified Organic Farmland Acreage, Livestock Numbers, and Farm Operations.

79. Who Are Smithfield Foods? *Ecologist*, December 2003–January 2004, pp. 48–53.

80. Ibid., p. 51.

81. This Little Piggy Went to the Global Market, Worldwatch Institute, 2004, available at worldwatch.org/pubs/goodstuff/meat/.

82. Large US Cattle Lots Threaten Water, Food-Sierra Club, *Hay and Forage Grower*, August 13, 2002, available at hayandforage.com.

83. Who Are Smithfield Foods? pp. 48–53.

84. Ibid., p. 49.

85. Ibid.

86. Ibid.

87. Animal Welfare, Sustainable Table, n.d., available at sustainabletable.org/issues/animalwelfare/.

88. Vander Wal, 44 States Regulate Odors on Hog Farms.

89. L. Silver, Why GM Is Good for Us, *Newsweek*, March 20, 2006, p. 51.

90. L. Lai et al., Generation of Cloned Transgenic Pigs Rich in Omega-3 Fatty Acids, *Nature Biotechnology*, April 2006, pp. 435–436.

91. United States Sheep Industry, American Sheep Industry Association, 2004, available at sheepusa.org.

92. Sheep and Wool: Overview, U.S. Department of Agriculture, 2003, available at ers.usda.gov/Briefing/Sheep/.

93. E. Nieves, The Job No Americans Want Isn't Getting Any Easier, *Washington Post*, April 3, 2006, p. A2.

94. Table 29: Sheep and Lambs-Inventory, Wool Production, and Number Sold: 2002 and 1997, 2002 Census of Agriculture.

95. Briefing Room, Sheep and Wool: Overview. *USDA Economic Research Service*, n.d., available at ers.usda.gov/Briefing/Sheep/.

96. Ibid.

Chapter Eight

1. A. P. McGinn, Blue Revolution: The Promises and Pitfalls of Fish Farming, *Worldwatch*, March–April 1998, p. 13.

2. World Fish Production, 1950–2003, Earth Policy Institute, 2005; B. Halweil, Aquaculture Pushes Fish Harvest Higher, in *Vital Signs, 2005*, ed. L. Starke (New York: Norton), pp. 26–27.

3. D. Pauly and R. Watson, Counting the Last Fish, *Scientific American*, July 2003, pp. 34–39; J. Larsen, Fish Catch Leveling Off, *San Diego Earth Times*, March 2003, available at sdearthtimes.com/et0203/et0203s13.html.

4. Seafood Consumption Reaches Record Levels in 2004, National Oceanographic and Atmospheric Administration, 2005, available at noaanews.noaa.gov/stories2005/s2531.htm.

5. Shrimp Leads Record Gain in Seafood Consumption, National Fisheries Institute, 2004, available at nfi.org/?a=news&b=News%20Releases&year=&x=3679.

6. Halweil, Aquaculture Pushes Fish Harvest Higher, in *Vital Signs*, ed. L. Starke, New York, Norton, 2005), p. 26.

7. G. Hardin, The Tragedy of the Commons, *Science*, December 13, 1968, pp. 1243–1248; G. Hardin, Extensions of "The Tragedy of the Commons," *Science*, May 1, 1998, pp. 682–683; J. Burger and M. Gochfeld, The Tragedy of the Commons, *Environment*, December 1998, pp. 4–13, 26–27.

8. J. L. Jacobson and A. Rieser, The Evolution of Ocean Law, *Scientific American Quarterly*, Fall 1998, pp. 100–105.

9. Ibid.; United Nations Convention on the Law of the Sea (A Historical Perspective), United Nations, 2006, available at un.org/Depts/los/convention_agreements/convention_historical _perspective.htm; O. R. Young et al., Solving the Crisis on Ocean Governance, *Environment*, May 2007, pp. 20–32.

10. D. Pauly and J. Maclean, *In a Perfect Ocean: The State of Fisheries and Ecosystems in the North Atlantic Ocean* (Washington, D.C.: Island Press, 2003), p. 16.

11. Fish on the Move as North Sea Hots Up, *New Scientist*, May 21, 2005, p. 21.

12. United Nations Convention on the Law of the Sea; United Nations Convention on the Law of the Sea, Wikipedia, 2005, available at wikipedia.org/wiki/United_Nations _Convention_on_the_Law_of_the_Sea.

13. J. Kettlewell, Ocean Census Discovers New Fish, BBC News, October 23, 2003, available at news.bbc.co.uk/2/hi/science/nature/3210544.stm.

14. C. Safina, The World's Imperiled Fish, *Scientific American Quarterly*, Fall, 1998, p. 59.

15. J. Raloff, Empty Nets, *Science News*, June 4, 2005, p. 360.

16. Ibid.

17. Ibid.

18. Ibid.

19. C. F., Protecting the World's Oceans, *International Herald Tribune*, May 24, 2004, p. 7.

20. Pauly and Watson, Counting the Last Fish, p. 35.

21. Larsen, Fish Catch Leveling Off.

22. Big-Fish Stocks Fall 90 Percent Since 1950, Study Says, *National Geographic News*, May 15, 2003, available at news.nationalgeographic.com/news/2003/05/0515_030515_fishdecline .html.

23. A. C. Revkin, Conservation as the Catch of the Day for Trawlnets, *New York Times*, July 29, 2003, p. F3.

24. J. Eilperin, Oceans Have Fewer Kinds of Fish, *Washington Post*, July 29, 2005, p. A3; E. Williamson, Scientists Try to Count Fish in the Sea, *Washington Post*, April 10, 2006, p. B1; J. Eilperin, Wave of Marine Species Extinctions Feared, *Washington Post*, August 24, 2005, p. 1.

25. R. A. Myers and B. Worm, Rapid Worldwide Depletion of Predatory Fish Communities, *Nature*, May 15, 2003, pp. 280–283.

26. W. Broad and A. C. Revkin, Has the Sea Given Up Its Bounty? *New York Times*, July 29, 2003, p. F1, F2.

27. B. Halweil, Aquaculture Pushes Fish Harvest Higher, in *Vital Signs 2005*, ed. L. Starke (New York: Norton), p. 27.

28. Pauly and Maclean, *In a Perfect Ocean*, p. 88.

29. Science News This Week, *Science News*, April 30, 2005, p. 278.

30. Safina, The World's Imperiled Fish, p. 60.

31. T. Hayden, Fished Out, *U.S. News & World Report*, June 9, 2003, p. 41.

32. Fewer Shark Attacks, But Fewer Sharks, Too, *International Herald Tribune*, January 29, 2004, p. 4; International Shark Attack File, Graph of Worldwide Trends in Shark Attacks over the Past Century, Florida Museum of Natural History, 2005, available at flmnh.ufl.edu/ fish/sharks/statistics/Trends.htm; J. Whitty, The Fate of the Ocean, *Mother Jones*, March– April 2006, p. 39; J. Simpson, Disaster in the Deep, available at sciencenow.sciencemag.org/ cgi/content/full/2007/330/3; J. Eilperin, Fish Story's New Reality Is That Man Bites Shark, *New York Times*, May 27, 2007.

33. Halweil, Aquaculture Pushes Fish Harvest Higher, p. 27.

34. Hayden, Fished Out, pp. 40–41.

35. World Fish Restoration a Global Challenge, UN Report Warns, 2005, available at ens-newswire.com/ens/mar2005/2005–03–07–03.asp.

36. Underwater World: Atlantic Cod, Fisheries and Oceans Canada, 2004, available at dfo-mpo.gc.ca/zone/underwater_sous-marin/atlantic/acod_e.htm.

37. Pauly and Maclean, *In a Perfect Ocean*, p. 16.

38. D. MacKenzie, Protected Fish Plundered in "Unintentional" Raids, *New Scientist*, October 1, 2005, p. 9.

39. J. Eilperin, Off the Cape, the Cod Continue to Dwindle, *Washington Post*, August 17, 2005, p. A2.

40. Where Have All the Salmon Gone? *Economist*, April 30, 2005, p. 48.

41. Shrinking Salmon, 2001, available at pbs.org/wgbh/evolution/library/10/3/l_103_02 .html.

42. S. Leahy, Let the Big Fish Go to Save the Species, *New Scientist*, June 25, 2005, p. 11; J. Raloff, Empty Nets, *Science News*, June 4, 2005, pp. 360–362.

43. J. McCrae, Swordfish, 1994, available at oregonstate.edu/odfw/devfish/sp/sword.html; Background: Pacific Halibut, n.d., Fisheries Management, available at pcouncil.org/halibut/halback.html; T. Bougher, Bluefin Tuna: Biology, n.d. people.cornellcollege.edu/t-bougher/geo105/biology.html; Strait Talk, Species at Risk Profile: Rockfish. Georgia Strait Alliance, 2002, available at georgiastrait.org/?q=node/414.

44. Hayden, Fished Out, p. 42.

45. Ibid.

46. Safina, The World's Imperiled Fish, p. 60; We're Eating an Entire Food Chain, *New Scientist*, February 18, 2006, p. 21.

47. American Fisheries Society Lists Marine Stocks at Risk of Extinction, 2000, available at fisheries.org.

48. B. Harder, Sea Change, *Science News*, April 24, 2004, p. 259.

49. More Fish in the Sea, *New Scientist*, June 19, 2004, p. 5.

50. Pauly and Maclean, *In a Perfect Ocean*, p. 38.

51. Seafood Supply and U.S. Trade, Department of Commerce, U.S. Bureau of the Census, 2004, available at globefish.org/files/ustrade_218.pdf; USA: Seafood Consumption Reaches Record Levels in 2004, 2005, available at fishupdate.com/news/fullstory.php/aid/3332/USA.

52. H. Johnson, Top 10 U.S. Consumption by Species Chart, National Fisheries Institute, 2005, available at aboutseafood.com/media/top_10.cfm.

53. Livestock Floor the Environment with Estrogen, *Environmental Science and Technology*, July 1, 2004, p. 241A.

54. S. Jobling et al., Wild Intersex Roach (*Rutilus rutilus*) Have Reduced Fertility, *Biology and Reproduction*, August 2002, pp. 515–524, available at biolreprod.org/cgi/content/abstract/67/2/515; Hormones in the Water Devastate Wild Fish, *New Scientist*, May 26, 2007, p. 16.

55. Sewage and Intersex Fish, Live Science, 2005, available at livescience.com/animalworld/ap_061114_intersex_fish.html; D. A. Fahrenthold, Male Bass across Region Found to Be Bearing Eggs, *Washington Post*, September 6, 2006, p. A1; A. Underwood, Rivers of Doubt, *Newsweek*, June 11, 2007, pp. 56–57.

56. Up Front, *Ecologist*, October 2006, p. 9; P. Montague, Modern Environmental Protection—Part 4, Environmental Research Foundation, September 7, 2000, available at rachel.org/BULLETIN/index.cfm?St=4.

57. "Intersex" Fish Found Off Calif. Coast, USA Today, November 15, 2005, available at usatoday.com/news/nation/2005–11–15–calif-fish_x.htm.

58. "Intersex" Fish Found Off California Coast; Want a Sex Change? Buy Some Suncream! *Ecologist*, April 2006, p. 12.

59. J. R., Happy Fish? *Science News*, December 16, 2006, p. 398; J. R., Could Prozac Muscle Out Mussels? *Science News*, December 2, 2006, p. 366; J. R., Sharks, Dolphins Store Pollutants, *Science News*, December 2, 2006, p. 366; G. Peterson, Water Worries, *E Magazine*, July–August 2007, pp. 16–20.

60. A. Pearson, "Safe" Heavy Metals Hit Fish Senses, *New Scientist*, April 7, 2007, p. 12.

61. Johnson, Top 10 U.S. Consumption by Species Chart; A. Pimental, Seafood Consumption Reaches Record Levels in 2004, National Oceanographic and Atmospheric Administration, 2005, available at noaanews.noaa.gov/stories2005/s2531.htm.

62. R. Blaur and S. Diaby, U.S. and World Shrimp Trade: Trends in Production, Imports, and Exports, National Marine Fisheries Service, 2004. 257, 3, Slide 3; 271, 2 slide 2, available at fas.usda.gov/ffpd/Fishery_Products_Presentations/fish-presentations.htm.

63. Southern Shrimp Alliance Welcomes Department of Commerce Findings Regarding Shrimp Dumping, available at ktvotv3.com/global/story.asp?s=2715474&ClientType =Printable.

64. Johnson, Top 10 U.S. Consumption by Species Chart.

65. China, Vietnam and Brazil Increase Share of US Shrimp, 2003, available at eurofish.dk/indexSub.php?id=1560&easysitestatid=-1093594546.

66. B. Halweil and D. Nierenberg, Watching What We Eat, in *State of the World 2004*, ed. L. Starke (New York: Norton, 2004), p. 92.

67. Ibid.

68. Crabs and Shellfish, 2005, available at chesapeakebay.net/info/crabshell.cfm.

69. R. Blauer, U.S. and World Shrimp Trade: Trends in Production, Imports, and Exports, USDA Foreign Agricultural Service, 2006.

70. J. Weinberg, Atlantic Surfclam, National Oceanographic and Atmospheric Administration, 2000, available at nefsc.noaa.gov/sos/spsyn/iv/surfclam/.

71. Why There Really Is Sauce in an Oyster, *New Scientist*, March 26, 2005, p. 26.

72. Weinberg, Atlantic Surfclam.

73. Bay Scallop, n.d., available at environmentaldefense.org/page.cfm?tagID=15815.

74. P. M. Halpin, Sea Scallop, Seafood Watch, 2005, Monterey Bay Aquarium.

75. C. Dean, Lobster Boom and Bust, *New York Times*, August 9, 2005, p. F1.

76. I. Polansky, Oysters, Oysters, Oysters! 1998, available at globalgourmet.com/food/egg/egg0298/oysters.html; I. Polansky, Oysters, available at Soupsong.com/foyster.html.

77. D. Brown, *The Seduction Cookbook* (New York: Innova Publishing, 2005).

78. T. Batten, Oyster, n.d., available at ocean.udel.edu/mas/seafood/oyster.html.

79. J. Wharton, American Oyster, 2005, available at chesapeakebay.net/info/American _oyster.cfm.

80. Batten, Oyster.

81. Chesapeake Bay Oyster Catch, 1880–2003, Worldwatch Institute, 2004; T. Horton, Why Can't We Save the Bay? *National Geographic*, June 2005, p. 36.

82. Wharton, American Oyster.

83. J. Eilperin, World's Fish Supply Running Out, Experts Warn, *Washington Post*, November 3, 2006, p. A1.

84. World Fish Production, 1950–2003, Earth Policy Institute, 2005, available at earth-policy.org/Indicators/Fish/Fish_data.htm; A. Neori et al., Blue Revolution Aquaculture,

Environment, April 2007, pp. 36–43; B. Halweil, Is More Fish Farming a Good Thing? Worldwatch Institute, February 23, 2007, available at worldwatch.org/node/4932/print.

85. J. Marra, When Will We Tame the Oceans? *Nature*, July 14, 2005, p. 175.

86. R. L. Naylor et al., Effect of Aquaculture on World Fish Supplies, *Nature*, June 29, 2000, p. 1018.

87. K. Betts, Will Fish Farms Feed the World? *Environmental Science and Technology*, December 1, 2003, pp. 429A–430A.

88. *Ecologist*, November 2006, p. 12.

89. The Promise of a Blue Revolution, *Economist*, August 9–15, 2003, p. 19.

90. G. Felcyn, Shrimp Trade Case Threatens U.S. Exports, Shrimp Task Force, 2004, available at citac.info/shrimp/press_releases/2004/trade_case_07_02.htm; H. Josupeit, An Overview on the World Shrimp Market. *Globefish*, 2004 Available at globefish.org/dynamisk .php4?id=2269.

91. World Fish Production, 1950–2003, Earth Policy Institute, 2005, available at earth-policy.org/Indicators/Fish/Fish_data.htm.

92. D. R. Montgomery, Geology, Geomorphology, and the Restoration Ecology of Salmon, *GSA Today* [Geological Society of America], November 2004, p. 4.

93. A. P. McGinn, Blue Revolution: The Promises and Pitfalls of Fish Farming, *Worldwatch*, March–April 1998, p. 13.

94. The Promise of a Blue Revolution, p. 20.

95. J. Eilperin, Fish Farming's Bounty Isn't without Barbs, *Washington Post*, January 24, 2005, p. A1.

96. J. Tibbetts, Eating Away at a Global Food Source, *Environmental Health Perspectives*, April 2004, p. A286.

97. A. Gosline, How Long Can We Keep Fishing to Feed Fish? *New Scientist*, October 2, 2004, p. 16; Halweil, Is More Fish Farming a Good Thing?.

98. The Promise of a Blue Revolution, p. 21.

99. E. Stoddard, Eating Fish: Good for Heart, Bad for Environment? Scientific American .com, August 10, 2007, available at sciam.com/print_version.cfm?articleID=5081474F-E7F2–99DF-3F5AD9A.

100. A. Krebiehl, Developing an Appetite for Seafood, *Financial Times*, 2005, available at ft.com/cms/s/0/e38784fe-7c41–11d9–8992–00000e2511c8.html.

101. Stoddard, Eating Fish.

102. Ibid.

103. The Promise of a Blue Revolution, p. 21.

104. Stoddard, Eating Fish.

105. E. Gies, Sustainable Seafood, *E Magazine*, November–December 2005, p. 42.

106. R. L. Naylor, J. Eagle, and W. L. Smith, Salmon Aquaculture in the Pacific Northwest, *Environment*, October 2003, p. 30.

107. Farmed Salmon, *Ecologist*, October 2005, p. 30.

108. Naylor et al., Effect of Aquaculture on World Fish Supplies, p. 1016.

109. Q. Schiermeier, Fish Farms' Threat to Salmon Stocks Exposed, *Nature*, October 23, 2003, p. 753.

110. Ibid.

111. Farmed Salmon, p. 32.

112. Salmon Scam: Consumer Reports Analysis Reveals that Farm-Raised Salmon Is Often Sold as "Wild," Consumer Reports.org, August 2006, consumerreports.org/cro/food/food-shopping/meats-fish-protein-foods/mislabeled-salmon/salmon-8-06/overview/0608_salmon _ov.htm.

113. The Promise of a Blue Revolution, p. 20.

114. S. Doughton, Research Fuels Fear of Gene-Altered Fish, *Seattle Times*, June 8, 2004, available at seattletimes.nwsource.com/html/localnews/2001950789_genefish08m.html.

115. Ibid.

116. R. L. Naylor, J. Eagle, and W. L. Smith, Salmon Aquaculture in the Pacific Northwest, *Environment*, October 2003, p. 31.

117. J. Blythman, Who Needs a Traditional Fishmonger? *Ecologist*, September 2004, p. 46.

118. Naylor, Eagle, and Smith, Salmon Aquaculture in the Pacific Northwest, p. 31.

119. R. A. Hites, Global Assessment of Organic Contaminants in Farmed Salmon, *Science*, January 9, 2004, p. 226; R. Edwards, A Fishy Tale of Salmon, Dioxins and Food Safety, *New Scientist*, January 17, 2004, p. 8.

120. E. Hood, Are Farmed Salmon Fit Fare? *Environmental Health Perspectives*, April 2004, p. A274.

121. J. C. Ryan, Canadian Fish Farms Spread Disease to Wild Salmon, *Worldwatch*, May–June 2003, p. 7.

122. A New Way to Feed the World, *Economist*, August 9–15, 2003, p. 9.

123. J. Owen, Shrimp's Success Hurts Asian Environment, Group Says, *National Geographic News*, December 20, 2004, available at news.nationalgeographic.com/news/2004/06/0621 _040621_shrimpfarm.html.

124. Ibid.

125. S. Pratt, Fish Advisories on the Rise, *Geotimes*, November 2004, p. 14.

126. S. Levine, Who'll Stop the Mercury Rain? *U.S. News & World Report*, April 5, 2004 [U.S. edition], p. 70; Gillette, B., Mercury Rising, *E Magazine*, July–August 1998, p. 42.

127. M. Janofsky, E. P. A. Says Mercury Taints Fish across U.S., *New York Times*, August 25, 2004, p. A21.

128. S. Steingraber, How Mercury-Tainted Tuna Damages Fetal Brains, In These Times, 2004, available at inthesetimes.com/article/1787/.

129. E. Sadler, untitled, *New Scientist*, March 11, 2006, p. 14; Burros, M. High Mercury Levels are Found in Tuna Sushi. New York Times, January 23, 2008.

130. R. Renner, Mercury Woes Appear to Grow, *Environmental Science and Technology*, April 15, 2004, p. 144A.

131. Agency Updates Advice to Pregnant and Breastfeeding Women on Eating Certain Fish, Food Standards Agency, 2003, available at foodstandards.gov.uk/news/pressreleases/tuna_mercury.

132. Pratt, Fish Advisories on the Rise, p. 15; Mercury in Tuna, Consumer Reports.org, July 2006, available at http://www.consumerreports.org/cro/babies-kids/child-safety/food/mercury-in-tuna/tuna-safety/overview/0607_tuna_ov.htm.

133. S. Fields, Great Lakes Resource at Risk, *Environmental Health Perspectives*, March 2005, p. A169; K. Schott, RFK Jr. Stumps for Cleaner Environment at UW-La Crosse, *La Crosse Tribune*, November 3, 2005, available at lacrossetribune.com/articles/2005/11/03/news/03rfkjr.prt.

134. High Levels of Mercury in Lake Fish, *Chemical and Engineering News*, August 9, 2004, p. 22.

135. S. Squires, Benefits of Fish Exceed Risks, Studies Find, *Washington Post*, October 18, 2006, p. A14; Touting the Benefits of Fish, *New York Times*, October 19, 2006, p. A26; N. Bakalar, Nutrition: Study Questions Limits on Fish in Pregnancy, *New York Times*, February 27, 2007, p. F1.

136. Fish Threat to Human Health, *Ecologist*, September 2006, p. 10.

Chapter Nine

1. Tomatoes: Fruit or Vegetable, 2002, available at cookinglouisiana.com/Articles/Tomatoes-Veg-or-Fruit.htm.

2. G. Lucier, S. Pollack, M. Ali, and A. Perez, Fruit and Vegetable Backgrounder, *USDA Economic Research Service*, 2006, p. 7.

3. U.S. per Capita Fruit and Vegetable Consumption, 1976–2003, University of California, n.d., available at postharvest.ucdavis.edu/datastorefiles/234–66.pdf; G. Lucier, Vegetable Consumption Away from Home on the Rise, *Amber Waves*, September 2003; G. Lucier and A. Jerardo, Vegetables and Melons Situation and Outlook Yearbook, USDA, Economic Research Service, 2006.

4. G. Danekas, Missouri Agricultural Statistics Service, 2005, available at agebb.missouri.edu/mass/immrel/bull16e.htm; P. Hollis, U.S. Fall Vegetable Acreage Increases, Southeast Farm Press, 2005, available at southeastfarmpress.com/mag/farming_us_fall_vegetable/.

5. S. Pollack, Consumer Demand for Fruit and Vegetables: The U.S. Example, in Changing Structure of Global Food Consumption and Trade, p. 50, USDA Economic Research Service, 2001.

6. S. Pollack and A. Perez, Fruit and Tree Nuts Situation and Outlook Yearbook, USDA, Economic Research Service, 2005, available at ers.usda.gov/publications/FTS/yearbook05/FTS2005.pdf.

7. Highlight: Fresh-Market Fruit Production, USDA, Economic Research Service, Fruit and Tree Nuts Outlook/FTS-302, 2003, available at ers.usda.gov/publications/fts/jan03/fts-302.pdf.

8. Fruit and Tree Nuts Outlook, Highlight: Fresh-Market Fruit Production, USDA Economic Research Service, Fruit and Tree Nuts Outlook/FTS-302, 2003, p. 16.

9. S. Rawlins, Number of Farms Dedicated to Fruit Production Continues to Drop, American Farm Bureau Federation, available at fb.com/issues/analysis/Fruit_Production.html.

10. Ibid.

11. G. Lucier, Briefing Room, Vegetables and Melons: Background, USDA Economic Research Service, 2000.

12. Ibid.

13. Tomatoes: Fruit or Vegetable, p. 24.

14. Reality Check, *Newsweek*, February 5, 2007, p. 7.

15. Facts about Farmworkers, National Center for Farmworker Health, n.d., available at ncfh.org/factsheets.php.

16. Y. Sarig, J. F. Thompson, and G. K. Brown, Alternatives to Immigrant Labor? 2000, Center for Immigration Studies, available at cis.org/articles/2000/back1200.html.

17. Ibid.

18. Y. Sarig, J. F. Thompson, and G. K. Brown, Alternatives to Immigrant Labor? Center for Immigration Studies, 2000, available at cis.org/articles/2000/back1200.html.

19. Antioxidants, Healthy Eating Club, n.d., available at healthyeatingclub.org/info/articles/vitamins/antioxidants.htm; Fighting Heart Disease with Fruit, BBC News, March 2, 2001, available at news.bbc.co.uk/1/hi/health/1196255.stm.

20. Ibid.

21. Ibid.

22. Apples and More, University of Illinois, n.d., available at urbanext.uiuc.edu/apples/.

23. S. Taylor, Research Gives Apple Products an All-Around Thumbs Up, Apple Products Research and Education Council, 2004, appleproducts.org/pr19.html; Why "An Apple a Day" Is Still Good Advice, Vermont Apples, available at vermontapples.org/nutrition.html.

24. Antioxidants, 2002, Healthy Eating Club, available at healthyeatingclub.org/info/articles/vitamins/antioxidants.htm.

25. S. P. Seitz, McCafferty, and J. Kimberly, Research Suggests That Nutrients in Apples and Apple Juice Improve Memory and Learning, Apple Products Research and Education Council, 2004, available at appleproducts.org/pr21.html.

26. Good Fruit Grower, 2007, available at goodfruit.com/issues.php?article=1506&issue=55.

27. 2004 U.S. Fruit and Vegetable Outlook, Arizona State University, 2004.

28. 2004 Fruit and Vegetable Outlook, 2004, available at nfapp.east.asu.edu/Outlook02/Grapes.htm.

29. 2004 U.S. Fruit and Vegetable Outlook.

30. W. D. Gubler, Organic Grape Production in California, available at landwirtschaft-bw.info/servlet/PB/menu/1043205_l1/index.html.

31. Pesticides in Imported Grapes, *Environmental Working Group Food News*, n.d., available at foodnews.org.

32. H. Brunke, Commodity Profile with an Emphasis on International Trade: Grapes, University of California, Agricultural Marketing Res Center, 2002, available at agmrc.org/fruits/info/ccpgrapes.pdf.

33. C. S. Smith, *Wine: California, Global*, Rural Migration News, University of California, 2005, available at migration.ucdavis.edu/rmn/more.php?id=1069_0_5_0.

34. Varietals, n.d., available at aboutwines.com/home/reference/varietals/main.html.

35. U.S. Strawberry Industry, 1970–2004, USDA Economic Research *Service*, 2005, available at usda.mannlib.cornell.edu/MannUsda/viewDocumentInfo.do?documentID=1381.

36. The U.S. and World Situation: Strawberries, USDA Foreign Agricultural Service Horticulturaland Tropical Products Division, 2003.

37. Strawberry Information and Facts, Columbia4Kids, n.d., available at columbia4kids.com/articles/strawberry.php.

38. Strawberries and More, University of Illinois, n.d., available at urbanext.uiuc.edu/strawberries/.

39. S. Arnot, Berry Berry Strawberry, *Sauce Magazine*, n.d., available at saucecafe.com/article/5/15/1.

40. The U.S. and World Situation: Strawberries, USDA Foreign Agricultural Service Horticultural and Tropical Products Division, 2003.

41. S. Pollack and A. Perez, Fruit and Tree Nuts Outlook/FTS-305, USDA Economic Research Service, 2003, p. 12.

42. R. Hartnell, Berries Earn More Health Honors, *Vital Choices*, September 20, 2004, available at imakenews.com/vitalchoiceseafood/e_article000305777.cfm; Blueberries: The World's Healthiest Foods, n.d., available at whfoods.org/genpage.php?tname=foodspice&dbid=8#summary.

43. G. Lucier and A. Jerardo, *Vegetables and Melons Situation and Outlook Yearbook*, USDA Economic Research Service, 2006; Briefing Room: Potatoes, USDA, n.d.

44. C. Plummer, Briefing Room: Potatoes: Background, *USDA Economic Research Service*, 2004, available at ers.usda.gov/Briefing/Potatoes/background.htm.

45. Ibid.

46. Ibid.

47. S. L. Hafez and P. Sundararaj, Soil Fumigation for Nematode Management, 2005, available at potatogrower.com/?pageID=09&featureID=25.

48. Report Card: Pesticides in Produce, Environmental Working Group, 2005, available at foodnews.org/pdf/EWG_pesticide.pdf..

49. When Potatoes Become Human, *Ecologist*, June 2005, p. 8.

50. Ibid.

51. A. Shin, Outbreaks Reveal Food Safety Net's Holes, *Washington Post*, December 11, 2006, p. A1; E. Schlosser, Has Politics Contaminated the Food Supply? *New York Times*, December 11, 2006, p. A27; Reckless with Food Safety, *New York Times*, December 12, 2006, p. A30; Food for Thought, *Nature*, February 15, 2007, pp. 683–684; E. Williamson, FDA Was Aware of Dangers to Food, *Washington Post*, April 23, 2007, p. A1.

52. E. Williamson, FDA Was Aware of Dangers to Food, *Washington Post*, April 23, 2007, pp. A1.

53. Barrionuevo, Food Imports Often Escape Scrutiny, *New York Times*, May 1, 2007, pp. C1, C6.

54. Ibid.

55. R. Weiss, Tainted Chinese Imports Common, *Washington Post*, May 20, 2007, p. 1; D. Barboza, China Finds Poor Quality on Its Store Shelves, *New York Times*, July 5, 2007, p. C1, C2.

56. A. E. Cha, Farmed in China's Foul Waters, Imported Fish Treated with Drugs, *Washington Post*, July 6, 2007, p. A1.

57. R. Weiss, Tainted Chinese Imports Common, *Washington Post*, May 20, 2007, pp. A1.

58. Ibid.

59. Barrionuevo, Food Imports Often Escape Scrutiny.

60. Veganism, Wikipedia, n.d., available at wikipedia.org/wiki/Vegan.

61. How Many Vegetarians Are There? Vegetarian Resource Group, 2000, available at vrg.org/nutshell/poll2000.htm.

62. R. Corliss, Should We All Be Vegetarians? *Time Magazine*, October 14, 2002, p. 59.

63. R. Berry, *Famous Vegetarians and Their Favorite Recipes* (New York: Pythagorean Books, 1993).

64. J. Siegel, Smart Kids Said More Likely to Become Vegetarians as Adults, *Jerusalem Post*, December 15, 2006, p. 8.

65. Vegetarian Diets, American Dietetic Association, 2003, available at eatright.org/cps/rde/xchg/ada/hs.xsl/advocacy_933_ENU_HTML.htm.

66. Ibid.

67. Ibid.

68. Veganism.

69. Corliss, Should We All Be Vegetarians? pp. 56–64.

70. How to Win an Argument with a Meat Eater, n.d., available at vegsource.com/how_to _win.htm.

71. Plummer, Briefing Room.

72. Now, It's Not Personal, *Worldwatch*, July–August 2004, pp. 12–19.

73. Brett, Vegetarianism, n.d., available at alumnus.caltech.edu/~brett/whyveg.html.

74. Think You Can Be a Meat-Eating Environmentalist? People for the Ethical Treatment of Animals, advertisement, n.d.

75. M. Prewitt, Cheap Burgers in Paradise—History of the Hamburger, n.d., available at mcspotlight.org/media/press/restaurant_news.html.

76. People for the Ethical Treatment of Animals. [advertisement], n.d.

77. Veganism.

78. J. Motavalli, The Case against Meat, *E Magazine*, January–February 2002, p. 28; F. Pearce, Earth: The Parched Planet, *New Scientist*, February 25, 2006, p. 35.

79. Ibid., p. 29.

80. L. Else, The True Cost of Meat, *New Scientist*, August 14, 2004, pp. 42–45.

81. B. Friedrich, Britain's Environment Agency: Go Vegetarian to Stop Climate Change, *Huffington Post*, June 23, 2007, available at huffingtonpost.com/bruce_friedrich/britains-environment-age_b_53454.html.

82. Siegel, Smart Kids Said More Likely to Become Vegetarians as Adults.

83. Slaughterville Snubs Veggie Name Change, *Ecologist*, April 2004, p. 9.

84. S. Deneen, Body of Evidence: Were Humans Meant to Eat Meat? *E Magazine*, January–February 2002, p. 33; Plant Eaters vs. Meat Eaters, *Living Vegetarian*, n.d., available at jtcwd.com/vegie/plant_or_meat_eaters.html; Humans Are Naturally Plant-Eaters and Not Meat-Eaters, n.d., available at jtcwd.com/vegie/chart.html.

85. Plant Eaters vs. Meat Eaters, n.d., Living Vegetarian, available at jtcwd.com/vegie/plant_or_meat_eaters.html.

86. History and Origin of the Potato, Sun Spiced, n.d., available at sunspiced.com/phistory.html.

87. Plant Eaters vs. Meat Eaters.

Chapter Ten

1. P. Chek, You Are What You Eat—Processed Foods, 2002, available at chekinstitute.com/printfriendly.cfm?select=42.

2. Ibid.

3. Ibid.

4. P. Lempert, Inside the Battle for Your Supermarket's Shelves, MSNBC, 2004, available at msnbc.msn.com/id/4955805/.

5. Elements of Food Processing Methods and Equipment, Ohio State University, 1999, available at class.fst.ohio-state.edu/FST401/Information/Elements-Food-Processing.html.

6. Career Guide to Industries, 2006–2007, Food Manufacturing, Bureau of Labor Statistics, U.S. Department of Labor, 2005, available at bls.gov/oco/cg/cgs011.htm.

7. Ibid.

8. Ibid.

9. Ibid.

10. What Meat Means, *New York Times*, February 6, 2005, p. 12.

11. Killing for a Living: How the Meat Industry Exploits Workers, n.d., available at goveg.com/workerRights.asp?pf=true.

12. E. Schlosser, *Fast Food Nation* (New York: HarperCollins, 2002), p. 197.

13. E. Schlosser, *Fast Food Nation: The Dark Side of the All-American Meal* (New York: Penguin Books, 2002), p. 197.

14. E. Schlosser, Slow Food, *Ecologist*, April 2004, p. 41.

15. Ibid.

16. M. Pollan and E. Schlosser, Modern Meat, 2002, available at pbs.org/wgbh/pages/frontline/shows/meat/slaughter/slaughterhouse.html.

17. Blood, Sweat, and Fear, *Human Rights Watch*, New York, 2005, available at hrw.org/reports/2005/usa0105.

18. Consumer Reports Finds 71 Percent of Store-Bought Chicken Contains Harmful Bacteria, press release, Consumers Union, February 23, 1998.

19. P. S. Mead et al., Food-Related Illnesses and Death in the United States, Centers for Disease Control, 1999, available at cdc.gov/ncidod/eid/vol5no5/mead.htm.

20. G. Volke, E. W. Allen, and M. Ali, Wheat Backgrounder, USDA Economic Research Service, 2005, available at ers.usda.gov/Publications/whs/dec05/whs05K01; D. Karp, Puff the Magic Preservative: Lasting Crunch, But Less Scent, *New York Times*, October 25, 2006, p. F1.

21. The Food Commission Guide to Food Additives, Food Commission [UK], 2004, available at foodcomm.org.uk/additives_poster.pdf.

22. Mead et al., Food-Related Illness and Death in the United States.

23. Food Additives, Food Standards Australia New Zealand, n.d., available at foodstandards .gov.au/foodmatters/foodadditives.cfm.

24. Ibid.

25. M. Burros, Which Cut Is Older? (It's a Trick Question), *New York Times*, February 21, 2006, p. A12.

26. Food Additives.

27. V. Gilman, Food Coloring, *Chemical and Engineering News*, August 25, 2003, p. 34.

28. FDA and Monosodium Glutamate (MSG), Food and Drug Administration, 1995, available at cfsan.fda.gov/~lrd/msg.html.

29. Monosodium Glutamate, Wikipedia, 2005, available at wikipedia.org/wiki/Monosodium _glutamate.

30. Food Processing and Nutrition, Government of Victoria, Australia, 2005, available at betterhealth.vic.gov.au/bhcv2/bhcarticles.nsf/pages/Food_processing_and_nutrition?open; G. Gopalan, B. V. Rama Sastri, and S. C. Balasubramanian, Cooking and Preserving Nutrition, 2000, available at webhealthcentre.com/general/diet_nutrition_cooknut.asp.

31. Faux Food: Where Have All Our Nutrients Gone? Available at health.msn.com/fitness/ articlepage.aspx?cp-documentid=100164969.

32. D. Reid, Milk and Dairy, n.d., available at hps-online.com/troph9.htm.

33. Milk, Wikipedia, n.d., available at en.wikipedia.org/wiki/Milk.

34. M. Miller, Cheese. Agricultural Marketing Resource Center, 2005, available at agmrc .org/agmrc/commodity/livestock/dairy/dairy+cheese.htm.

35. Ibid.; Butter and Cheese Production: Wisconsin, 2000–2004, *State of Wisconsin Agricultural Statistics*, 2005, available at nass.usda.gov/Statistics_by_State/Wisconsin/Publications/ Annual_Statistical_Bulletin/index.asp.

36. Miller, Cheese.

37. Ibid.

38. Food and Agriculture Industry: Natural and Processed Cheese, 1997, available at epa .gov/ttn/chief/ap42/ch09/final/c9s06–1.pdf; N. Willman and C. Willman, Home Cheese-making, 2005, available at cheeselinks.com.au/products.html; Cheese: Development of Structure, 2004, available at cip.ukcentre.com/cheese4.htm; Cheese, Fresh, 2000, available at wholehealthmd.com/print/view/1,1560,FO_275,00.html.

39. Why Does Swiss Cheese Have Holes in It? 2002, available at ask.yahoo.com/20020610 .html.

40. Butter Manufacture, University of Guelph, n.d., available at foodsci.uoguelph.ca/ dairyedu/butter.html; Butter, Wikipedia, n.d., available at wikipedia.org/wiki/Butter.

41. History of Margarine, National Association of Margarine Manufacturers, available at margarine.org/historyofmargarine.html.

42. B. Worthington-Roberts, Milk, *Encarta*, 2006, available at encarta.msn.com/text _761562453_0/Milk.html; Outline of Ice Cream Manufacture, N. E. M. Business Solutions, n.d., available at cip.ukcentre.com/icream.htm; Ice Cream, Wikipedia, n.d., available at en.wikipedia.org/wiki/Ice_cream.

43. Ice Cream, *Chemical and Engineering News*, November 8, 2004, p. 51.

44. It's a Wrap—What's New in the Area of Packaging? *Food Today*, 2005, available at eufic.org/gb/food/pag/food33/food333.htm; Food Packaging, National Food Processors Association, n.d., available at gmabrands.com; The Safety of Plastic Food Packaging, American Plastics Council, 1999, available at plasticsinfo.org.

45. Food Packaging, The Site, n.d., available at thesite.org.uk/healthandwellbeing/ fitnessanddiet/food/foodpackaging.

46. To Replace Oil, U.S. Experts See Amber Waves of Plastic; American Crops Could Be Used in Place of Many Products' Petroleum Base, Some Scientists Say, Truth About Trade and Technology, 2005, available at truthabouttrade.org/print.asp?id=4063.

47. Food Packaging.

48. Understanding Plastic Film, American Plastics Council, 1996, p. 4.

49. M. Knopper, The Perils of Plastic, *E Magazine*, September–October 2003, pp. 40–41; E. Jardina and D. Fischer, What Can I Do? *Inside Bay Area*, 2005, available at insidebayarea .com/bodyburden/ci_2603026; H. L. Obidzinski, Wrapping It Up, University of Buffalo, 1999, available at eng.buffalo.edu/Courses/ce435/PlasticWrap/plasticwrap.html.

50. B. Trivedi, The Hard Smell, *New Scientist*, December 16, 2006, pp. 36–38; M. W. Pressler, Appealing to the Senses, *Washington Post*, February 19, 2006, p. F1; T. B. Rivedi, The Hard Smell, *New Scientist*, December 16, 2006, pp. 36–39; N. Blackburn, A Sweet Smell— Israeli Scientists Enhance Food with the Scent of Flowers, *Israel 21c*, April 22, 2007, available at israel21c.net.

51. Food Packaging.

52. M. Locke, Edible Packaging Could Reduce Plastic Food Packaging, *USA Today*, December 16, 2004, available at organicconsumers.org/foodsafety/packaging122004.cfm; Indestructible Bag Ban, *New Scientist*, October 25, 2005, p. 7; J. Marshall, Packaging Unwrapped, *New Scientist*, April 7, 2007, pp. 37–41.

53. M. L. Gavin, Deciphering Food Labels, 2005, available at kidshealth.org/parent/food/ general/food_labels.html.

54. Quoted in F. Forencich, Is It Food? *GoAnimal*, 2005, available at goanimal.com/ newsletters/2005/food_food_product/food_products.html.

55. S. Davidson, Free Food: Packaging Labels Contain Errors, *Live Science*, 2004, available at livescience.com/strangenews/041229_free_food.html.

56. F. Hall, AFTC Testifies in Senate on Ag Transport, Energy Issues, Current Trucking Industry News, 2005, available at truckinginfo.com/news/news-detail.asp?news_id=55817.

57. N. Blisard and H. Stewart, *Food Spending in American Households*, USDA Economic Research Service Information Bulletin Number 23, 2007.

58. Ibid.

59. T. Dunkle, Is Your Weight Killing You? DietPower, 2006, available at dietpower.com/features/overweight_risks.php.

60. P. Lempert, Inside the Battle for Your Supermarket's Shelves, MSNBC, 2004, available at msnbc.msn.com/id/4955805/.

61. The Biggest Cheese, 2002, available at images.forbes.com/images/forbes/2002/0415/130chart3_426_400.gif.

62. Kraft Brands North America, n.d., available at kraft.com/brands.

63. Lempert, Inside the Battle for Your Supermarket's Shelves.

64. J. Mercola, Four Ways Junk Food Marketing Targets Your Kids, 2003, available at mercola.com/2003/nov/26/junk_food_marketing.htm.

65. Ibid.

66. C. E. Mayer, Groups to Use Consumer-Protection Laws in Suit against Kellogg, Viacom, *Washington Post*, January 19, 2006, p. D3.

67. Ibid.

68. SpongeBob, Kellogg Get the Big Squeeze, January 23, 2006, available at commercialfreechildhood.org/news/spongebobsqueeze.htm.

69. E. Assadourian, When Good Corporations Go Bad, *Worldwatch*, May–June 2005, p. 18.

70. E. Assadourian, The Role of Stakeholders, *Worldwatch*, September–October 2005, p. 22.

71. Food and Nutrition Information Center, USDA, 2005, available at nal.usda.gov/fnic/Fpyr/pyramid.html.

Chapter Eleven

1. L. McShane, American Chestnut Sets Record, Crowned Hot Dog Champ, *Washington Post*, July 5, 2007, p. A2.

2. Takeru Kobayashi, International Federation of Competitive Eating, n.d., available at ifoce.com/eaters.php?action=detail&sn=22.

3. Joey Chestnut, International Federation of Competitive Eating, n.d., available at ifoce.com/eaters.php?action=detail&sn=106.

4. Cookie Jarvis, International Federation of Competitive Eating, n.d., available at ifoce.com/eaters.php?action=detail&sn=9.

5. Eric Booker, International Federation of Competitive Eating, n.d., available at ifoce.com/eaters.php?action=detail&sn=14.

6. Sonya Thomas, International Federation of Competitive Eating, n.d., available at ifoce.com/eaters.php?action=detail&sn=20.

7. D. Sharp, Sonya Thomas Sets New Record at Lobster-Eating, August 13, 2005, available at boston.com/news/local/maine/articles/2005/08/13/Sonya_thomas_sets_new.

8. D. Kadlec, The Marriage of Gluttony and Sport, *Time Magazine*, April 26, 2004, p. 19; P. Hurley, Competitive Eating Is Not a Sport, 2005, available at useless-knowledge.com/1234/aug/article098.html; Sharp, Sonya Thomas Sets New Record at Lobster-Eating Contest; L. Christie, Getting Paid to Eat, 2005, available at money.cnn.com/2005/07/01/pf/eating_to_live/.

9. C. Zimmer, Underground Gourmet: Mole Sets a Speed Record, *New York Times*, February 8, 2005, pp. F1, F2.

10. M. D. Myers, Definition of Obesity, 2004, available at weight.com/definition.asp?page=1.

11. Body Mass Index, Wikipedia, n.d., available at en.wikipedia.org/wiki/Body_mass_index.

12. Overweight and Obesity: Defining Overweight and Obesity, Centers for Disease Control and Prevention, n.d., available at cdc.gov/nccdphp/dnpa/obesity/defining.htm.

13. Body Mass Index, Wikipedia.

14. C. W. Schmidt, Obesity: A Weighty Issue for Children, *Environmental Health Perspectives*, October 2003, p. A702.

15. Body Mass Index.

16. Overweight and Obesity.

17. D. Grady, The Secret Lives of Fat Cells, *International Herald Tribune*, July 12, 2004, p. 12.

18. D. Grady, Fat: The Secret Life of a Potent Cell, *New York Times*, July 6, 2004, pp. F1, F6; Myers, Definition of Obesity.

19. Obesity in the U.S., n.d., available at obesity.org; V. Reitman, Diet or Die, *Jerusalem Post*, Up Front insert section, January 20, 2006, p. 22.

20. R. Longley, Americans Getting Taller, Bigger, Fatter, Says CDC, n.d., available at usgovinfo.about.com/od/healthcare/a/tallbutfat.htm.

21. Ibid.

22. F as in Fat: How Obesity Policies Are Failing in America—2006, Trust for America's Health, available at rwjf.org/research/researchdetail.jsp?ia=138&id=2974; C. Newman, Why Are We So Fat? *National Geographic*, August 2004, p. 52; National Health Interview Survey, National Center for Health Statistics, 2006, available at cdc.gov/nchs/about/major/nhis/released200603.htm.

23. Overweight and Obesity in U.S. Cities, American Obesity Association, 2005, available at obesity.org; no author given, Obesity in America. Time, September 10, 2007, p. 10.

24. E. Marshall, Public Enemy Number One: Tobacco or Obesity? *Science*, May 7, 2004, p. 804.

25. R. Winslow and R. Rundle, Obese American Teens Seek Stomach Surgery, *Wall Street Journal*, October 7, 2003, p. A1.

26. F as in Fat; S. A. Hearne, L. M. Segal, P. J. Unruh, M. J. Earls, and P. Smolarcik,Trust for America's Health, 2004, available at healthyamericans.org/reports/obesity/.

27. N. Hellmich, Obesity Surges among the Affluent, *USA Today*, May 3, 2005, p. 1A.

28. Obesity in the U.S.

29. The Overweight Pet, 2006, available at thepetcenter.com/imtop/overweight.html; The Nation's Pets are Living Large … Too Large, available at purina.com/science/research/PetsLivingLarge.aspx.

30. R. Rivenburg, Letting Out the Seams, *Jerusalem Post*, February 19, 2006, p. 15.

31. C. Newman, Why Are We So Fat? *National Geographic*, August 2004, p. 52.

32. Fat Nation Fights Back—Sort Of, *U.S. News & World Report*, July 1, 2002, p. 4.

33. K. Pallarito, Plumping Up Profits, *U.S. News & World Report*, December 20, 2004, p. 41.

34. Newman, Why Are We So Fat? p. 52; Ambulances Brought into Service for XL Patients, 2006, available at ems1.com.

35. W. St. John, One Size Does Not Fit All, So Funeral Industry Adapts to the Trend in Obesity, *International Herald Tribune*, September 29, 2003, p. 5.

36. Ibid.; L. Tanner, Obesity Epidemic Hits Child Safety Seats, MSNBC.com, April 3, 2006, available at msnbc.msn.com/id/12122112/.

37. St. John, One Size Does Not Fit All, p. 5.

38. Ibid.

39. R. Lalasz, Will Rising Childhood Obesity Decrease U.S. Life Expectancy? Population Reference Bureau, available at prb.org/Articles/2005/WillRisingChildhoodObesityDecrease USLifeExpectancy.aspx; J. Raloff, Inflammatory Fat, *Science News*, February 28, 2004, pp. 139–140; D. Grady, Fat: The Secret Life of a Potent Cell, *New York Times*, July 6, 2004, pp. F1, F6; N. Seppa, Put Down That Fork, *Science News*, January 14, 2006, p. 21.

40. F as in Fat; S. Perrine, Low-Carb Craze Threatens to Head Off Track, *USA Today*, April 13, 2004, p. 23A.

41. Obesity, *U.S. News & World Report*, December 6, 2004, pp. 53–54; E. Ross, Study Links Midlife Obesity, Dementia, *Columbus Dispatch*, April 29, 2005, p. A5; American Dietetic Association, Why Worry Now? Obesity: A National Epidemic, *Newsweek*, March 22, 2004.

42. A. Underwood and J. Adler, What You Don't Know about Fat, *Newsweek*, September 20, 2004, p. 57.

43. The Shape of Things to Come, *Economist*, December 13, 2003, p. 5.

44. C. Power, Big Trouble, *Newsweek*, August 11, 2003, p. 44.

45. F as in Fat; S. A. Hearne, L. M. Segal, P. J. Unruh, M. J. Earls, and P. Smolarcik, Trust for America's Health, 2004, available at healthyamericans.org/reports/obesity/.

46. A Fact Sheet from the National Diabetes Education Program, n.d., available at ndep.nih .gov/diabetes/youth/youth_FS.htm.

47. N. R. Kleinfeld, Diabetes and Its Awful Toll Quietly Emerge as a Crisis, *New York Times*, January 9, 2006, pp. 1, 18–19.

48. Life Expectancy Hits Record High, National Center for Health Statistics, 2005, available at cdc.gov/nchs/pressroom/05facts/lifeexpectancy.htm.

49. R. Lalasz, Will Rising Childhood Obesity Decrease U.S. Life Expectancy? Population Reference Bureau, 2005, available at prb.org/Articles/2005/WillRisingChildhoodObesity DecreaseUSLifeExpectancy.aspx; R. Stein, Baby Boomers Appear to Be Less Healthy Than Parents, *Washington Post*, April 20, 2007, p. A1.

50. M. D. Lemonick, How We Grew So Big, *Time Magazine*, August 9, 2004, pp. 48, 49; Study: Child Obesity Expected to Soar Worldwide, 2006, MSNBC, available at msnbc.msn .com/id/11694799/print/1/displaymode/1098/.

51. A. Stone, The All-You-Can-Eat Gene, *Discover*, October 2004, p. 17.

52. K. De Seve, Food Addiction, *ScienCentral News*, May 11, 2004, available at sciencentral .com/articles/view.php3?article_id=218392245.

53. S. Velu, Taste the Difference, *ScienCentral News*, 2002, available at http://www .sciencentral.com/articles/view.php3?language=english&type=article&article_id=218391842.

54. J. Kaiser, Mysterious, Widespread Obesity Gene Found Through Diabetes Study, *Science*, April 13, 2007, p. 185; Gene Makes Some People Fatter, *New Scientist*, April 21, 2007, p. 16.

55. J. Kaiser, Mysterious, Widespread Obesity Gene Found through Diabetes Study, *Science*, April 13, 2007, p. 185.

56. A. Spake, Rethinking Weight, *U.S. News & World Report*, February 9, 2004, p. 53.

57. How Some People Manage to Stay Thin in a Fat, Fat World, *Tucson Citizen*, September 11, 2004, available at tucsoncitizen.com/breaking/091104globesity.html.

58. J. Raloff, Still Hungry? *Science News*, April 2, 2005, p. 216.

59. G. Kolata, Hormone Holds a Clue in Fight against Obesity, *International Herald Tribune*, April 3–4, 2005, p. 2.

60. How Some People Manage to Stay Thin in a Fat, Fat World.

61. Raloff, Still Hungry? p. 216.

62. J. Gramza, Infectious Obesity, *ScienCentral News*, 2001, available at sciencentral .com/articles/view.php3?article_id=218391620; R. R. Britt, Virus May Contribute to Obesity, MSNBC.Com, August 20, 2007, available at msnbc.msn.com/id/20359961/print/1/displaymode/1098/.

63. N. Angier, Study Finds Signs of Elusive Pheromones in Humans, *New York Times*, March 12, 1998, p. A22.

64. R. Stein, Way to Shrink, Grow Fat Is Found, *Washington Post*, July 2, 2007, p. A1.

65. R. M. Henig, Fat Factors, *New York Times Magazine*, August 13, 2006, pp. 31–32, 52–56; M. Bajzer and R. J. Seeley, Obesity and Gut Flora, *Nature*, December 21–28, 2006, pp. 1009–1010; R. E. Ley, P. J. Turnbaugh, S. Klein, and J. I. Gordon, Human Gut Microbes Associated with Obesity, *Nature*, December 21–28, 2006, pp. 1022–1023.

66. Overweight and Obesity: Contributing Factors, Centers for Disease Control and Prevention, 2005, available at cdc.gov/nccdphp/dnpa/obesity/contributing_factors.htm; J. Pytnam, J. Allshouse, and L. S. Kantor, U.S. per Capita Food Supply Trends: More Calories, Refined Carbohydrates, and Fats, *Food Review*, USDA, Economic Research Service, Winter 2002, p. 8.

67. P. Taylor, C. Funk, and P. Craighill, Eating More; Enjoying Less, Pew Research Center, 2006, available at pewresearch.org/pubs/309/eating-more-enjoying-less.

68. Obesity Policy and the Law of Unintended Consequences, *Amber Waves*, 2005, available at ers.usda.gov/AmberWaves/June05/Features/ObesityPolicy.htm.

69. USDA, *Agriculture Fact Book 2001–2002* (Washington, D.C.: U.S. Government Printing Office, 2002).

70. C. Newman, Why Are We So Fat? *National Geographic*, August 2004, p. 49; F. Wooldridge, Food and Environment: Part 26—Next Added 100 Million Americans, 2006, available at americanchronicle.com/articles/27686.

71. K. Mulvihill, Healthwatch: Americans Load Up on Junk Food, 2004, available at cbs5 .com/news/local/2004/06/02/HealthWatch:_Americans_Load_Up_on_Junk_Food.html; Scientists Target Soda as Main Cause of Obesity, MSNBC, 2006, available at msnbc.msn.com/id/ 11686974/.

72. Schmidt, Obesity, p. A705.

73. J. de la Roca, A Far Eastern Perspective, *Jerusalem Post Special Supplement*, March 17, 2006, p. 30.

74. C. Gorman, How to Eat Smarter, *Time Magazine*, December 8, 2003, pp. 66–67.

75. W. Duffy, Sweet Death, *Ecologist*, November 2003, p. 44.

76. J. Putnam, J. Allshouse, and L. S. Kantor, U.S. per Capita Food Supply Trends: More Calories, Refined Carbohydrates, and Fats, *Food Review*, Winter 2002, p. 3.

77. Duffy, Sweet Death, p. 46.

78. USDA, *Agriculture Fact Book 2001–2002*.

79. Duffy, Sweet Death, pp. 44–53.

80. Sweet Smell of Excess, *Ecologist*, November 2003, pp. 42–43.

81. A. L. Kelly, Can We Downsize? *Los Angeles Times*, April 5, 2004, p. F1.

82. Sweet Smell of Excess, pp. 42–43.

83. Kelly, Can We Downsize? pp. F1, F4; D. Leonhardt, The Economics of Behaving against Our Own Best Interests, *International Herald Tribune*, May 2, 2007, p. 12.

84. Down with Supersizing, *International Herald Tribune*, March 6–7, 2004, p. 8.

85. D. M. Cutler, E. L. Glaeser, and J. M. Shapiro, Why Have Americans Become More Obese? National Bureau of Economic Research, 2003, p. 21.

86. Be Afraid. Be Very Afraid, *International Herald Tribune*, December 21, 2004, p. 8.

87. L. Krantz and S. Sveum, *The World's Worsts* (New York: HarperCollins, 2005), pp. 195–196.

88. A. Spake, Building Illness, *U.S. News & World Report*, June 20, 2005, p. 54.

89. J. Mercola, U.S. Food Industry Comes under Scrutiny, 2003, available at mercola.com/ 2003/apr/5/food_industry.htm.

90. C. Watkins, Can Obese People Sue the Food Industry—and Win? *Inform*, August 2004, pp. 505–507.

91. M. Warner, Salads or No, Cheap Burgers Revive McDonald's, *New York Times*, April 19, 2006, pp. A1, C4.

92. Untitled, *Reader's Digest*, December 2003, p. 26.

93. D. Levitsky, The More You Have on Your Plate, the More You Overeat, 2004, available at newswise.com/articles/view/507424.

94. T. Dunkle, Is Your Weight Killing You? DietPower, 2006, available at dietpower.com/ features/overweight_risks.php.

95. Y. Ben Ami, McDonald Has No Farm, *Haaretz* [Israeli English-language newspaper], December 19, 2003, p. B7; M. Khan, The Dual Burden of Overweight and Underweight in Developing Countries, Population Reference Bureau, 2006, available at http://www.prb.org/Articles/2006/TheDualBurdenofOverweightandUnderweightinDevelopingCountries.aspx.

96. N. Rigby, Advertising to Children, *Ecologist*, April 2004, p. 16.

97. E. Olson, Kids Getting Steady Diet of Junk Food Advertising, *International Herald Tribune*, March 30, 2007, p. 13.

98. A. Martin, Kellogg Scales Back Sugary Ads, *International Herald Tribune*, June 15, 2007, p. 14; A. Martin, Kellogg to Phase Out Some Food Ads to Children, *New York Times*, June 14, 2007, p. C1, C2.

99. A. Spake, Learning about Fat, *U.S. News & World Report*, October 11, 2004, p. 35.

100. J. Wiecha, Soundbites, *New Scientist*, April 8, 2006, p. 11.

101. A. Spake, Thinner Schools, *U.S. News & World Report*, December 15, 2003, p. 64.

102. Ibid.

103. Fighting Obesity in Schools, *International Herald Tribune*, January 17–18, 2004, p. 6; A. Shin, Removing Schools' Soda Is Sticky Point, *Washington Post*, March 22, 2007, p. D3.

104. The Soda Scourge, *Washington Post*, May 11, 2006, p. A26.

105. D. Kiley, A Food Fight over Obesity in Kids, *Business Week Online*, September 30, 2004, available at businessweek.com/bwdaily/dnflash/sep2004/nf20040930_0110_db035.htm; J. E. Caplan et al., How Bill Put the Fizz in the Fight against Fat, *Time*, May 16, 2006, pp. 27–29; Sack, K. Schools Found Improving on Nutrition and Fitness. New York Times, October 20, 2007, p. A10.

106. Obesity Report Cards, Environmental Health Perspectives, September 2004, p. A735.

107. Kiley, A Food Fight over Obesity in Kids; M. Burros and M. Warner, Bottlers Agree to a School Ban on Sweet Drinks, *New York Times*, May 4, 2006, pp. A1, A22.

108. A. Waters, Eating for Credit, *New York Times*, February 24, 2006, p. A23.

109. Limits Placed on Fat in Public School Meals, *International Herald Tribune*, March 5, 2004, p. 5; Of Taters and Tots, *Science News*, February 18, 2006, pp. 109–110.

110. P. Thomas, A Big Fat Problem, *Ecologist*, December 2006, pp. 33–43; Enter the Obesogen, *Economist*, February 24, 2007, p. 83; Fattening Toxins, *Worldwatch*, May–June 2007, p. 7; G. Stemp-Morlock, Exploring Developmental Origins of Obesity, *Environmental Health Perspectives*, May 2007, p. A242.

111. C. Brownlee, Buff and Brainy, *Science News*, February 25, 2006, p. 124.

112. C. Lambert, The Way We Eat Now, 2004, available at sophists.org/article-print-300.html; P. J. Wart, Amish Lifestyle Shows Activity Key to Fitness, Vanderbilt University, 2005, available at vanderbiltowc.wellsource.com/dh/contentasp?ID=1593.

113. Wart, Amish Lifestyle Shows Activity Key to Fitness.

114. When Work Is a Workout, *International Herald Tribune*, January 15, 2004, p. 8.

115. D. Grady, New Weight-Loss Focus: The Lean and the Restless, *New York Times*, May 24, 2005, pp. F1, F6.

116. Sack, K. Schools Found Improving on Nurition and Fitness, New York Times, October 20, 2007.

117. D. Suzuki, Suburban Sprawl Is Bad for People and the Planet, Environmental News Network, 2003, available at enn.com/news/2003-09-16/s_8319.asp.

118. C. W. Schmidt, Obesity: A Weighty Issue for Children, *Environmental Health Perspectives*, October 2003, p. A703.

119. Fatter and Gassier, *Worldwatch*, January–February 2007, p. 7.

120. C. W. Schmidt, Sprawl: The New Manifest Destiny? *Environmental Health Perspectives*, August 2004, p. A625.

121. Spake, Thinner Schools.

122. Schmidt, Sprawl.

123. Ibid.

124. How Some People Manage to Stay Thin in a Fat, Fat World, *Tucson Citizen*, September 11, 2004, available at tucsoncitizen.com/breaking/091104globesity.html.

125. T. Dunkle, Is Your Weight Killing You? DietPower, 2006, available at dietpower.com/features/overweight_risks.php.

126. M. Carmichael, Stronger, Faster, Smarter, *Newsweek*, April 9, 2007, pp. 30–35; C. Brownlee, Buff and Brainy, *Science News*, February 25, 2006, p. 122.

127. NPD Reports Americans Eat Differently 20 Years Later, NPD Group, 2005, available at npd.com/press/releases/press_051006.html; Americans Are More Accepting of Heavier Bodies, 2006, available at msnbc.msn.com/id/10807526/print/1/displaymode/1098/.

128. A. Spake, Rethinking Weight, *U.S. News & World Report*, February 9, 2004, p. 52.

129. Y. Garcia, What about Large Women's Clothes? Hu.mtu.edu/~pjsotiri/subjects/courses/shop/clothing.html.

130. G. Kolata, Genes Take Charge, and Diets Fall by the Wayside, *New York Times*, May 8, 2007, p. F1.

131. NPD Reports Americans Eat Differently 20 Years Later.

132. G. Kolata, Low Calories, Empty Promises, *International Herald Tribune*, January 6, 2005, p. 10; D. Butler, Slim Pickings, *Nature*, March 18, 2004, pp. 252–254.

133. N. Bakalar, Nutrition: An Upside to Hard Times, *New York Times*, October 9, 2007.

134. K. M. Reese, Mouth Device Fights Obesity, *Chemical and Engineering News*, June 14, 2004, p. 80.

135. S. Brink, America's Expanding Waistline, *U.S. News & World Report*, October 27, 2003, p. 71.

136. G. Gardner and B. Halweil, The World Pays a Heavy Price for Malnutrition, *International Herald Tribune*, March 9, 2000, p. 9; N. Singer, Do My Knees Look Fat to You? *New York Times*, June 15, 2006, pp. G1, G3; D. K., What to Expect of Liposuction, *U.S. News and World Report*, June 18, 2007, pp. 56–57.

137. Grady, Fat, pp. F1, F6.

138. *Ecologist*, September 2005, p. 10.

139. Laser Melts Fat Away, *New Scientist*, April 15, 2006, p. 7.

140. M. Santora, Obese Teenagers Turning to Surgery, *International Herald Tribune*, November 27–28, 2004, p. 5.

141. M. Freudenheim, A Boom in Surgery to Shrink the Stomach, *International Herald Tribune*, August 30–31, 2003, p. 9; D. Grady, Hazards Found in Surgery for Obese, *International Herald Tribune*, May 6, 2004, p. 11.

142. B. Strelsand, Weighing the Risks, *U.S. News & World Report*, March 27, 2006, p. 62; R. Pear, Obesity Surgery Often Leads to Complications, Study Says, *New York Times*, July 24, 2006, p. A13.

143. A Staple in Time, *Economist*, March 27, 2004, p. 51; Does Weight Stay Off? Consumer Reports.org, February 2006, available at consumerreports.org.

144. Stomach Stapling Leads to Longer Lives, MSNBC.com, August 22, 2007, available at msnbc.msn.com/id/20395008/.

Chapter Twelve

1. J. I. Nassauer, Agricultural Landscapes in Harmony with Nature, in *Fatal Harvest: The Tragedy of Industrial Agriculture*, ed. A. Kimbrell (Washington, D.C.: Island Press, 2002), p. 189.

2. D. Nierenberg and B. Halweil, Cultivating Food Security, in *State of the World 2005*, ed. L. Starke (New York: Norton), pp. 62–77; L. Horrigan, R. S. Lawrence, and P. Walker, How Sustainable Agriculture Can Address the Environmental and Human Health Harms of Industrial Agriculture, *Environmental Health Perspectives*, May 2002, pp. 445–456; D. Tilman, K. G. Cassman, P. A. Matson, R. Naylor, and S. Polasky, Agricultural Sustainability and Intensive Production Practices, *Nature*, August 8, 2002, pp. 671–677; M. Margolis, Crisis in the Cupboard, *Newsweek*, June 9, 2003, pp. 50–54; B. Halweil, Where Have All the Farmers Gone, *Worldwatch*, September–October 2000, pp. 12–28; K. A. Annan, Sustaining Our Future, *Environment*, October 2000, pp. 25–30.

3. N. Wirzba, ed., *The Art of the Commonplace: The Agrarian Essays of Wendell Berry* (Washington, D.C.: Shoemaker and Hoard, 2002), p. x.

4. T. Jones, The Scoop on Dirt, *E Magazine*, September–October 2006, p. 34.

5. Pesticides in Ground Water, *Pesticides News*, no. 21, September 1993, p. 18, available at pan-uk.org/pestnews/issue/pn21/pn21p18a.htm.

6. News Briefs, *Environmental Science and Technology*, July 1, 1998, p. 307A.

7. C. M. Cooney, Urban and Farm Runoff Still a National Problem, *Environmental Science and Technology*, June 23, 2004, available at pubs.acs.org/subscribe/journals/esthag-w/2004/jun/policy/cc_runoff.html.

8. J. Robbins, How to Win an Argument with a Meat Eater, 2006, available at vegsource.com/how_to_win.htm.

9. F. Pearce, Thirsty Meals That Suck the World Dry, *New Scientist*, February 1, 1997, p. 7.

10. D. Pauly and J. Maclean, *In a Perfect Ocean* (Washington, D.C.: Island Press, 2003).

11. Ibid., pp. 91–120; Sustaining America's Fisheries and Fishing Communities, Environmental Defense, 2007, available at environmentaldefense.org/article.cfm?contentID=6109.

12. R. W. Kates, Ending Hunger: Current Status and Future Prospects. *Consequences* 2:2 (1996): 3–11; G. Conway, Food for All in the 21st Century, *Environment*, January–February

2000, pp. 9–18; Filling the World's Belly, *Economist*, December 13, 2003, pp. 3–5; F. M. Lappé, J. Collins, and P. Rosset, *World Hunger: Twelve Myths*, 2nd ed. (New York: Grove Press, 1998); V. Smil, *Feeding the World: A Challenge for the 21st Century* (Cambridge, Mass.: MIT Press, 2000).

13. Consumption Statistics, Centers for Disease Control, n.d., available at 5aday.com/html/research/consumptionstats.php.

14. F. Kuchler, E. Golan, J. N. Variyam, and S. R. Crutchfield, Obesity Policy and the Law of Unintended Consequences, *Amber Waves*, USDA Economic Research Service, June 2005, available at ers.usda.gov/AmberWaves/June05/Features/ObesityPolicy.htm.

15. Ibid.

16. E. Rosenthal, Europeans Take Aim at Junk Food Advertisers, *International Herald Tribune*, January 6, 2005, pp. 1, 3.

17. A. A. Leiserowitz, R. W. Kates, and T. M. Parris, Do Global Attitudes and Behaviors Support Sustainable Development? *Environment*, November 2005, p. 28.

18. M. Bryant, Human Nature, Mating Motives May Lead to Murder, Book Theorizes, University of Texas, 2005, available at utexas.edu/opa/news/2005/05/psychology23.html.

19. C. Larson, The End of Hunting? *Washington Monthly*, January–February 2006, available at washingtonmonthly.com/features/2006/0601.larson.html.

20. J. Kettlewell, Farm Animals "Need Emotional TLC," BBC News, n.d., available at http://news.bbc.co.uk/2/hi/science/nature/4360947.stm.

21. Wronged, *Economist*, August 27, 2005, p. 12.

22. Capital Punishment in the United States, Wikipedia, 2008.

23. T. Regan, *Empty Cages* (Lanham, Md.: Rowman & Littlefield, 2004).

24. P. Singer, *Animal Liberation* (New York: HarperCollins, 1975); Regan, *Empty Cages*.

25. A. Shin, Outbreaks Reveal Food Safety Net's Holes, *Washington Post*, December 11, 2006, p. A1; Reckless with Food Safety, *New York Times*, December 12, 2006, p. A30; E. Schlosser, Has Politics Contaminated the Food Supply? *New York Times*, December 11, 2006, p. A27.

Additional Readings

References to pages in *Time*, *Newsweek*, and *U.S. News & World Report* refer to the International Edition. Page numbers may be different in the edition for readers in the United States.

A Better Road to Hoe: The Economic, Environmental, and Social Impact of Sustainable Agriculture. St. Paul, Minn.: Northwest Area Foundation, 1994.

Ackerman, J. How Altered? *National Geographic*, May 2002, pp. 34–50.

Altieri, M. A. Agroecology in Action. 2000. Available at cnr.berkeley.edu/~agroeco3/modern_agriculture.html.

August, M., M. Cooper, D. Bjerklie, L. McLaughlin, W. Cole, A. Daruvalla, J. Israely, J. Perry, J. Ressner, and A. Smith. Should We All Be Vegetarians? *Time*, October 14, 2002, pp. 56–64.

Barstow, C. *The Eco-Foods Guide*. Gabriola Island, B.C.: New Society Publishers, 2002.

Berry, W. *The Unsettling of America*. San Francisco: Sierra Club Books, 1997.

Berry, W. Death of the American Family Farm. *Progressive*, April 2002.

Blatt, H. *America's Environmental Report Card: Are We Making the Grade?* Cambridge, Mass.: MIT Press, 2004.

Blayney, D. P. The Changing Landscape of U. S. Milk Production. USDA Economic Research Service, 2002. Available at ers.usda.gov.

Compa, L. *Blood, Sweat, and Fear*. New York: Human Rights Watch, 2004.

Blythman, J. Permanent Global Summertime. *Ecologist*, September 2004, pp. 18–32.

Brady, N. C., and R. R. Weil. *The Nature and Properties of Soils*, 13th ed. Upper Saddle River, N.J.: Prentice Hall, 2001.

Brownell, K. D., and K. B. Horgen. *Food Fight: The Inside Story of the Food Industry, America's Obesity Crisis, and What We Can Do About It*. New York: McGraw-Hill, 2003.

Brownlee, C. Arbiter of Taste. *Science News*. volume 168. December 3, 2005, p. 356.

Brush, S. B. *Farmers, Bounty: Locating Crop Diversity in the Contemporary World*. New Haven, Conn.: Yale University Press, 2004.

Butzke, C. E., and L. F. Bisson. Wine. *Encarta Online Encyclopedia*, 2005. Available at encarta.msn.com/encyclopedia_761576868/Wine.html.

Career Guide to Industries, 2006–2007. Food Manufacturing. Washington, D.C.: Bureau of Labor Statistics, U.S. Department of Labor, 2005.

Casa, K. The Changing Face of Farming in America. *National Catholic Reporter*, February 12, 1999, p. 1.

Cloud, J. My Search for the Perfect Apple. *Time*, March 12, 2007, pp. 37–42.

Clover, C. *The End of the Line.* London: Ebury Press, 2005.

Coleman, F. C., W. F. Figueira, J. S. Ueland, and L. B. Crowder. The Impact of United States Recreational Fisheries on Marine Fish Populations. *Science*, September 24, 2004, pp. 1958–1959.

Critser, G. *Fatland: How Americans Became the Fattest People in the World.* Boston: Houghton Mifflin, 2003.

Cummings, C. H. Trespass. *Worldwatch*, January–February 2005, pp. 24–35.

Cummins, R., and B. Lilliston. *Genetically Engineered Food: A Self-Defense guide for Consumers*, 2nd ed. New York: Marlowe & Company, 2000.

Cut the Fat. *Consumer Reports*, January 2004, pp. 12–16.

Cutler, D. M., E. L. Glaeser, and J. M. Shapiro. Why Have Americans Become More Obese? National Bureau of Economic Research Working Paper 9446, 2003.

Dean, C. Lobster Boom and Bust. *New York Times*, August 9, 2005, p. F1.

Dimitri, C., and L. Oberholtzer. *Market-Led Versus Government-Facilitated Growth: Development of the U.S. and EU Organic Agricultural Sectors.* Washington, D.C.: USDA Economic Research Service, 2005.

Dudley, K. M. *Debt and Dispossession: Farm Loss in America's Heartland.* Chicago: University of Chicago Press, 2000.

Duram, L. A. *Good Growing: Why Organic Farming Works.* Lincoln, Neb.: Bison Books, 2005.

Egan, T. Amid Dying Towns of Rural Plains, One Makes a Stand. *New York Times*, December 1, 2003.

Ellstrand, N. C. *Dangerous Liaisons: When Cultivated Plants Mate with Their Wild Relatives.* Baltimore: Johns Hopkins University Press, 2003.

Epstein, E. Roots. *Scientific American*, May 1973, pp. 48–58.

Ervin, D. E., C. F. Runge, E. A. Graffy, W. E. Anthony, S. S. Batie, P. Faeth, T. Penny, and T. Warman. A New Strategic Vision. *Environment*, July–August 1998, pp. 8–15, 35–40.

Farabee, M. J. Photosynthesis. 2001. Available at emc.maricopa.edu/faculty/farabee/BIOBK/BioBookPS.html.

Farm Animal Waste and the Clean Water Dilemma. *Farm Sanctuary News*, Spring 1998. Available at http://farmsanctuary.org/newsletter/news_cleanwater.htm.

Finn, C. Temperate Berry Crops. In *Perspectives on New Crops and New Uses*, edited by J. Janick. Alexandria, Va.: ASHS Press, 1999, pp. 324–334.

Frequently Asked Questions About Sustainable Agriculture. Union of Concerned Scientists, 2001. Available at ucsusa.org/food_and_environment/sustainable_food/questions-about-sustainable-agriculture.html.

Fruit and Tree Nuts Outlook. Highlight: Fresh-Market Fruit Production. Economic Research Service/FTS-302, USDA, January 30, 2003.

Gardner, B. U. S. Agriculture in the Twentieth Century. *EH.NET Encyclopedia*. Available at http://eh.net/encyclopedia/article/gardner.agriculture.us.

Gillham, O. *The Limitless City*. Washington, D.C.: Island Press, 2002.

Gilman, V. Food Coloring. *Chemical and Engineering News*, August 25, 2003, p. 34.

Glasser, J. A. Broken Heartland. *U.S. News & World Report*, May 7, 2001, pp. 16–20.

Gluckman, P., and M. Hanson. *Mismatch: Why Our World No Longer Fits Our Bodies*. New York: Oxford University Press, 2006.

Goff, S. A., and J. M. Salmeron. Back to the Future of Cereals. *Scientific American*, August 2004, pp. 26–33.

Gorman, C. How to Eat Smarter. *Time Magazine*, December 8, 2003, pp. 60–69.

Graham-Rowe, D., and B. Holmes. The World Can't Go On Living Beyond Its Means. *New Scientist*, April 2, 2005, pp. 8–10.

Granatstein, D., and E. Kirby. Current Trends in Organic Tree Fruit Production. Washington State University CSANR Report no. 4, May 2002.

Gratzer, W. *Terrors of the Table*. New York: Oxford University Press, 2005.

Hails, R. S. Assessing the Risks Associated with New Agricultural Practices. *Nature*, August 8, 2002, pp. 685–688.

Halweil, B. Where Have All the Farmers Gone? *Worldwatch*, September–October 2000, pp. 12–28.

Halweil, B. *Eat Here*. New York: Norton, 2004.

Halweil, B. Can Organic Farming Feed Us All? *Worldwatch*, May–June 2006, pp. 18–24.

Halweil, B. Still No Free Lunch: Nutrient Levels in U.S. Food Supply Eroded by Pursuit of High Yields. Organic Center, 2007.

Halweil, B., and D. Nierenberg. Watching What We Eat. In *State of the World*, edited by L. Starke. New York: Norton, 2004, pp. 68–95.

Hayden, T. Fished Out. *U.S. News & World Report*, June 9, 2003, pp. 38–45.

Hearne, S. A., L. M. Segal, P. J. Unruh, M. J. Earls, and P. Smolarcik. F as in Fat. Trust for America's Health, 2004. Available at healthyamericans.org.

Higgins, A. Why the Red Delicious No Longer Is. *Washington Post*, August 3, 2005, p. A1.

Hilborn, R., T. A. Branch, B. Ernst, et al. State of the World's Fisheries. *Annual Review of Environment and Resources* 28 (2003): 359–399.

Hill, J. O., Wyatt, H. R., and Peters, J. C. Modifying the Environment to Reverse Obesity. In *Essays of the Future of Environmental Health Research*, edited by J. Goehl. Bethesda, Md.: National Institutes of Health, 2005.

Hillel, D., ed. *Encyclopedia of Soils in the Environment*. St. Louis, Mo.: Elsevier, 2004.

History and Origin of the Potato. N.d. Available at sunspiced.com/phistory.html.

Horrigan, L., R. S. Lawrence, and P. Walker. How Sustainable Agriculture Can Address the Environmental and Human Health Harms of Industrial Agriculture. *Environmental Health Perspectives*, May 2002, pp. 445–456.

How to Win an Argument with a Meat Eater. N.d. Available at animalsvoice.com/PAGES/invest/meat.html.

Huang, J., C. Pray, and S. Rozelle. Enhancing the Crops to Feed the Poor. *Nature*, August 8, 2002, pp. 678–684.

Huang, P. M. Soil Mineral—Organic Matter—Microorganism Interactions: Fundamentals and Impacts. In *Advances in Agronomy*, vol. 82, edited by D. L. Sparks. Orlando, Fla.: Academic Press, 2004, pp. 391–472.

Ikerd, J. Economic Fallacies of Industrial Hog Production. 2001. Available at www.ssu.missouri.edu/faculty/jikerd/papers/EconFallacies-Hogs.htm.

Ikerd, J. Sustainable Farming: Reconnecting with Consumers. University of Missouri, 2002. Available at www.ssu.missouri.edu/faculty/jikerd/papers/HawaiiSA.html.

In the Battle of the Bulge, More Soldiers Than Successes. Pew Research Center, April 26, 2006. Available at pewresearch.org/social/pack.php?PackID=10.

Karlen, D. L., S. S. Andrews, and J. W. Doran. Soil Quality: Current Concepts and Applications. In *Advances in Agronomy*, vol. 74, edited by D. L. Sparks. Orlando, Fla.: Academic Press, 2001, pp. 1–41.

Kimbrell, A., ed. *Fatal Harvest: The Tragedy of Industrial Agriculture*. Washington, D.C.: Island Press, 2002.

Kindhart, J. D. Blueberries. USDA, 1994.

Kuepper, G., and L. Gegner. Organic Crop Production Overview. National Sustainable Agriculture Information Service, 2004.

Lambert, C. The Way We Eat Now. N.d. Available at sophists.org/article-print-300.html.

Lappe, F. M., J. Collins, and P. Rosset. *World Hunger: Twelve Myths*, 2nd ed. New York: Grove Press, 1998.

Lawrence, F. Chicken. *Ecologist*, September 2004, pp. 40–45.

Lemonick, M. D. How We Grew So Big. *Time*, August 9, 2004, pp. 40–51.

Logue, A. W. *The Psychology of Eating and Drinking*, 3rd ed. New York: Routledge, 2004.

Meat Packaging Materials. USDA Food Safety and Inspection Service, 2000. Available at fsis.usda.gov/OA/pubs/meatpack.htm.

Midkiff, K. *The Meat You Eat*. New York: St. Martin's Press, 2004.

Miller, H. I., and G. Conko. *The Frankenfood Myth: How Protest and Politics Threaten the Biotech Revolution*. Westport, Conn.: Praeger, 2004.

Mitchell, S., and C. Christi. *Fat Is Not Your Fate*. New York: Simon & Schuster, 2005.

Montgomery, D. R. Geology, Geomorphology, and the Restoration Ecology of Salmon. *GSA Today* [Geological Society of America], November 2004, pp. 3–12.

Montgomery, D. R. *The Erosion of Civilizations*. Berkeley: University of California Press, 2007.

Motavalli, J. Across the Great Divide. *E Magazine*, January–February 2002, pp. 34–39.

Motavalli, J. The Case Against Meat. *E Magazine*, January–February 2002, pp. 26–32.

Motavalli, J. Rights from Wrongs. *E Magazine*, March–April 2003, pp. 26–33.

Naylor, R. L. Agriculture and Global Change. In *Earth Systems: Processes and Issues*, edited by W. G. Ernst. Cambridge: Cambridge University Press, 2000, pp. 462–475.

Naylor, R. L., J. Eagle, and W. L. Smith. Salmon Aquaculture in the Pacific Northwest: A Global Industry with Local Impacts. *Environment*, October 2003, pp. 18–39.

Naylor, R. L., R. J. Goldberg, J. H. Primavera, N. Kautsky, M. C. M. Beveridge, J. Clay, C. Folke, J. Lubchenco, H. Mooney, and M. Troell. Effect of Aquaculture on World Fish Supplies. *Nature*, June 29, 2000, pp. 1017–1024.

Nestle, M. *Food Politics*. Berkeley: University of California Press, 2002.

Newman, C. Why Are We So Fat? *National Geographic*, August 2004, pp. 46–61.

Nierenberg, D. *Happier Meals: Rethinking the Global Meat Industry*. Worldwatch Paper 171, Worldwatch Institute, 2005.

Nizeyimana, E. L., G. W. Petersen, and E. D. Warner. Tracking Farmland Loss. *Geotimes*, January 2002, pp. 18–21.

Noncitrus Fruits and Nuts, 2004 Preliminary Summary. National Agricultural Statistics Service, USDA, January 2005.

Now, It's Not Personal. *Worldwatch*, July–August 2004, pp. 12–19.

Oberholtzer, L., C. Dimitri, and C. Greene. Price Premiums Hold on as U.S. Organic Produce Market Expand. Economic Research Service, USDA, May 2005.

Okie, S. *Fed Up: Winning the War against Childhood Obesity*. Washington, D.C.: Joseph Henry Press, 2005.

Oliver, M. A. Soil and Human Health: A Review. *European Journal of Soil Science*, December 1997, pp. 573–592.

Pauly, D., V. Christensen, S. Guenette, T. J. Pitcher, U. R. Sumaila, C. J. Walters, R. Watson, and D. Zeller. Towards Sustainability in World Fisheries. *Nature*, August 8, 2002, pp. 689–695.

Pauly, D., and J. Maclean. *In a Perfect Ocean: The State of Fisheries and Ecosystems in the North Atlantic Ocean*. Washington, D.C.: Island Press, 2003.

Pauly, D., and R. Watson. Counting the Last Fish. *Scientific American*, July 2003, pp. 34–39.

Pearce, F. The Famine Fungus. *New Scientist*, April 26, 1997, pp. 32–36.

Perez, A., B.-H. Lin, and J. Allshouse. Demographic Profile of Apple Consumption in the United States. Economic Research Service/USDA, Fruit and Tree Nuts S&O/FTS-292, September 2001.

Perry, J., D. Banker, and R. Green. Broiler Farms' Organization, Management, and Performance. Agriculture Information Bulletin no. 748, 1999. Available at ers.usda.gov/publications/aib748/.

Pfeiffer, D. A. Eating Fossil Fuels, 2004. Available at vivelecanada.ca/article.php?story=20040611151023655&mode=print.

Pierzynski, G. M., J. T. Sims, and G. F. Vance. *Soils and Environmental Quality*, 3rd ed. Boca Raton, Fla.: Lewis Publishers, 2005.

Pollan, M. No Bar Code. *Mother Jones*, May–June 2006, pp. 36–45.

Pollan, M. Unhappy Meals. *New York Times*, January 28, 2007.

Pool, R. *Fat*. New York: Oxford University Press, 2001.

Pretty, J. Agroecology in Developing Countries: The Promise of a Sustainable Harvest. *Environment*, November 2003, pp. 8–20.

Pringle, P. *Food, Inc.: The Promises and Perils of the Biotech Harvest*. New York: Simon & Schuster, 2003.

Raloff, J. The Ultimate Crop Insurance. *Science News*, September 11, 2004, pp. 170–172.

Raloff, J. Food Colorings. *Science News*, January 8, 2005, pp. 27–29.

Rap Sheet on Animal Factories. Sierra Club, 2002.

Regan, T. *Empty Cages: Facing the Challenge of Animal Rights*. Lanham, Md.: Rowman & Littlefield, 2003.

Rich, D. K. Sustainable Agriculture Is More Than Organic Methods. *San Francisco Chronicle*. July 3, 2004, p F1.

Salmon Farms and Hatcheries. *Environment*, April 2004, pp. 40–45.

Sampson, R. N. *The Ethical Dimension of Farmland Protection: Farmland, Food and the Future*. Ankeny, Iowa: Soil Conservation Society of America, 1979.

Sanchez, P. A., and M. S. Swaminathan. Cutting World Hunger in Half. *Science*, January 21, 2005, pp. 357–359.

Sarig, Y., J. F. Thompson, and G. K. Brown. Alternatives to Immigrant Labor? Center for Immigration Studies, 2000. Available at cis.org/articles/2000/back1200.html.

Schlosser, E. *Fast Food Nation*. New York: Houghton Mifflin, 2001.

Schubert, R. Farming's New Feudalism. *Worldwatch*, May–June 2005, pp. 10–15.

Shape of Things to Come. *Economist*, December 13, 2003, pp. 1–16.

Shaw, T. *Everything I Ate: A Year in the Life of My Mouth*. New York: Reed Elsevier, 2005.

Shute, N. Better Safe Than Sorry. *U.S. News & World Report*, May 28, 2007, pp. 67–70.

Singer, M. J., and D. N. Munns. *Soils: An Introduction*, 5th ed. Upper Saddle River, N.J.: Prentice Hall, 2001.

Singer, P. *Animal Liberation*. New York: HarperCollins, 1975.

Smil, V. *Feeding the World: A Challenge for the Twenty-first Century*. Cambridge, Mass.: MIT Press, 2000.

Soil Association. *Seeds of Doubt: North American Farmers' Experiences of GM Crops*. Bristol, U.K.: Soil Association, 2002.

Soil Erosion and Conservation. 2000. Available at seafriends.org.nz/enviro/soil/erosion.htm.

Spake, A. Rethinking Weight. *U.S. News & World Report*, February 9, 2004, pp. 50–56.

Spanier, A. M., F. Shahidi, T. H. Parliament, C. Mussinian, C.-T. Ho, and E. T. Contis, eds. *Food Flavor and Chemistry*. Royal Society of Chemistry, 2005.

Stengel, P., and S. Gelin. *Soil: Fragile Interface*. Enfield, N.H.: Science Publishers, 2003.

Stewart, A. *The Earth Moved*. Chapel Hill, N.C.: Algonquin Books of Chapel Hill, 2004.

Stockdale, E. A., N. H. Lampkin, M. Hovi, et al. Agronomic and Environmental Implications of Organic Farming Systems. In *Advances in Agronomy*, vol. 70, edited by D. L. Sparks. Orlando, Fla.: Academic Press, 2001, pp. 261–327.

Taming the Wild Strawberry. Vegetarians in Paradise. N.d. Available at vegparadise.com/highestperch45.html.

Taylor-Davis, S., and M. B. Stone. Food Processing and Preservation. MSN Encarta, 2005. Available at encarta.msn.com/text_761560675_0/Food_Processing_and_Preservation.html.

Tibbetts, J. Eating Away at a Global Food Source. *Environmental Health Perspectives*, April 2004, pp. A282–A291.

Tilman, D., K. G. Cassman, P. A. Matson, and S. Polasky. Agricultural Sustainability and Intensive Production Practices. *Nature*, August 8, 2002, pp. 671–677.

Trewavas, A. Malthus Foiled Again and Again. *Nature*, August 8, 2002, pp. 668–670.

Tudge, C. *So Shall We Reap*. New York: Penguin Books, 2003.

Understanding Plastic Film. American Plastics Council, 1996.

Underwood, A., and J. Adler. You Know About Fat. *Newsweek*, September 20, 2004, pp. 54–60.

United States Agriculture: Overcoming Barriers to Sustainability. N.d. Available at geologyandgeography.vassar.edu/saed/ingraham.htm.

Vegetarian Diets. American Dietetic Association, 2003. Available at eatright.org/Public/GovernmentAffairs/92_17084.cfm.

What Is Sustainable Agriculture? N.d. Available at www.sarep.ucdavis.edu/concept.htm.

Whitty, J. The Fate of the Oceans. *Mother Jones*, March–April 2006, pp. 32–48.

Wicks, D. Humans, Food, and Other Animals: The Vegetarian Option. In *A Sociology of Food and Nutrition*, 2nd ed., edited by J. Germov and L Williams. New York: Oxford University Press, 2004, pp. 263–287.

Wiebe, K. Linking Land Quality, Agricultural Productivity, and Food Security. Agricultural Economic Report no. 823, U.S. Department of Agriculture, 2003.

Williams, M. J. Are High Seas and International Marine Fisheries the Ultimate Sustainable Management Challenge? *Journal of International Affairs*, Fall–Winter 2005, pp. 221–234.

Winston, M. L. *Travels in the Genetically Modified Zone*. Cambridge, Mass.: Harvard University Press, 2002.

Wirzba, N., ed. *The Art of the Commonplace: The Agrarian Essays of Wendell Berry*. Washington, D.C.: Shoemaker and Hoard, 2002.

Young, O. R. Taking Stock: Management Pitfalls in Fisheries Science. *Environment*, April 2003, pp. 25–33.

Index